Structure, Dynamics,
Interactions and Evolution
of
Biological Macromolecules

The cover design presents a simplified view of the structure of theλcro-protein-DNA complex as suggested from X-ray crystallographic studies. For the DNA, the orange circles correspond to the phosphates and the smaller yellow circles correspond to the bottom of the major groove. For the cro protein each circle corresponds to an α-carbon position.

(Kindly supplied by Pr. B. W. Matthews,
Institute of Molecular Biology, University of Oregon, Eugene, U.S.A.)

Structure, Dynamics, Interactions and Evolution of Biological Macromolecules

Proceedings of a Colloquium held at Orléans, France
on July 5-9, 1982
to Celebrate the 80th Birthday of Professor Charles Sadron

edited by

CLAUDE HÉLÈNE

*Centre de Biophysique Moléculaire, CNRS, Orléans
and Muséum National d'Histoire Naturelle, Paris*

D. REIDEL PUBLISHING COMPANY
DORDRECHT : HOLLAND / BOSTON : U.S.A.
LONDON : ENGLAND

Library of Congress Cataloging in Publication Data

Main entry under title:

Structure, dynamics, interactions and evolution of biological macro-
     molecules.

     "The colloquium was sponsored by the Centre National de la
Recherche Scientifique and the Muséum National d'Histoire Naturelle"—
Pref.
     1. Molecular biology–Congresses.   2.   Macromolecules–
Congresses.   3.   Chemical evolution–Congresses.   I.   Hélène, C.
II.   Sadron, Charles.   III.   Centre national de la recherche scientifique
(France)   IV.   Muséum national d'histoire naturelle (France)
QH506.S837   1983   574.8'8   82-24105
ISBN-13:978-94-009-7054-0   e-ISBN-13:978-94-009-7052-6
DOI: 10.1007/978-94-009-7052-6

Published by D. Reidel Publishing Company,
P.O. Box 17, 3300 AA Dordrecht, Holland.

Sold and distributed in the U.S.A. and Canada
by Kluwer Boston Inc.,
190 Old Derby Street, Hingham, MA 02043, U.S.A.

In all other countries, sold and distributed
by Kluwer Academic Publishers Group,
P.O. Box 322, 3300 AH Dordrecht, Holland.

D. Reidel Publishing Company is a member of the Kluwer Group.

This book is dedicated to

Professor Charles SADRON

on the occasion of his 80th birthday

Professor Charles Sadron lecturing during the
"Symposium" (Greek name for "banquet") held
in Orléans on July 1982.

CONTENTS

II. PROTEIN STRUCTURE AND DYNAMICS

III - EVOLUTION OF BIOLOGICAL MACROMOLECULES

PREFACE

A Colloquium was held in Orléans on July 5-9, 1982, to celebrate the 80th birthday of Professor Charles SADRON. This meeting
was devoted to the "Structure, Dynamics, Interactions and Evolution
of Biological Macromolecules", research areas whose development in
France owes so much to Pr. Sadron's efforts. The Colloquium was
sponsored by the Centre National de la Recherche Scientifique and the
Muséum National d'Histoire Naturelle. This book is a collection of the
plenary lectures delivered during this scientific meeting together with
several contributions from former collaborators of Pr. Charles Sadron.

Charles Sadron was born in 1902 at Châteauroux, near the
center of France. After highschool in his native town he moved to the
University of Poitiers where he was graduated in Physics and then obtained the Agrégation degree in 1926. Appointed as a highschool teacher
in Strasbourg, he devoted his spare time to research in magnetism
under the direction of Pr. P. Weiss. He got his Ph.D. from the University of Strasbourg in 1932 and then did a post-doctoral work at
Caltech with Pr. Von Karman. Back to France in 1934, he was appointed as Associate Professor at the University of Strasbourg in 1937. His
interest in colloidal suspensions and polymers dates back to these times when he started to study the morphology of chain polymers.

When the second world war started, Charles Sadron moved to
Clermont-Ferrand with other members of the University of Strasbourg.
He and several other members of his community then unwillingly made
a second and unexpected "post-doctoral" stay for two years at the concentration camp of Buchenwald and then in Dora, under the "supervi-
sion" of W. Von Braun. Pr. Charles Sadron fortunately survived and
resumed his activities at Strasbourg as Professor of Physics. In 1947
he started a research laboratory (Centre d'Etudes de Physique Macro-
moléculaire) devoted to the study of the physico-chemical properties
of macromolecules. The development of these research activities led
the CNRS to create in Strasbourg a research institute, the "Centre de
Recherches sur les Macromolécules (CRM) in 1954. Pr. Charles
Sadron was of course the first director of this Institute. He was already
interested in biological macromolecules and several research groups
at the CRM chose this new orientation.

C. Hélène (ed.), Structure, Dynamics, Interactions and Evolution of Biological Macromolecules, xi–xii.
Copyright © 1983 by D. Reidel Publishing Company.

In 1961, a Chair of Biophysics was created for Pr. Charles Sadron at the Muséum National d'Histoire Naturelle where he introduced this new branch of Science and created a new laboratory. He then convinced the CNRS that the creation of a research Institute devoted to Molecular Biophysics was timely. In 1967, the "Centre de Biophysique Moléculaire" (CBM) opened in Orléans ; several research groups with complementary expertise in Physics, Chemistry and Biology gathered in this new Institute under the dynamic stimulus of Pr. Charles Sadron. When Pr. Sadron retired in 1974, this Institute had acquired an international reputation and its main scientific orientations on protein and nucleic acid conformations and interactions are still actively pursued. Since 1974, Pr. Sadron has remained very active, coming to the laboratory to write, discuss with friends and former collaborators and keep himself informed about new developments in the life sciences.

The areas which have been covered by the lectures of the Colloquium are those in which Pr. Charles SADRON has been involved during the last three decades in the three laboratories he has created. The organization of the Colloquium and the publication of the present book have been inspired by the spirit of Professor Charles SADRON to whom all participants are dedicating their contribution.

Many chapters of this book deal with recent developments in the field of nucleic acid and protein structures, their dynamic state in solution and their interactions to form multimacromolecular complexes. Problems related to the origin of life, prebiotic processes and evolution are dealt with in several chapters. One of the contributors to this last part of the Symposium, Pr. F. Egami, did not come to Orléans. In his last letter sent on May 20, 1982, he explained why he would no longer be able to come to Orléans ; on July 1, he sent the manuscript of the lecture he would have liked to dedicate to Pr. Sadron. Pr. Egami died on July 17, 1982. Science has lost a great man and we will all miss this friend ; but Pr. Egami's enthousiasm and kindness will always remain in our memories.

Claude HELENE

September 15, 1982.

ACKNOWLEDGEMENTS

The Colloquium on "STRUCTURE, DYNAMICS, INTERAC-
TIONS AND EVOLUTION OF BIOLOGICAL MACROMOLECULES" held
in Orléans on July 5-9, 1982, was sponsored by the Centre National de
la Recherche Scientifique and the Muséum National d'Histoire Naturelle.
Support was also obtained from the "Association pour le Développement
des Sciences Biophysiques", the "Institut de Recherches Internatio-
nales Servier", and from Beckman France, Rhône-Poulenc Santé and
Virax S.A.

The local organization of the Colloquium has been a collective
enterprise involving many members of the Centre de Biophysique Moléculaire. I would like to thank all of those who have participated in
making this meeting a success. Special mention should be made of
Mrs. Jeannine Florian who - as usual - took care of everything, and
Mrs. Christiane David who not only participated in the organization but
also did a magnificent job of typing - or retyping - many of the manus-
cripts reproduced in the present book.

I - NUCLEIC ACIDS . STRUCTURES, DYNAMICS
AND ASSOCIATIONS WITH PROTEINS

LEFT-HANDED DNA IN CHEMICAL AND BIOLOGICAL SYSTEMS

Alexander Rich
Department of Biology
Massachusetts Institute of Technology
Cambridge, Massachusetts 02139 U.S.A.

DNA can have both right-handed and left-handed double helical conformations. This idea has old roots and was considered shortly after the formulation of the double helix by Watson and Crick in 1953 (1). In this early period, attention was directed toward the question of whether or not the X-ray diffraction patterns of DNA fibers were consistent with the Watson and Crick formulation. Efforts spanning over a decade produced a great deal of experimental evidence which in essence stated that it was possible to modify the initial Watson-Crick right-handed helix in a number of details so that one could produce a model consistent with the experimental X-ray data (2). It was obvious to workers in the field, however, that although DNA fiber analysis generated physical data in agreement with the Watson-Crick structure, they did not unambiguously prove the structure. The X-ray diffraction patterns were limited, and far from atomic resolution. Because of the limited resolution the patterns had to be interpreted and a number of assumptions made. One assumption was that the structure was perfectly regular with an asymmetric unit consisting of one nucleotide. However, the agreement was reasonable and this led to the general adoption of the right-handed B-DNA conformation as the predominant form in biological systems. Considerable attention was paid to the fact that under certain conditions alternative conformational variants were seen. The most important of these was the A-DNA pattern which originally was produced by air-dried fibers (3). The greater wealth of diffraction data seen in the A pattern made it possible to get more information. The relative ease with which it slipped into the B pattern reinforced the idea that these were two conformational variants; it was found that both patterns could be fitted by a right-handed double helix.

Although there was discussion of the possibility of left-handed helices, especially shortly after the double helix formulation, there were very few systematic studies of left-handed helices published during the next decade or so. Left-handed double helical DNA regions might be anticipated to facilitate replication fork movement and chain

C. Hélène (ed.), Structure, Dynamics, Interactions and Evolution of Biological Macromolecules, 3–21.

separation during DNA replication. Helical flexibility as a mechanism
to either generate structural recognition signals along the DNA molecule or regulate superhelical topology was not discussed extensively.
It is only relatively recently that there has been increased emphasis
on alternative conformational models for DNA (4). Part of the stimulus
came from the development of the idea that DNA might not exist as a
continuously coiled double helix but that it might adopt a side-by-side
conformation (5, 6). In the face of suggestions of this type, it was
clear that the fiber diffraction studies with their limited information
content could no longer be responsive to the question of what detailed
conformations DNA is capable of accommodating.

Single Crystal Analysis

The recent development of chemical methods for synthesizing DNA
made it possible to produce oligonucleotides in amounts great enough
for crystallization experiments. Unlike fiber diffraction patterns,
single crystals can diffract to a very high resolution and they produce
very large numbers of reflections or experimental measurements. This
in turn makes it possible to solve the structures and observe fine details of DNA conformation in which very little or no interpretation is
required. In an atomic resolution ($<1.0Å$) electron density map, every
atom is seen and no interpretation is required to obtain bond angles,
distances, ring pucker, etc. Studies of this type carry with them the
promise of resolving issues dealing with fine details of DNA conformation.

For many years there have been single crystal X-ray studies of nucleic acid bases, nucleosides and monomeric nucleotides. Several single crystal diffraction studies have been carried out in which the purine-pyrimidine base pairs have been shown to have variety of hydrogen
bonding (7). However, the first visualization of the double helix at
atomic resolution occurred almost ten years ago with the solution of
the structures of rApU (8, 9) and rGpC (10, 11) which were found to
form right-handed double helical fragments of the type which had been
anticipated for RNA molecules. This work on ribonucleotide fragments
laid the groundwork for a more extended subsequent analysis of deoxyo-
ligonucleotides.

We have collaborated with J. H. van Boom and his colleagues in
Leiden University who have developed methods for synthesizing oligonucleotides. In an attempt to work with tetramers, van Boom synthesized
all four of the self-complementary deoxynucleotide tetramers containing
guanine and cytosine residues. It seemed likely that these bases might
produce stable oligonucleotide double helical fragments which could
then be crystallized. Three of the four were crystallized and one of
them, d(CpGpCpG), crystallized with apparent ease. This suggested that
it might be worth exploring the hexamer d(CpGpCpGpCpG). It was
discovered that crystals could be formed readily using magnesium and
spermine as cations (12). Furthermore, these crystals diffracted to
$0.9Å$ resolution. It became clear that solution of this crystal struc-

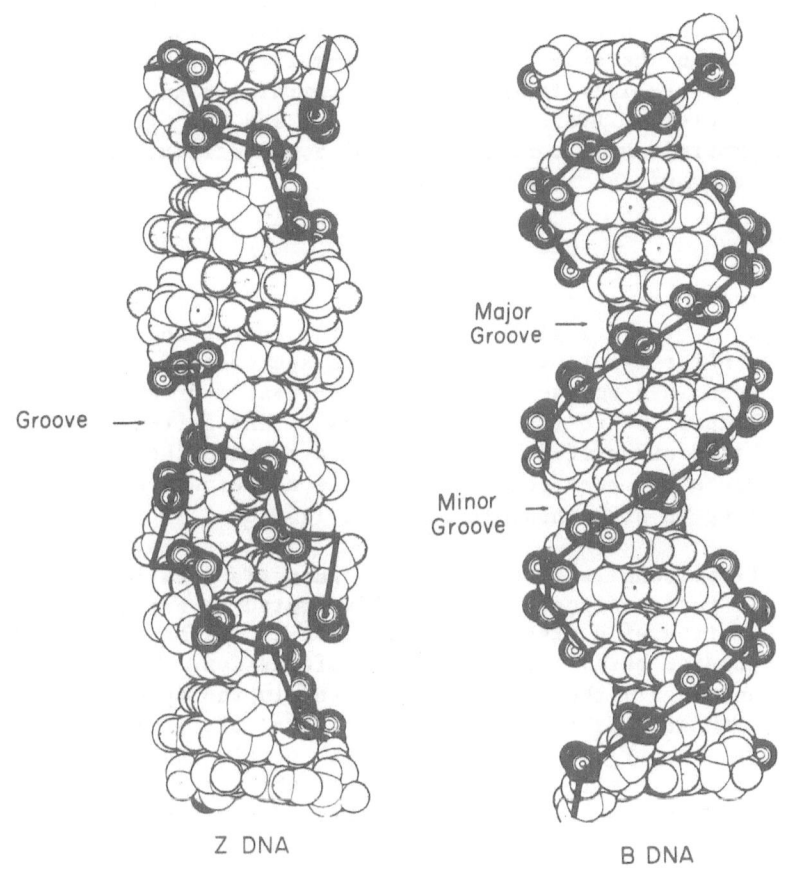

Fig. 1. Van der Waals drawings showing side views of Z–DNA and B–DNA.
The irregularity of the Z–DNA backbone is illustrated by the
heavy lines which go from phosphate to phosphate residues
along the chain. This includes positions in Z–DNA where the
phosphate residues are missing in the crystal structure but
would be occupied in a continuous double helix. The groove
in Z–DNA is quite deep, extending to the axis of the double
helix.   In contrast, B–DNA has a smooth line connecting the
phosphate groups and there are two grooves, neither one of
which extends to the helix axis of the molecule.

ture would make it possible to visualize not only the details of the double helix but would also yield information about hydration and ions.

The structure was solved using the multiple isomorphous replacement method. Various heavy metal cation derivatives were obtained in the crystals to solve the structure. The asymmetric unit of the crystal had one double-helical fragment of DNA containing six base pairs, two spermine molecules, a hydrated magnesium ion and 62 water molecules. The asymmetric unit contained almost 5,000 Daltons and thus it was as large as some of the smaller proteins.

Left-Handed Z-DNA

The resultant structure was quite unusual (12). It was a left-handed helix. However, it was not a simple left-handed helix, but one in which the alternating purine-pyrimidine sequence of the oligonucleotide was used in a unique way. A van der Waal's drawing is shown in Figure 1 in comparison to right-handed B-DNA. The left-handed helix has the phosphates arranged in a zig-zag array; hence the name Z-DNA. The same Z structure has been found in crystals of tetramers d(CpGpCpG) (13, 14) as well as in other hexamer crystals (15).

The unusual backbone is associated with guanosine residues which have an unusual conformation. This is illustrated in Figure 2, which shows the conformation of deoxyguanosine as seen in Z-DNA and B-DNA. In Z-DNA the guanine bases adopt the syn conformation. The anti conformation is found in all the bases in B-DNA and in the cytosine residues in Z-DNA. NMR studies in solution have shown that purine residues can adopt the syn and anti conformations with equal ease (16). An early analysis of the rotational barriers about the glycosyl bonds suggested that purines can rotate more easily than pyrimidines into the syn conformation (17).

Another difference is found in the pucker of the sugar ring. The guanosine in Z-DNA (Fig. 2) has the C3' endo conformation (which is the preferred conformation for ribonucleotides) while in B-DNA the C2' endo conformation is adopted for all nucleotides. However, the cytidine residues in Z-DNA have the C2' endo conformation of the sugar. The asymmetric unit in Z-DNA is thus a dinucleotide consisting of a cytidine residue in one conformation and a guanosine residue in a different conformation.

One consequence of these differences in conformation is that the bases in Z-DNA have a different orientation relative to the sugar phosphate backbone than they do in B-DNA. In essence, the bases must "flip over" in order to go from one form to the other. This is illustrated in Figure 3 where the bases are drawn as flat boards with shading on one side. When a Z-DNA segment is introduced into the molecule, the bases in that segment are inverted.

In addition, Z-DNA has only one groove while the B-DNA helix has

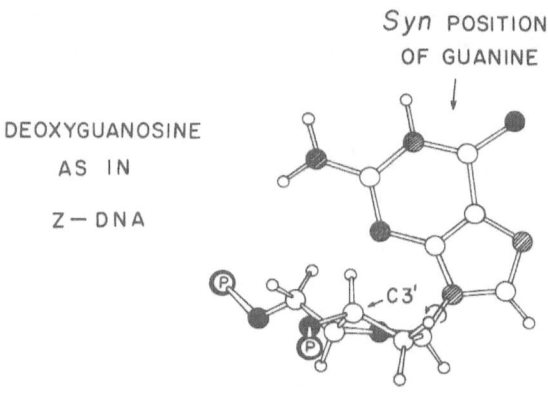

C3' endo Sugar Pucker

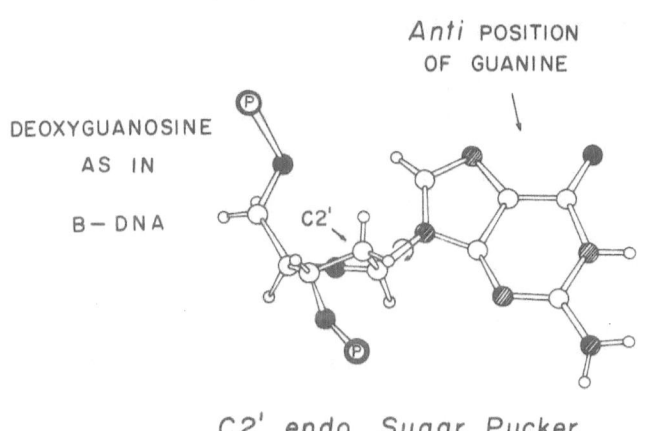

C2' endo Sugar Pucker

Fig. 2.   Conformation of deoxyguanosine in B-DNA and in Z-DNA.   The
          sugar is oriented so that the plane defined by C1'-O1'-C4' is
          horizontal. Atoms lying above this plane   are   in   the   endo
          conformation.   The   C3'   is endo in Z-DNA while in B-DNA the
          C2' is endo. These different ring puckers are associated
          with   significant   changes   in the distance between the phos-
          phorus atoms. In addition, Z-DNA has guanine in the syn po-
          sition, in contrast to the anti position in B-DNA. A curved
          arrow around the glycosyl carbon-nitrogen bond indicates the
          site of rotation.

two (Figure 1). The single groove in Z-DNA is analogous to the minor
groove of B-DNA. The bases which form the concave major groove surface
of B-DNA form the convex outer surface of Z-DNA. This change in conformation is principally associated with the <u>syn</u> conformation of guanosine which has the effect of transforming a concave major groove surface into a convex outer part of Z-DNA.

The Z-DNA helix has 12 base pairs per turn of the helix with a
helical pitch of 44.6 Å. This is in contrast to the 10 base pairs per
turn occupying a distance of 34 Å in right-handed B-DNA. The diameter
of the Z-DNA helix is only 18 Å compared to the 20 Å found in B-DNA.
In Z-DNA it was possible to assign precise parameters associated with
the structure almost immediately on its discovery. The reason is that
the crystal diffracted to a resolution of 0.9 Å, i.e. atomic resolution, so every atom was visualized in the electron density map. In the
crystal lattice, the hexamer Z-DNA fragments are stacked upon each
other in such a manner that they make a structure in which the base
pairs have a stacking which runs continuously through the crystal parallel to the <u>c</u> axis, and around it the sugar-phosphate chains form a
continuum except that they are interrupted by the absence of a phosphate group every six residues. The structure in the crystal has
enough regularity so that it is visualized as a continuous helix (Fig.
1).

Figure 4 shows an end view of Z-DNA using space-filling atoms.
This shows a cross-section of three base pairs as viewed down the axis.
The helix twists in a clockwise direction moving toward the reader.
Thus, on the left side the three phosphate groups are visible while
only one is seen on the right. It can be seen that the groove in Z-DNA
is fairly deep, and the axis is found very close to the O2 oxygen of
the cytosine residue. In this diagram the imidazole ring of guanine is
shown projecting on the outside of the molecule where it is readily accessible to other substances.

Since purine nucleotides can adopt the <u>syn</u> conformation more readily than pyrimidine nucleotides, it is reasonable to believe that the Z
conformation will be favored in sequences with alternating purines and
pyrimidines. Fiber X-ray diffraction data indicates that the Z conformation is adopted by a DNA polymer containing alternating cytosine and
adenine on one strand and thymine and guanine on the other strand (18).
Although the geometrical requirements are satisfied using an
adenine-thymine base pair as well as a guanine-cytosine base pair, the
AT base pair will be somewhat less stable. This is related to the fact
that there is a water molecule in the groove of Z-DNA to which the N2
amino group of guanine is hydrogen bonded, and this water in turn is
hydrogen bonded to the phosphate of guanosine. The <u>syn</u> conformation is
stabilized by this bridging water molecule. Although adenine can adopt
the <u>syn</u> conformation, it does not have an amino group in the 2 position, so this bridging water molecule cannot be positioned there. This
would undoubtedly contribute to some destabilization for AT base pairs.

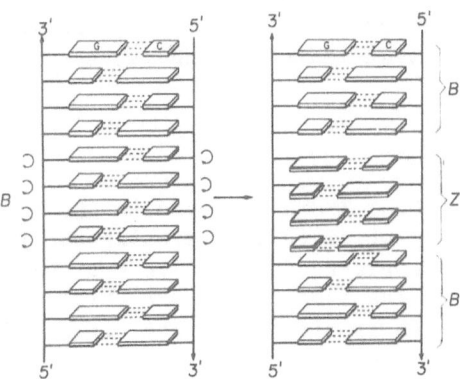

Fig. 3. A diagram illustrating part of the conformational change if a
four base pair segment of B-DNA were converted into Z-DNA.
This conversion is accomplished by inversion of the Z-DNA
bases relative to those in B-DNA. This inversion is illustrated by shading one surface of the bases. All of the dark
shaded areas are at the bottom in B-DNA. In the segment of
Z-DNA, however, four of them are turned upwards, as indicated
by the curved arrows. Rotation of the guanine residues about
the glycosyl bond produces deoxyguanosine in the <u>syn</u> conformation while for cytidine residues, both cytosine and deoxy-
ribose are inverted.

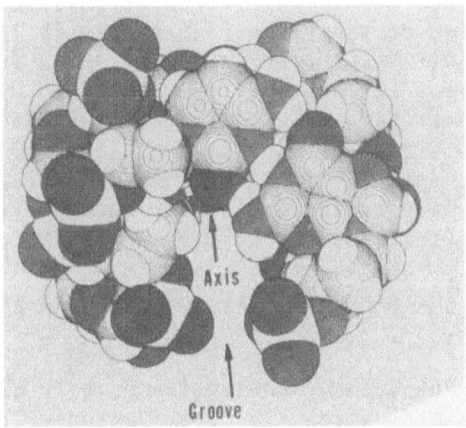

Fig. 4. A van der Waals drawing of a fragment of Z-DNA as viewed down
the axis of the helix. Three base pairs are shown, and the
deep groove is seen which extends almost to the axis of the
molecule. In these three base pairs the groove rotates
clockwise toward the reader. For that reason, three phosphates are visible on the left and only one on the right.
The N7 and C8 atoms of guanine are near the outer wall of the
molecule. A solid black dot indicates the axis of the molecule.

Z-DNA can also form in oligonucleotides containing A-T base pairs. This has been shown in the crystal structure of the octamer d(CGCATGCG) which adopts the same crystal lattice and structure as d(CGCGCGCG) (19).

Solution Studies of Z-DNA

The first evidence about Z-DNA is found in the work of Pohl and Jovin (20) (1972). They observed that raising the salt concentration of poly(dG–dC) to 4 M NaCl produced a near inversion of the circular dichroism spectrum. In the Z-DNA structure (Figure 1), the phosphate groups on opposite sugar phosphate chains are closer together than they are across the minor groove of B-DNA. This suggests that the structure is likely to be stabilized by increasing the concentration of cations. Thus it was reasonable to believe that the high salt form of poly(dG–dC) is Z-DNA. Conclusive proof of this was established by using the fact that the Raman spectra of the high salt and the low salt forms of poly(dG–dC) differ from each other in a significant manner (21). Comparison of the Raman spectrum of the Z-DNA hexamer crystals reveals that they have a spectrum identical to that of the high salt form of poly(dG–dC) and quite different from that of the low salt form (22). Thus the high salt form of poly(dGdC) is identical to Z-DNA.

In a poly(dG–dC) solution there is an equilibrium between B-DNA and Z-DNA, where the equilibrium constant is determined by the environment. The midpoint for the conversion is 2.5 M Na^+ or 0.7 M Mg^{++} (20). Evidence for the existence of an equilibrium is obtained from the fact that the original solution of the hexamer which crystallized as Z-DNA was a low salt solution (12) and there is no evidence for Z-DNA formation under those conditions as judged by the circular dichroism. The crystals themselves must have been nucleated by a minor component of Z-DNA in the low-salt solution and as the crystals grew, the equilibrium converted all of the material into the Z form. Recognition of this equilibrium is a useful step toward developing an understanding of the effect of various modifications of the polymer on the ability to form Z-DNA. It is also likely that there is a similar equilibrium in native DNA, which is strongly influenced by base sequence as well as methylation, as discussed below.

Another example in which the conformation of poly(dG–dC) is influenced by chemical modification is bromination. Bromination of poly(dG–dC) in a high salt solution yields a low salt stable form of the Z polymer (23). Bromination of approximately a third of the guanine residues on the C8 position and somewhat less of cytosine on the C5 position is adequate to stabilize the Z conformation in physiological salt solution. It is likely that this stabilization of Z is partly associated with the presence of the bulky bromine atom which favors the guanine syn conformation rather than anti, and partly due to its presence on cytosine C5. At this level of bromination, the circular dichroism is completely reversed so that it looks similar to the high salt form of poly(dG–dC).

The major conclusion reached from these studies is that the equilibrium between the B form and Z form of poly(dG-dC) is strongly influenced by cations and by the substituents which are attached to the polymer. A number of these substituents profoundly influence the equilibrium between the two forms and favor the Z-DNA conformation.

Methylation Stabilizes Z-DNA

One of the principal modifications of the nucleic acids, especially in higher eukaryotes, is methylation of cytosine on the 5 position where the cytosine is followed by a guanine. Extensive studies have been carried out which have shown that methylation of DNA is generally associated with turning off transcription; conversely, demethylation is associated with the initiation of transcription or the expression of the gene (24, 25). One of the features apparent in the structural (12) work was the fact that CG sequences play an important role in the Z conformation. Accordingly, it is reasonable to ask to what extent methylation might modify the distribution between right-handed B-DNA and left-handed Z-DNA. Behe and Felsenfeld (26) have addressed this problem directly by synthesizing a polymer, poly(dG-m^{5}dC), which is fully methylated. They found that methylation of the cytosine residue on the 5 position had a profound effect in altering the equilibrium between B-DNA and Z-DNA in solution. If one has poly(dG-dC) and poly(dG-m^{5}dC) in 50 mM NaCl, both molecules are in the right-handed B-DNA conformation. In order to convert poly(dG-dC) to left-handed Z-DNA, the magnesium concentration must be raised to 760 mM (20). However, for the fully methylated polymer, 0.6 mM is adequate to convert it to Z-DNA. There has been a decrease by three orders of magnitude in the amount of magnesium ions needed to stabilize Z-DNA if the polymer is methylated.

We have recently solved the structure of a methylated hexamer, (m^{5}dC-dG)$_3$ (27). Figure 5 compares these two helices. The overall form of the methylated Z-DNA molecule is similar to that seen in the unmethylated molecule. This result is reasonable in view of the fact that antibodies raised against non-methylated Z-DNA can also recognize Z-DNA formed by the methylated polymer (28). However, there have been some alterations in the geometry of the molecule, principally associated with the fact that the methyl group is very close to the carbon atoms C1' and C2' of the adjacent guanosine residue. This has made small changes in the helix.

The stabilization of Z-DNA by methylation is due to two factors. One of these is that the methyl group fills a vacancy or depression on the surface of Z-DNA. This is shown in Figure 5, in which the arrow near the unmethylated structure points to a depression at the side of the molecule; in the methylated structure the depression is filled by the methyl group. The methyl group is in van der Waal's contact with the imidazole ring of guanine immediately above it in Figure 5, and also with the C1' and C2' carbon atoms of the same guanosine. In effect, these form a small hydrophobic patch on the surface of the molec-

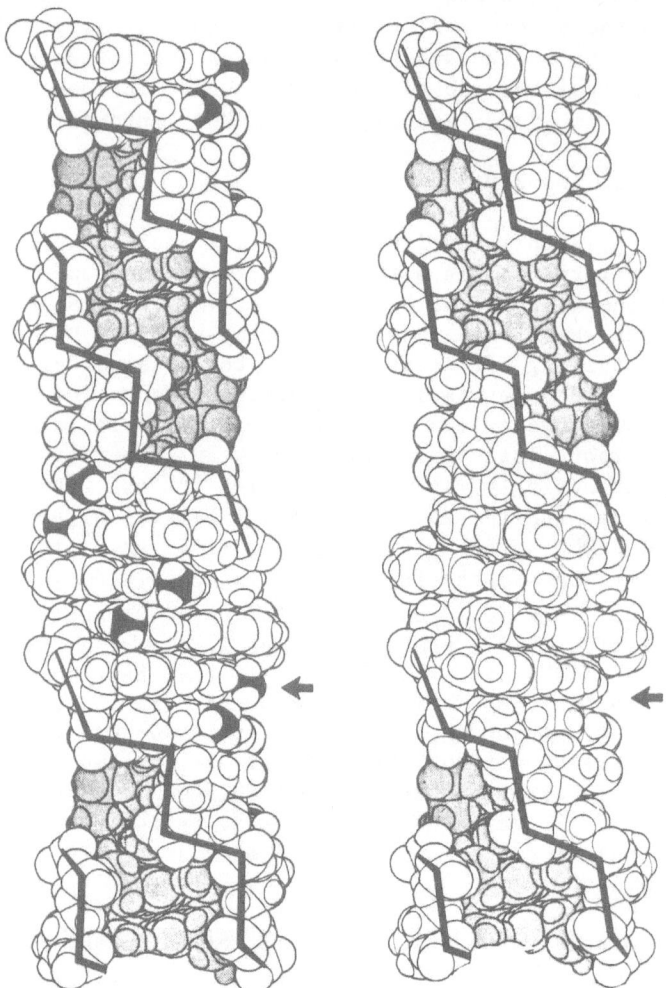

Fig. 5. Two van der Waal's drawings showing the structure of Z-DNA in
both its unmethylated and methylated forms as determined in
single crystal structures of $(C-dG)_3$ and $(m^5dC-dG)_3$ respectively. The groove in the molecule is shown by the shading.
The black zig-zag line goes from phosphate group to phosphate
group to show the arrangement of the sugar phosphate backbone. The methyl groups on the C5 position of cytosine in
the methylated structure are shaded black. The arrow illustrates that a depression found on the surface of the unmethylated structure is filled by the methyl groups. The methyl
group which is in close contact with the imidazole ring of
guanosine above it and the C1 and C2' carbon atoms of the
sugar ring.

ule.   The methyl group fills a depression in the molecule which would
otherwise be filled by water molecules.   This is in contrast to the situation   in   B-DNA where the methyl group projects out into the solvent
from the major groove of the double helix (27).   The effect of methylation is thus two-fold: one, to destabilize B-DNA by interposing a hydrophobic group into the water; and secondly,   to   stabilize   Z-DNA   by
closing off the access of water molecules into a hydrophobic region and
thereby allowing the molecule to be stabilized by hydrophobic bonding.
The   structural studies thus account for the considerable change in the
equilibrium between B-DNA and Z-DNA.   It also provides a structural rationale for why methylation of DNA might favor the formation of small
segments of Z-DNA.

Z-DNA in Biological Systems

     After the discovery of Z-DNA, opinions were mixed as to its re-
levance to biology.   It was clear that Z-DNA could form only under specialized conditions that would stabilize it.   Because of increased
phosphate-phosphate repulsion forces the left-handed helical structure
is inherently less stable than B-DNA in a physiological salt solution.
In view of the equilibrium between the two forms, it is clear that a
number of components in the environment could stabilize Z-DNA.

     It has been shown that Z-DNA can exist in the same molecule with
B-DNA.   This was shown by cloning oligo(dG-dC) into plasmids and then
isolating segments containing both the plasmid DNA and the oligo(dG-dC)
(29, 30).   It could be shown that raising the salt concentration
resulted in forming Z-DNA in the segment cloned into the plasmid DNA.
However, the plasmid itself largely remained in the form of B-DNA.

     Antibodies have been used to demonstrate the existence of Z-DNA in
vivo.   To carry this out, left-handed brominated poly(dG-dC) was used
as an antigen for injection into rabbits.   The brominated DNA polymer
was complexed to methylated bovine serum albumin, which is a positively
charged molecule.   It was found that this polymer was a strong immunogen and produced antibodies specific against Z-DNA (23).   These antibodies would not react with native or denatured DNA, brominated poly dG,
brominated poly dG.poly dC, E. coli DNA and a variety of other DNA or
RNA molecules. The antibody appears to have a specificity solely
directed toward Z-DNA.   It was shown that bromination was not essential
for producing the antibody, since unmodified poly(dG-dC) complexed to
methylated bovine serum albumin would also direct production of antibodies against Z-DNA although at a reduced titre. This is probably associated with the formation of local segments of Z-DNA on the protonated
surface of methylated bovine serum albumin.

     Antibodies against nuclear components have been found in certain
pathological   states.   In the autoimmune disease systemic lupus
erythematosus a variety of antibodies have been identified against
various nucleic acid components. We have detected anti Z-DNA antibodies in MRL mice, a strain of mice which is known to develop a

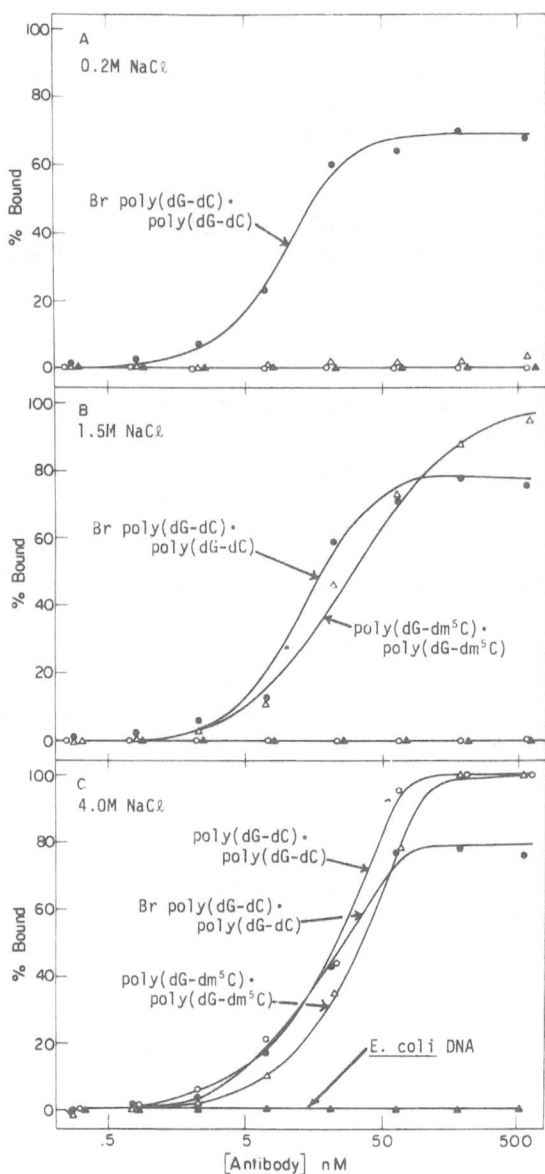

Fig. 6. Specificity of affinity-purified anti-Z-DNA antibody as measured by direct binding to ^{3}H-brominated
poly(dG-dC)·poly(dG-dC), ^{3}H-poly(dG-dm[5]dC)·poly(dG-m[5]dC),
[3]H-poly(dG-dC)·poly(dG-dC) and ^{3}H-labelled E. coli DNA in
0.2 M NaCl (A), 1.5 M NaCl (B), and 4.0 M NaCl (C).

lupus-like disease. Over 20 mice were examined and all of the mice
with the disease contained serological reactivity against Z-DNA, many
of them with very high titre (23). This provided some evidence suggesting the existence in the mouse nucleus of Z-DNA as immunogen.
Further work on humans with this disease also resulted in the detection
of antibodies specific for Z-DNA (31).

The specificity of the rabbit antibodies against Z-DNA could be
further demonstrated by the discovery that these antibodies would combine with Z-DNA even in high salt solutions. This suggested that the
antibodies were not solely recognizing the charged phosphate groups but
might also be directed toward the stacked purines and pyrimidines forming the outer convex wall of Z-DNA. The specificity is illustrated in
Figure 6, which shows direct binding of the antibody at various titres
to four different radioactive nucleic acids (28). At 0.2 M NaCl it can
be seen that the antibody precipitates the brominated polymer, while at
1.5 M NaCl it also precipitates only the polymer methylated on cytosine
C5. From other studies it was known that the brominated polymer is
fully in the Z conformation in 0.2 M NaCl (23) while in 1.5 M salt the
methylated polymer is also in the Z conformation (Behe and Felsenfeld,
1981). However, it is only in 4 M NaCl--the environment in which it is
known to be in the form of Z-DNA (22)--poly(dG-dC) itself is also bound
by the antibody. At that elevated salt concentration there is no binding to E. coli DNA. Figure 6 shows that the antibody combines with a
substrate only when the conformation of the substrate is in the form of
Z-DNA.

This reaction has been used as the basis for purifying the antibodies from rabbit serum. The serum is combined with poly(dG-dC) in 4 M
NaCl where the polymer adopts the Z-DNA conformation. The antibody is
then precipitated as an immune complex from which the purified antibodies can be isolated (28). These antibodies as well as antibodies from
the whole serum can be used to look for Z-DNA in biological systems.

Z-DNA in Drosophila Polytene Chromosomes

The polytene chromosomes of Drosophila melanogaster are found in
salivary glands of third instar larvae. These are interphase chromosomes, active in both transcription and replication. The alignment of
the thousand or more chromatids making up a polytene chromosome is so
precise that a characteristic band-interband pattern is created which
reflects the extent of coiling of the individual chromatid fibers. The
banding pattern of each chromosome is essentially constant from nucleus
to nucleus, so it must reflect basic features of chromatid structure.
This chromosome has been used to look for Z-DNA with the technique of
indirect immune fluorescence. In this technique the anti Z-DNA antibody is added to the chromosomes once the nucleus has been squashed and
the chromosomes fixed to a glass slide. Then a secondary antibody
(goat anti-rabbit-IgG) is added which has attached to it a fluorescent
dye. Illuminating the chromosomes in the exciting wavelength and observing it at the emitting wavelength of the fluorescent dye make it

possible to identify those parts of the chromosome to which the anti-Z-DNA antibody is bound. Figure 7 shows a photomicrograph of a polytene chromosome under phase microscopy and fluorescence microscopy. Under phase microscopy it can be seen that the chromosome is made up of a series of dark bands with the lighter interband regions between them. Under fluorescent light a number of fluorescent segments can be seen along the chromosome which vary somewhat in intensity. Close comparison of these photographs reveals that the fluorescence is mostly in the interband region of the chromosome rather than the band region (28). However, in puffs the band region fluoresces. The majority of the DNA is found in the band region, but nonetheless it is only the interband which fluoresces under the experimental conditions employed for this fixing and staining. The intensity of staining of various interbands seems to vary by as much as a factor of 10. This intensity variation is reproducible and is a function of the position on the chromosome. This conclusion is reinforced by the fact that in regions of homologous nonpairing, where the chromosome has split into two parts (asynapsis), identical staining patterns are found in the separated homologous regions.

There are about 5,000 band-interband systems in the Drosophila genome, and on the average they contain 30 kilobases of DNA (32). The amount of DNA in the interband has been variously estimated as 5% or 30% of the total DNA. Electron microscopic studies of the interband regions reveal that it is made of fibrillar structures some 60Å in width (32). This is to be contrasted with the densely granular organization of the band regions which undoubtedly contain large clumps of nucleosomes. The band-interband system appears to be the unit of transcription and possibly even replication. In the immunofluorescence experiments a number of attempts were made to vary the method of fixing and preparing the chromosomes for visualization (28). The method used for fixing the chromosomes involves the removal of large amounts of chromosomal proteins, including histones and other proteins (33). It cannot be stated that failure to stain the band region is a consequence of the fact that there is no Z-DNA in there. It is possible that even though a large fraction of the protein has been removed, nonetheless it may still block access to the band regions. Changing the method of fixing may change the staining patterns.

Under physiological conditions the Z-DNA conformation is somewhat less stable than the B conformation and therefore has to be stabilized. There are at least four ways of maintaining the Z conformation in biological systems. 1) Supercoiling. Z-DNA twists the double helix in a left-handed mode, opposite to the right-handed B-DNA conformation. Because of this one can interchange strongly negative supercoiled DNA for segments of Z-DNA. It has been demonstrated that negatively supercoiled plasmids form Z-DNA, as shown by antibody binding (34). 2) Binding to proteins which are specific for the Z conformation. These may involve electrostatic interactions with basic residues. 3) Binding to specific ions. For example, spermine or spermidine stabilize crystals of Z-DNA so that they form with a regularity which yields an atom-

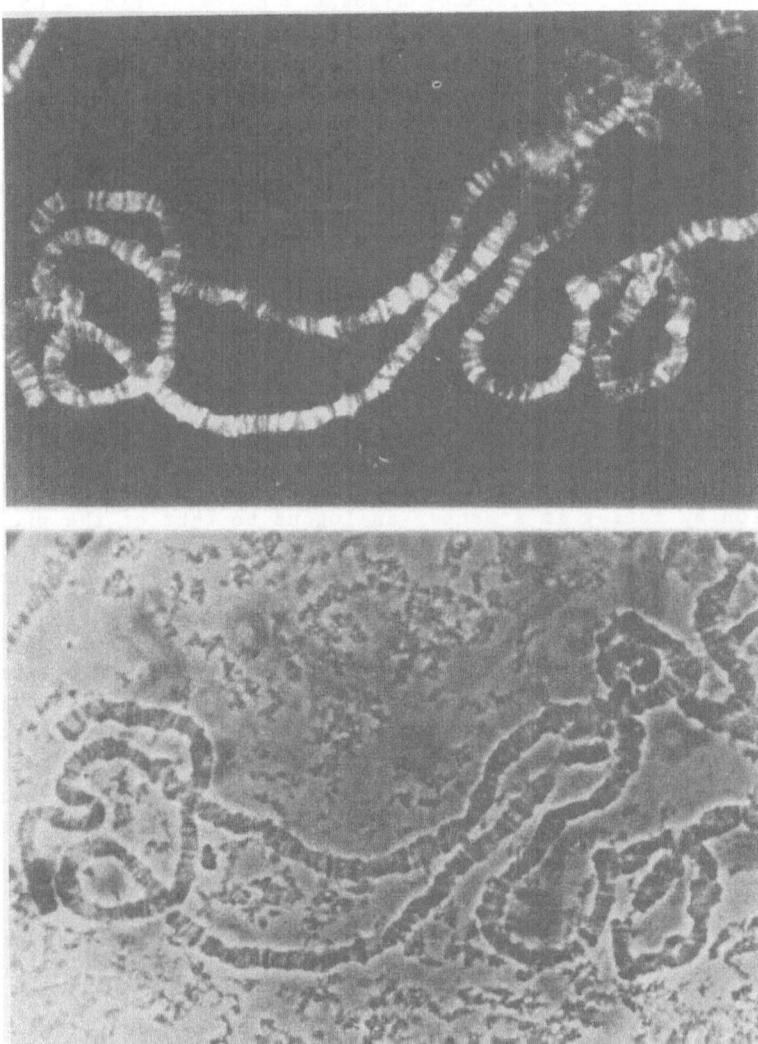

Fig. 7. Photomicrographs of <u>Drosophila</u> polytene chromosome squashes as visualized in (bottom) the phase contrast and (top) dark-field fluorescence. The phase microscopic picture shows the typical darker bands separated by lighter interband segments. The fluorescent photograph shows fluorescence largely in regions which can be identified as the interbands; however, in the band regions which are puffed, considerable fluorescence is found. Notice that the intensity of fluorescence varies in different interbands; this pattern is reproduced in different organisms.

ic resolution diffraction pattern. 4) Modification such as methylation
of cytosine in the 5 position as discussed above. This is not necessarily applicable to <u>Drosophila</u> DNA which appears not to be methylated
(35).

Does Z-DNA Have a Role in Regulating Gene Expression?

The effects of methylation on gene expression (24, 25) suggest
that Z-DNA may be involved, especially since methylation favors Z-DNA
stabilization (26, 27). Two models may be considered for the role of
Z-DNA in regulating gene expression. There may be segments of DNA on
the 5' side of transcription sites which have the potential for forming
Z-DNA as has been shown in experiments on <u>Xenopus</u> tRNA genes (36). It
has been shown that a small segment of alternating purines and pyrimidines just to the left of the site where transcription is initiated appears to block transcription in an <u>in vitro</u> system; further, removal of
the segment leads to transcription. If these sections of Z-DNA had
Z-DNA binding proteins attached to them, they could effectively block
transcription by acting in a manner analogous to that of the prokaryotic repressor which blocks the access of RNA polymerase to the transcription site.   This type of regulation we term <u>proximal regulation</u>,
since the controlling interaction is very close to the position where
transcription begins.   Regulation in this site would be influenced by
the extent to which the sequence of nucleotides is capable of forming a
Z-DNA segment. This could be one of the elements which determines the
strength of a promoter. A sequence with a greater tendency to form
Z-DNA might make a poor promoter, while the absence of such a sequence
might produce a strong promoter. The isolation of genes together with
their 5' flanking sequences which can be used as substrates for transcription systems either <u>in vivo</u> or <u>in vitro</u> should make it possible to
clarify this type of regulation.

One of the main features of Z-DNA, however, is that it twists the
double helix in the opposite sense of B-DNA. Negative supercoiling
stabilizes Z-DNA (34). However, if a large segment of Z-DNA then con-
verts back into B-DNA, it will impart considerable twist to the DNA and
further increase its negative supercoiling. A change in the supercoiling would be evident not only nearby but also at considerable distance
away from the DNA. Access of polymerase to a DNA strand is determined
in part by the level of supercoiling (37). Thus Z-DNA segments may be
used to determine the level of supercoiling over long stretches of DNA
and it may thus serve as a system for regulating the access of polymerase to a number of genes, including several which are further away.
We may call this <u>distal regulation</u>. It is possible that the Z-DNA in
the interband regions of the <u>Drosophila</u> polytene chromosome is an example of this type of regulation. The interband regions might function
as control elements which can form either Z- or B-DNA and thereby regulate the access of polymerases to the band-interband system. An extension of this idea suggests the possibility that this system could be
used as a method of regulating cell commitment. DNA is organized into
domains which are isolated in terms of their supercoiling. The domain

might not be committed or able to express its genes while the control region is in the form of Z-DNA. However, when it switches to B-DNA, the genes of the domain are then accessible to polymerases. They might then be considered as committed in the sense that transcription is allowed.

What we have outlined here is a speculative model of the role of Z-DNA involving the regulation of transcription of genes which are close to the Z site (proximal) or further away from the Z site (distal). The central event for proximal regulation is the presence of Z-DNA binding proteins which may block the access of polymerase to the site and therefore control transcription directly. For distal regulation, the key event is the control of the level of supercoiling, which in turn is responsible for controlling the access of polymerase to a number of genes.

We suggest that the conformation of the DNA may be used by nature in regulating gene expression. Of course, this represents a description of molecular machinery but it does not describe in detail those events which are in control of the Z-DNA forming ability of a particular segment of DNA. Such control must include four elements listed above--the extent of methylation of CpG sequences, the binding of Z-DNA proteins (which may have sequence-specific sites on them and thus be analogous to repressors), the degree of supercoiling, and the availability of specific ions, including especially the polyamines. It is clear that considerable further effort will be required until we understand the role of Z-DNA in biological systems. However, consideration of hypothetical models such as these are often useful in suggesting experiments.

Acknowledgments: This research was supported by grants from the National Institutes of Health, the National Aeronautics and Space Administration, and the American Cancer Society.

Bibliography

(1) Watson, J.D. and Crick, F.H.: Nature 171, 737 (1953).
(2) Arnott, S.: Progress in Biophysics and Molecular Biology, J.A.V.
    Butler and D. Noble, eds., Pergamon Press, New York, 265 (1970).
(3) Franklin, R.E., and Gosling, R.G.: Acta Crystallographica 6, 673
    and 8, 151 (1953).
(4) Hopkins, R.C.: Science 211, 289 (1981).
(5) Rodley, G.A., Scobie, R.S., Bates, R.H., and Lewitt, R.M.: Proc.
    Natl. Acad. Sci. USA. 73, 2959 (1976).
(6) Gupta, G., Bansal, M., and Sasisekharan, V.: Proc. Natl. Acad.
    Sci. USA 77, 6486 (1980).
(7) Voet, D. and Rich, A.: Progr. Nucl. Acid Res.   Mol.   Biol.
    10, 183 (1970).
(8) Rosenberg, J.M., Seeman, N.C., Kim, J.J.P., Suddath, F.L., Nicho-
    las, H.B. and Rich, A.: Nature 243, 150 (1973).
(9) Seeman, N.C., Rosenberg, J.M., Suddath, F.L.   Kim, J.J.P., and
    Rich, A.: J. Mol. Biol. 104, 109 (1976).
(10) Day, R.O., Seeman, N.C., Rosenberg, J.M., and Rich, A.:   Proc.
    Natl. Acad. Sci. USA 70, 849 (1973).
(11) Rosenberg, J.M., Seeman, N.C., Day, R.O., and Rich, A.: J. Mol.
    Biol. 104, 145 (1976).
(12) Wang, A.H.-J., Quigley, G.J., Kolpak, F.J., Crawford, J.L., van
    Boom, J.H., van der Marel, G., and Rich, A.: Nature 282, 680
    (1979).
(13) Drew, H., Takano, T., Tanaka, S., Itakura, K.   and Dickerson,
    R.E.: Nature 286, 567 (1980).
(14) Crawford, J.L., Kolpak, F.J., Wang, A.H.-J., Quigley, G.J., van
    Boom, J.H., van der Marel, G., and Rich, A.: Proc. Nat. Acad.
    Sci. USA 77, 4106 (1980).
(15) Wang, A.H.-J., Quigley, G.J., Kolpak, F.J., van der Marel, G.,
    van Boom, J.H., and Rich, A. Science 211, 171 (1981).
(16) Son, T.-D., Guschlbauer, W., and Gueron, M.:   J.   Am.   Chem.
    Soc. 94, 7903 (1972).
(17) Haschemeyer, A.E.V. and Rich, A.: J.   Mol.   Biol.   27, 369
    (1967).
(18) Arnott, S., Chandrasekarran, D.L., Birdsall, D.L., Leslie,
    A.G.W., and Ratliff, R.L.: Nature 283, 743 (1980).
(19) Fujii, S., Wang, A.H.-J., Quigley, G.J., Westerink, H., van Boom,
    J.H. and Rich, A: submitted for publication.
(20) Pohl, F.M. and Jovin, T.M. (1972). J. Mol. Biol. 67, 375
    (1972).
(21) Pohl, R.M., Ranade, A., and Stockburger, M.: Biochim. Biophys.
    Acta 335, 85 (1973).
(22) Thamann, T.J., Lord, R.C., Wang, A.H.-J., and Rich, A.:   Nucleic
    Acid Res. 9, 5443 (1981).
(23) Lafer, E.M., Møller, A., Nordheim, A., Stollar, B.D., and Rich,
    A.: Proc. Natl. Acad. Sci. USA 78, 3546 (1981).
(24) Razin, A. and Riggs, A.D. Science 210: 604 (1980).
(25) Ehrlich, M., and Wang, R.Y.-H.: Science 212, 1350 (1981).
(26) Behe, M., and Felsenfeld, G.: Proc. Natl. Acad. Sci. USA 77,

6468 (1981).

(27) Fujii, S., Wang, A.H.-J., van der Marel, G., van Boom, J.H. and Rich, A.: Nuc. Acids Res. (in press).

(28) Nordheim, A., Pardue, M.L., Lafer, E.M., Møller, A., Stoller, B.D., and Rich, A.: Nature 294, 417 (1981).

(29) Klysik, J., Studivant, S.M., Larson, J.E., Hart, P.A., and Wells, R.D.: Nature 290, 672 (1981).

(30) Peck, L.J., Nordheim, A., Rich, A. and Wang, J.C.: Proc. Nat. Acad. Sci. 79, 4560 (1982).

(31) Lafer, E.M., Valle, R.P.C., Møller, A., Nordheim, A., Schur, P., Rich, A. and Stollar, B.D.: J. Clin. Invest., in press.

(32) Beermann, W.: Developmental Studies on Giant Chromosomes, W. Beermann, ed., Springer-Verlag, New York, 1 (1972).

(33) Dick, C., and Johns, E.W.: Exp. Cell. Res. 51, 626 (1968).

(34) Nordheim, A., Lafer, E.M., Peck, L.J., Wang, J.C., Stollar, B.D., and Rich, A. Submitted for publication.

(35) Argyrakis, M.P., and Bessman, M.J.: Biochim. Biophys. Acta 72, 120 (1963).

(36) Hipskin, R.A. and Clarkson, S.G.: Cell, in press.

(37) Smith, G.R.: Cell 24, 599 (1981).

THEORETICAL STUDIES OF THE POLYMORPHISM AND MICROHETEROGENEITY OF DNA.

BERNARD PULLMAN, ALBERTE PULLMAN and RICHARD LAVERY
Institut de Biologie Physico-Chimique, Laboratoire de
Biochimie Théorique, associé au C.N.R.S., 13, rue Pierre
et Marie Curie, PARIS 75005, FRANCE.

Introduction

It becomes well recognized today that DNA may exist in a number of biologically significant allomorphic forms which differ among themselves in geometrical parameters and, what seems to be particularly important, in conformation. This polymorphism may be due to external factors (humidity, salt concentration etc...) or, as discovered recently, induced by variations in the sequence of the base pairs. This last possibility leads to the consideration of a possible regional heterogeneity (microheterogeneity) of native DNA as a function of the order of the base pairs along its helical axis. Such microheterogeneity could be of particular significance for the specificity of interaction or association of the nucleic acids with external reactants.

Besides the classical B conformation, particularly outstanding in view of their possible biological significance are the A and the recently discovered left-handed Z conformations of DNA.

The necessity of distinguishing the double template role of DNA, for DNA in replication and for RNA in transcription, suggests by itself the possibility of using for this sake two differents conformations, "the DNA replicases being compatible only with the B conformation and related transient forms, and the RNA synthetases only with A forms" (1). In fact, it is repeatedly supposed that the conformational transition of B-DNA into the RNA-like A form is required for normal transcription (1-3). It is known since some time that DNA-RNA hybrids adopt A forms (4, 5) (although a B-DNA like structure seems to be possible for the highly solvated state of hybrids possessing certains kinds of specific sequences (6)). Recently Selsing et al. (7) have prepared a polynucleotide complex $dG_n.rC_{11}dC_{16}$ which has covalently linked DNA-DNA duplex and RNA-DNA hybrid tracts, thus demonstrating unambiguously that two quite different conformations can exist as neighbours on a DNA helix. Selsing et al. (8) have also provided a stereochemically acceptable model for the junction of neighbouring A-DNA and B-DNA conformations in a nucleic acid duplex, in which the junction region comprises one base pair and two internucleotide linkages around this base pair and results in a "bend", a change in the direction of the helix axis of the molecule. The model implies that local segments of A-DNA could exist in an other-

C. Hélène (ed.), Structure, Dynamics, Interactions and Evolution of Biological Macromolecules, 23–44.

wise B-DNA helix, with two bend junction regions bordering each A-DNA
segment. Moreover their study indicates a correlation of the B $\longrightarrow$ A
transition with several features of the initiation of RNA transcription.
On the other hand, it is also known that DNA binding sites for regulatory
proteins are frequently conspicuously rich in AT base pairs (see e.g.
9). From that point of view it is interesting to recall that it has been
shown (10) that contrary to previous beliefs A-T rich and even A-T very
rich (95%) DNA's can adopt the A form and furthermore that the very
recent X-ray analysis of the crystal structure of the self complementary
octamer d(GGTATACC) indicated its existence in the A conformation (11).
This oligonucleotide contains the TATA fragment, a sequence ubiquitous
in promoter regions and involved in binding RNA polymerase as a prelude
to transcription. The A double helical conformation was also found in
the crystal of the d(^{I}CCGG) tetramer (12) and the comaprison of its
hydration scheme with that of a recently studied B-DNA self-complementary dodecamer d(CGCGAATTCGCG) (13) has thrown light on the possible
role of water structure in the B $\longrightarrow$ A transition (14).

      Concerning the left-handed Z-form of DNA, its discovery is too
recent, spectacular and produced too strong an impact on the whole field
of research on the polymorphism and microheterogeneity of nucleic acids
to need to be recalled here (for a review see 15). We would simply like
to underline some of the very recent developments related to this particularly fascinating structure.

      1) Although the discovery of the left handed Z-DNA refers ori-
ginally to the crystal structure of the self complementary d(CGCGCG)
hexamer, it has now been abundantly demonstrated by numerous techniques
such as infrared spectroscopy (16), laser Raman spectroscopy (17),
circular dichroïsm (18, 19), nuclear magnetic resonance (20-23) and
proton nuclear Overhauser effect (24) that poly (dG-dC) . poly (dG-dC)
in high salt solution is in the Z-DNA form. The possibility that the Z
form may occur with other alternating d(purine-pyrimidine)$_n$ sequences
seems substantiated by recent studies on the conformational transitions
of poly (dA-dC) . poly (dG-dT) induced by high salt or in ethanolic
solution (25, 26).

      2) The alternating polymer, poly (dG-m^{5}dC) . poly (dG-m^{5}dC)
undergoes the B $\longrightarrow$ Z transition at salt-concentrations much lower than
those required to induce this transition in the unmethylated polymer.
The substitution of a methyl group at C5 of cytosine allows in fact the
polymer to adopt the Z conformation under physiological conditions (27,
28). The same phenomenon occurs when poly (dG-dC) . poly (dG-dC) is
methylated at the N7 of guanine through the use of dimethylsulfate (29).
A very recent work by de Sande and Jovin (30) shows that the interconversion between the right and left helical form of poly (dG-dC) . poly
(dG-dC) occurs at low concentrations of Mg Cl$_2$ and EtOH acting together
in a highly synergistic manner (e.g. 0.4 mM Mg Cl$_2$ at 20% EtOH). The Z-
DNA produced (designated as Z*) has certain unique properties.

      The Z form of poly (dG-dC) . poly (dG-dC) is also stabilized
by the interaction of the guanine moieties with the carcinogen N-acetoxy-
N-acetyl-2-aminofluorene (31-34), mitomycin (35, 36) and by complexation
with dien-Pt (37). On the other hand, the polymer does not adopt the Z
conformation, even in high salt concentration, when modified by cis-Pt

and trans-Pt complexes (37), upon reaction with aflatoxin B_1 (38), or upon association with highly basic histones (39). Studies with $d(CG)_n$ oligomers (40) indicate that a minimum values of n = 3 is necessary to obtain a left-handed Z helix in high salt solution. When $d(CG)_3$ is associated with three alternating AT pairs $(d(AT)_3(GC)_3)$ the Z helix is unstable and the B conformation is obtained even at high salt concentration.

3) The important question as to the feasibility of the junction between the B- and Z-type conformations seems to have received a positive answer (41-46). Following some authors (43, 44) this junction should span a transition region of about 11 bp (base pairs) which are in an intermediate state and it is the Z_{II} conformation which is most likely to be found at this interface (41). Following others (47), it is possible that as few as 1-2 base pairs would be adequate to negotiate the B-Z junction, the interconversion being able to occur without breaking the hydrogen bonds between the paired bases, by concerted changes in the sugar-phosphate backbone bonds, coupled to the anti-syn interconversion of the glycosidic bonds. A theoretical analysis has also been carried out on the possibility of the B $\longrightarrow$ Z transition in torsionnally stressed DNA (42).

4) Finally concerning the biological significance of Z-DNA, data are steadily accumulating, indicating that it actually may have an important role. Thus :

  a) Stable left-handed DNA helices were successfully used to
     induce in rabbits antibodies that are specific for Z-DNA
     (48, 49).
  b) The antibodies to left-handed Z-DNA were shown to bind to
     interband regions of Drosophila polytene chromosomes (50).
  c) Left-handed DNA has been shown to be present in restriction
     fragments and a recombinant plasmid (43, 44).
  d) The Z^*-DNA variety of the Sande and Jovin (30) can function
     as template for the E. Coli polymerase and is involved in
     binding of various drugs.
  e) On the other hand while Z-DNA binds histones, the presence
     of this form prevents nucleosome formation (51). This
     result suggests that Z-DNA if present will serve to disrupt
     the normal chromatin structure.

These are the first demonstrations of the "in vivo" presence of Z-type zones in DNA. Following Rich they could be involved essentially in the regulation of gene expression (52).

Less information is available about the biological role of the remaining allomorphic forms of DNA. Following ref. 3 (see also 53, 54) there are two main conformations in chromatin, distributed into segments according to the regularities of chromatin structure. The DNA constituent of nucleosome particles seems to exist in a conformation close to the C-DNA and the linker DNA in the B-form. As to the alternating B form it was proposed by Klug and colleagues (55) in order to account for the specificity of action of the enzyme DNAase I on double stranded DNA polymers.

To this rapid review of the experimental evidence on DNA polymorphism and heterogeneity it may be useful to add the recent more

direct evidence of a non-uniform backbone conformation of DNA obtained
by ^{31}P NMR spectroscopy (56), of the sequence dependence of the helical
repeat of DNA in solution (57-59) and of the possibility of existence of
non standard forms, as shown recently to occur in double stranded ribonucleic acids (60).

In view of assessing the possible role of the different known
forms of DNA in terms of their specific reactivities we have undertaken
recently in our laboratory a detailed examination of their structure
with the double objective of evaluating quantitatively : 1) one of the
main variable geometrical features associated with the conformational
changes of DNA, namely the accessibility to reactive sites and 2) two of
the main electronic features associated with these changes namely their
molecular electrostatic potential and their electrostatic field, these
factors being expected to be among the most decisive ones in invluencing
the reactional possibilities of the nucleic acids. We have explored from
this point of view A-DNA (61), B-DNA (62 and references given therein),
"alternating" B-DNA (63), C-DNA (64), D-DNA (63) and Z-DNA (65, 66) and
present here a rapid, synthetic view of some of the results obtained
(see also 67).

Method

The techniques employed for calculating the accessibility (68),
the molecular electrostatic potential (62, 69) and the electrostatic
field (70) of nucleic acids have been described in detail in our previous
publications and we shall therefore not restate them here.
We would like, however, to :
1) indicate that the accessibilities (the accessible areas on
the van der Waals surfaces of target atoms in the nucleic acids) have
been evaluated with respect to a test sphere of radius 1.2 Å. It was
demonstrated that these accessibilities may be considered as corresponding to an attack by a water molecule through one of its hydrogen atoms
and as representing the upper limit of the atomic accessibilities, within the nucleic acids, towards molecular species (68).
2) Recall that five representations are available for describing the molecular potentials. These are :
a) Point potentials.
b) Plane potentials (isopotential maps, whose minima represent
the main site potentials at reactive centers, in particular
at the purine and pyrimidine bases)
c) Radial potentials (computed generally in planes perpendicular to the helical axis)
d) Line potentials (an extension of the radial potentials
across the whole width of DNA)
e) Surface envelope potentials (potentials on envelopes formed
by the intersection of spheres centered on each atom of DNA
with radii equal to the van der Waals radius of the atoms
concerned, multiplied, if desired, by a factor F ; for
practical reasons these are generally presented in the
form of their projection on a two-dimensional "window" (68)).
3) Indicate that, so far, two representations have been used
for describing the molecular field (70) :

a) In selected molecular planes.
b) On the molecular surface envelopes.
The model helices of the different DNAs studied represent each
one complete turn of the appropriate conformation. The input data for
the computations have been taken from the following original sources : A
and B-DNA (71), alternating B-DNA (72), C-DNA (73), D-DNA (74), Z_I and
Z_{II}-DNAs (75).

Results and discussion
        A) An essential distinctive structural feature of the allomor-
            phic forms.
        One of the essential distinctive features which differentiates
the various allomorphs of DNA and which has a decisive influence on
the distribution of their molecular electrostatic potential and field
and on the accessibility to their reactive sites is the position of the
helical axis with respect to the complementary base pairs, adenine-
thymine and guanine-cytosine (67). The situation concerning this position
is summarized schematically in figure 1 from which it is evident that
the DNAs may be divided into two groups : 1) one involving the B and C-
DNAs (and also the alternating B-DNA), in which the helical axis passes
through the pairs of hydrogen-bonded complementary bases and 2) the
other, involving A, D and Z-DNAs, in which the helical axis is situated
outside these pairs. In this last situation we may distinguish two possibilities, one, characteristic of A-DNA, in which the helical axis is
located in what is the conventional language is called the major groove
and the other, characteristic of D and Z-DNAs, in which this axis is
located in the minor groove.

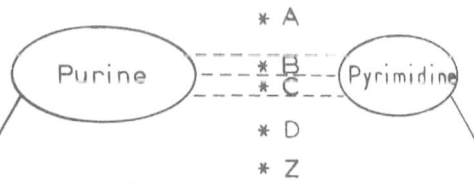

Major groove

* A
Purine ---- * B --- Pyrimidine
        * C
    * D
    * Z

Minor groove

Figure 1. The position of the helical axis with respect to the
base pairs in different conformers of DNA.

This displacement of the helical axis with respect to the base
pairs has a major consequence on the shape of the grooves of the nucleic

acids. The attention centered until recently essentially on B-DNA, considered and probably justly so as biologically the most significant form
of DNA, has accustomed us to believe that the nucleic acid double helix
is associated with two grooves which although referred to as "major" and
"minor" are nevertheless of rather comparable width and depth. In fact,
this particular situation is the consequence of the approximately central positioning of the helical axis with respect to the base-pairs in
this form of DNA. The displacement of the axis in the other forms has a
drastic influence on the nature of the grooves. Thus, the displacement
of the axis towards the major groove (or of the base pairs towards the
minor groove), as in A-DNA, renders the minor groove very shallow and
very wide, while the major groove becomes very deep but narrow. On the
contrary, the displacement of the helical axis into the minor groove (or
of the base pairs into the major groove) renders the major groove shallow and wide and the minor groove deep (and narrow in D-DNA, but narrow
or wide in Z-DNA following the type of phosphate considered). The situation is summarized schematically in fig. 2.

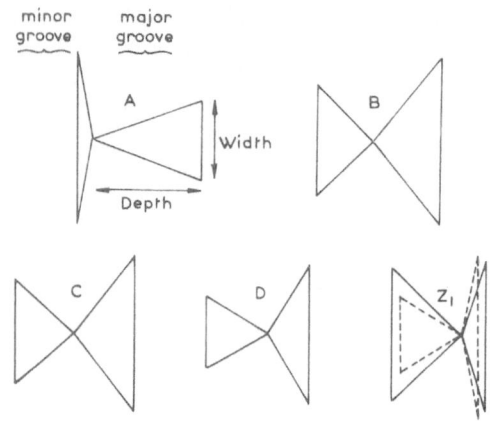

Figure 2. A schematic representation of the widths and depths
of the grooves in different conformers of DNA. In Z-DNA the
full lines refer to the GpC phosphates and the dotted lines to
the CpG phosphates.

B) Effect on the electrostatic molecular potential of the
grooves.
     One of the major consequences of the situation described above
concerns the distribution of the electrostatic molecular potential in
the grooves of the allomorphic forms. Although this distribution may and
has been obtained by a direct computation of the model helical turns of

the different DNA's, it may be advantageous from the analytical point of
view to study it stepwise and to consider thus first the electrostatic
molecular potential produced solely by the phosphate backbone of the
different allomorphs of DNA. This is justified because this distribution
represents numerically the principal component of the global potential.
Fig. 3 represents the potential due to the phosphate backbone alone
produced along a line (line potential) perpendicular to the helical axis
and passing through the center of it in the models studied (one helical
turn of the allomorphs) and bisecting the major (M) and minor (m) groove
of each allomorph (67). It is seen that for the related conformations of
B-, alternating B- and C-DNAs, in which the base pairs are located more
or less symmetrically about the helical axis, the mM line potential
exhibits a shallow curve with only a slight minimum on the side of the
minor groove. A very different situation prevails for A-DNA. In this
conformer, unique in having the base pairs largely displaced into the
minor groove, the mM line potential shows a strong asymmetry, with the
minimum inside the major groove, close to the backbone. The Z conformers,
exhibit an inverse effect. The displacement of their base pairs into the
major groove produces again a noticeable asymmetry of the phosphate
potential, but this time to the advantage of the minor groove.

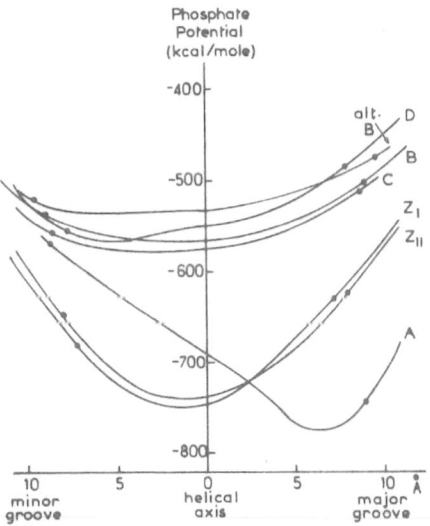

Figure 3. The potential due to the phosphate backbone along
the mM line in different conformers of DNA. The dots represent
the radii of the helices.

Table 1. Electrostatic molecular potential (kcal/mole)

BASE SITE		B-DNA (x)	B-DNA (xx)	A-DNA (xxx)	Z_I-DNA (xxxx)	Z_{II}-DNA (xxxx)	D-DNA (xxxx)	Alternating-B DNA (xxxx)	C-DNA (xxx)
G	N7	-683	-667	-787	-601	-646	-732		-696
	O6	-654	-645	-761	-600	-642	-708		-666
	C8	-630	-605	-699	-546	-563	-684		-632
	N3	-670	-669	-666	-658	-662	-768		-702
	N2	-623	-625	-623	-643	-627	-732		-651
A	N7	-650	-650	-748			-715	-621	-664
	C8	-610	-601	-670			-675	-579	-610
	N3	-669	-676	-663			-766	-653	-699
	N6	-600	-602	-682			-668	-586	-621
C	N4	-602	-616	-692	-550	-578	-669		-617
	O2	-645	-652	-638	-695	-607	-752		-663
T	O2	-663	-665	-655			-763	-642	-684
	O4	-612	-617	-725			-675	-607	-627

(x)   From complementary homopolymers poly(dG).poly(dC) and poly(dA).poly(dT), represented in figures by B.

(xx)   From alternating copolymers poly(dG-dC).poly(dG-dC) and poly(dA-dT).poly(dA-dT), represented in figures by B*.

(xxx) From complementary homopolymers.

(xxxx) From alternating copolymers.

Table 2. Accessibility (Å^2) (a).

BASE SITE	(x) B-DNA	(xx) B-DNA	(xxx) A-DNA	(xxxx) Z$_I$-DNA	(xxxx) Z$_{II}$-DNA	(xxxx) D-DNA	(xxxx) Alterna-ting-B DNA	C-DNA
G N7	4.1	3.	3.4	5.6	4.0	3.2		3.8
O6	2.7	2.5	2.3	3.4	2.6	3.4		2.5
C8	1.0	0.5	0.8	2.0	1.4	0.3		0.9
N3	0.1	0.2	0.9	0.	0.	0.		0.1
N2	0.	0.	0.3	0.3	0.6	0.		0.
A N7	2.6	2.5	2.7			2.6	2.6	2.5
C8	1.0	0.1	0.7			0.3	0.1	0.8
N3	0.7	0.9	2.1			0.3	1.1	1.0
N6	0.	0.1	0.1			0.1	0.1	0.1
C N4	0.2	0.4	0.1	0.	0.	0.2		0.2
O2	0.2	0.2	1.6	1.1	1.9	0.		0.1
T O2	0.9	1.1	2.8			0.3	1.1	1.0
O4	2.2	2.3	1.3			1.9	1.9	1.6

(a) The symbols (x) to (xxxx) have the same meaning as in table 1.

        This variable situation is bound, of course, to have a strong
effect on the global potential of the allomorphs. In order to be able to
give a more precise account of this overall situation, we need to complete the above description by including the contribution to the global
potential of the component generated by the base pairs. The essential
element to consider is the difference in the potential minima between
the major and minor groove side of the individual isolated G-C and A-T
base pairs. This difference (major groove minimum -minor groove minimum)
is found to be positive (8 kcal/mole) for the AT base pair and negative
(- 26 kcal/mole) for the GC base pair (67) (76). The asymmetry in the
distribution of the potential on the two sides of the base pairs is thus
small, but not negligeable. Whether it will have an effect on the overall
distribution of the potential in the grooves will depend on whether it
will be able to perturb the asymmetry due to the component of the potential produced by the phosphates. A detailed study (67) shows that in B-,
alternating B and C-DNAs, in which the contribution of this last component in the region of the base pairs presents only a very small variation,
the effect of the base-pairs is sufficient to produce the predominance,
from the point of view of the potential depth, of the minor groove in A-
T sequences and of the major groove in GC sequences. On the contrary,
this effect is unsufficient to influence the distribution of the potential in A- D- and Z-DNAs in which the overall groove potential is thus
independent of the base sequence and contains the deepest minimum in the
major groove in A-DNA and in the minor groove in Z-DNA. Explicit computations on the appropriate models of DNA confirm this situation which is
bound to have significant consequences for the interaction of the different forms of DNA, whether separated or interspersed within a continuous chain, with external reactants, in particular with non-intercala-
tive, medium or large ligands. We shall indicate in section D an example
of such a situation.
        Note that in all cases the deepest potentials are created in
the grooves of the DNA's. This is a general, major feature of the distribution of the potentials.
        C) The situation at the base active sites : potentials versus
            accessibilities.
        This is one of the major problems connected with the biochemi-
cal reactivity and biological functionning of DNA. Attention is centered
in particular on the nucleophilic atoms of the bases for which the
significance of the potential and its minima is the most straightforward.
Tables 1 and 2 present the results of explicit calculations of these two
fundamental quantities, potential and accessibility, for all such major
sites in the principal forms of DNA (67). A more convenient representation of the results (67), more suitable for direct utilisation, is probably the one given in figure 4 for the illustrative case of the C8 and
N2 atoms of guanine. It has the advantgae of containing simultaneously
the information relevant to both the potential and the accessibility of
these sites in the different types of DNA and it offers thus their instantaneous classification with respect to the two investigated properties.
The predictive information contained in figure 4 may be used for the
comparison of the different types of DNA whether in separate forms or
interspersed within, say, a predominantly B-DNA conformation. We shall

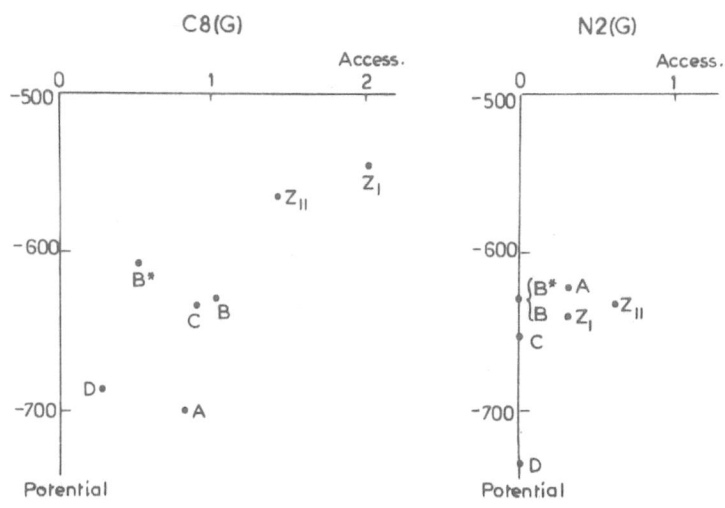

Figure 4. Accessibility (Å²) versus potential (kcal/mole) for the C8 and N2 atoms of guanine in the different allomorphic forms of DNA.

indicate in section D an example of the utility of such a quantitative, comparative and simultaneous study of the potentials and accessibilities of the nucleophilic sites of the bases.

D) <u>Field, by itself and versus potential.</u>

The computations of the electrostatic field associated with the nucleic acids are very recent. Results are available at present for B-DNA (70) as well as for A- and Z-DNA's (77). Because of the limits imposed on this paper we shall not dwell on this quantity here. We would just like to indicate for one example, which is typical of all the forms of DNA studied, one of the principal characteristics of the distribution of the field around these polymers and, at the same time, one of its essential differences with respect to the distribution of the potential.

The main result is illustrated in fig. 5 which presents in its central unit a diagrammatic representation of a helical turn of the B-DNA form of poly (dG) . poly (dC) (one recognizes the major groove in the lower part and the minor groove in the upper part of this diagram), in its left-hand side unit the distribution of the potential on the molecular surface envelope of this helical turn (following the technique presented in 62) and in its right hand side unit the corresponding distribution of the field. As we wish to restrict ourselves here only to general qualitative aspect of the results, we shall not present details of the values corresponding to the different degrees of shading in fig. 5 (these details may be found in ref. 70) but indicate simply that the potentials and the fields are the greater the deeper the shading.

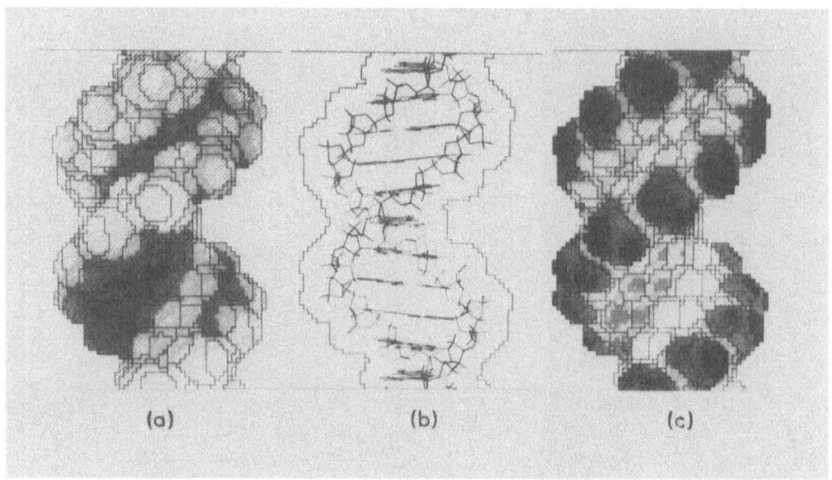

Figure 5. Poly (dG-dC) . poly (dG-dC) in the B-DNA form.
a) electrostatic potential on the surface envelope, b) a diagrammatic representation, c) electrostatic field on the surface envelope.

        The fundamental result to which we refered above is then imme-
diately visible : while the deepest potentials are observed for the
grooves of the DNA this particular case in the major groove, the strongest fields are located along the backbone around the phosphate groups.
This last result is typical of all the allomorphic forms studied and
represents thus a major feature of the distribution of the field,
clearly distinguishing it from the distribution of the potential.
        This situation has, of course, important consequence for the
study of the biochemical reactivity of the nucleic acid toward external
reactants (see section E).
        E) Exemples of application.
        The significance of accessibility to reactive sites of the
macromolecule for the biochemical behaviour of DNA seems, of course,
obvious. No reactivity can occur when a site is occluded and inaccessible.
A more delicate situation arises, however, when a number of competing
sites are partially and non-uniformly accessible. It certainly would be
a dangerous oversimplification to expect that the most accessible site
will always also be the most reactive one. The reason simply being that
reactions depend in general on more factors than accessibility alone. It
is surprising to observe how frequently people dealing with the reactivity of biomolecules and in particular biopolymers have tendancy to
forget this elementary truth.
        The significance of the molecular electrostatic potential and
of the electrostatic field for reactions, interactions or associations
involving B-DNA and its constituents has been abundantly demonstrated
(62) (70) (78). Because of lack of experimental data on their reactivity
a similar demonstration is much more difficult to provide for the remai-

ning forms of DNA and, a fortiori, for a comparative study of these
forms. A few examples may nevertheless be given.

Generally speaking it is obvious from its very definition that
the scalar molecular electrostatic potential of the polyanionic nucleic
acids is particularly appropriate for describing what is "felt" by an
electrophilic, charged species (a cation) approaching the macromolecule.
The interaction of these species constitutes in fact a system particularly
well suited to study with the aid of electrostatic potentials, because
the electrostatic term is likely to be dominant in their interaction
energy, at least at long or intermediate interaction distances. On the
other hand, the vectorial electrostatic field should be particularly
well adapted to describing what is "felt" by neutral dipolar molecules
approaching the nucleic acid. By the term "dipolar" we imply those
molecules for which a single center multipole expansion of the electron
density would be dominated by the dipole term. In such cases, the electrostatic interaction energy of these species may be approximated by the
scalar product of their dipole moments with the local field of the
macromolecule. A particularly important species of this sort is water.
Thus it is expected that a detailed knowledge of the field of macromolecules should provide a tool for investigating, in particular, their
zones of preferential hydration. These preferred domains of applicability
of these indices should be kept in mind when utilizing them for practical problems.

Because of limitation of space and because the problem of
hydration of the nucleic acid is such a big world in itself we shall
consider here only problems related to the electrostatic molecular potential (a discussion on the utilisation of the field for the study of
hydration may be found in ref. 70, 77 and 78).

We are going to consider two such problems :

1) As an example of the significance of the electrostatic
molecular potentials in the grooves of the surface envelopes of the
nucleic acids we may quote our recent demonstration of the role played
by the location of the surface potential minimum for the interaction of
different natural and synthetic nucleic acids with the oligopeptide
antibiotics netropsin (I) and distamycin A (II) (79). These antibiotics
bind to B-DNA, preferentially in the minor groove of AT rich sequences
and to poly(dA).poly(dT). They do not bind to A-DNA or to poly(dG).poly(dC)
(80-82). Their binding is thought to involve interactions between their
charged terminal groups and the phosphates of DNA and also hydrogen
bonds between their amide groups and the nucleophilic base atoms of DNA,
O2(T) and N3(A). This binding scheme is insufficient, however, to explain
by itself the very variable affinities of the antibiotics for nucleic
acids with different base-pair sequences or conformations. Moreover,
the elimination of the charged terminal guanidino group does not reduce
the binding of this netropsin derivative to poly(dA).poly(dT) (83) and,
further, the capacity to form hydrogen bonds with the O2(T) or N3(A)
atoms of the bases does not seem either to be indispensable for binding :
thus, the bis quaternary ammonium heterocycle III with no hydrogen bonding possibilities binds to the minor groove of the DNA helix as well as
the compounds listed in fig. 6, which altogether seem to behave similarly
to netropsin and distamycin A (84-87). A similar situation could possi-

bly occur with stilbamidine which is known to bind to DNA (85) (88)
although it lacks the OH group of the hydroxystilbamidine of fig. 6.

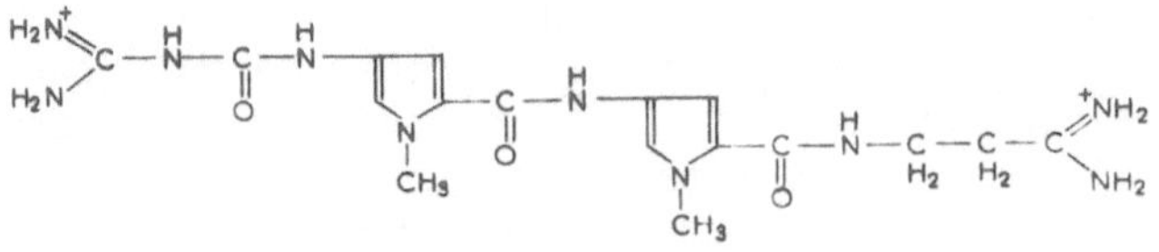

I. Netropsin

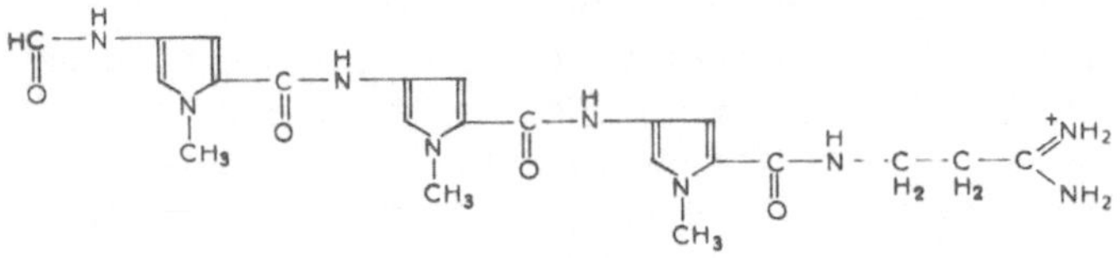

II. Distamycin A

III. A bisquaternary ammonium heterocycle with
no hydrogen bonding possibilities.

On the other hand, we have been able to show (79) that the
binding of the antibiotics correlates with the presence of the most
negative surface potential minimum in the minor groove of the nucleic
acids. Such is the case for poly(dA).poly(dT) with which the antibiotics
bind most strongly. It is also the case for poly (dA-dT) . poly (dA-dT),
which exists most probably in the alternating-B conformation. The binding with this sequence is somewhat weaker than with poly(dA).poly(dT)
and it is interesting to note that the minimum of potential in the minor

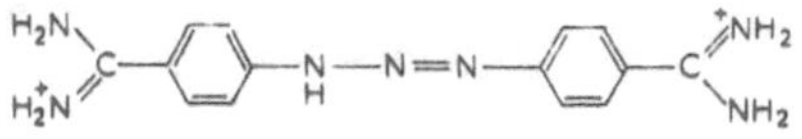

Berenil (a bisamidine)

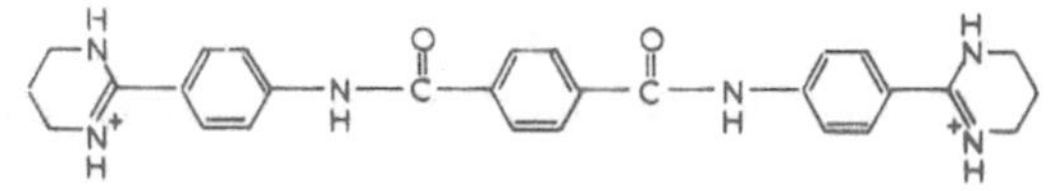

Terephtalanilide

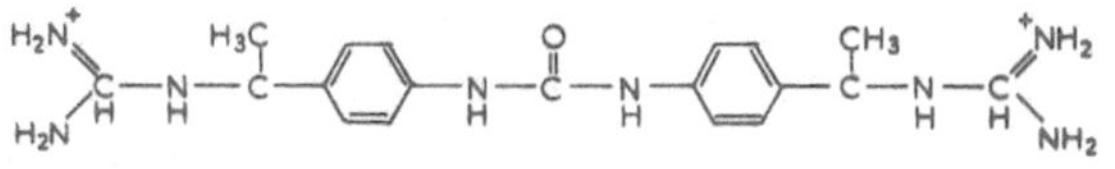

A bis (guanylhydrazone) : DDUG

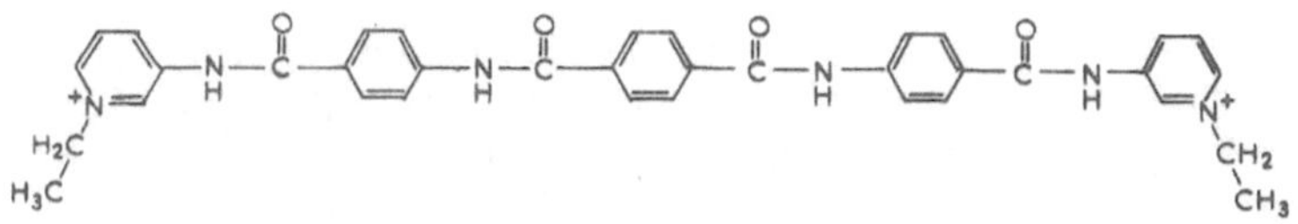

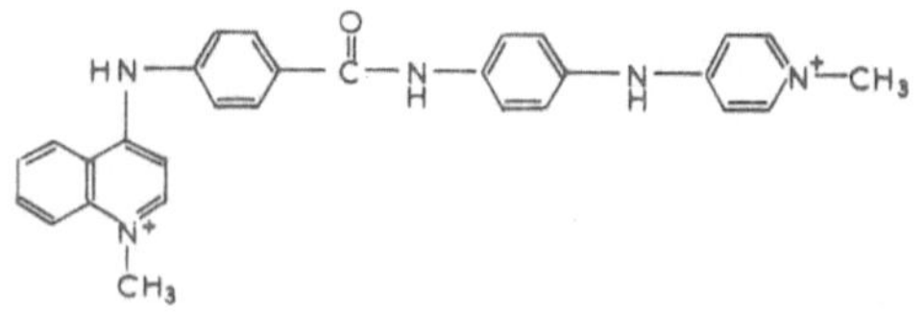

Bisquaternary ammonium heterocycles

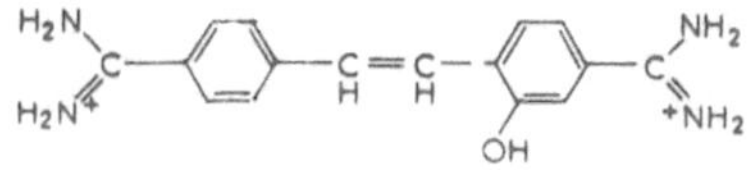

Hydroxystilbamidine

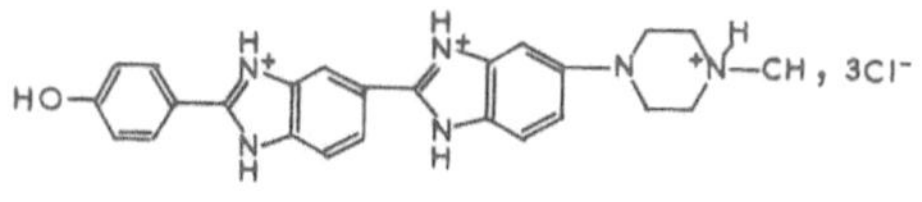

Hoechst 33258

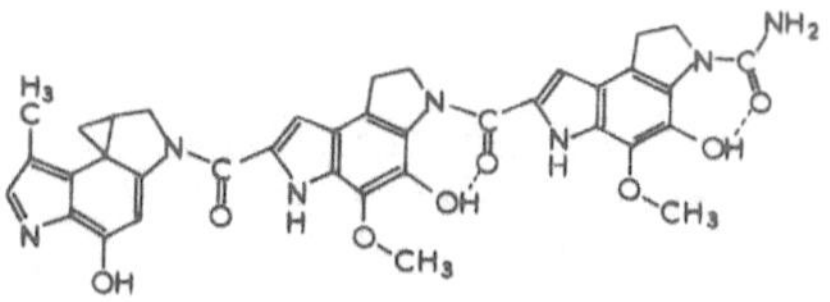

CC 1065

Figure 6. Drugs interacting preferentially with A-T sequences of DNA.

groove is also somewhat weaker in this alternating copolymer than in poly(dA).poly(dT). Inversely, the minor groove potentials are disfavoured in poly(dG.dC) and in A-DNA of any sequence and the antibiotics do not bind to these species.

A complementary argument in favour of the role of the molecular electrostatic potential in the interactions involving netropsin with nucleic acids may be found in the results of studies on the interaction of this antibiotic with the series $(dA)_n$. $(dT)_n$ (89) in which netropsin induces the formation of a base-paired duplex. While netropsin spans, upon interaction, about four adjacent base pairs, it is nevertheless observed that the efficiency of the interaction of this antibiotic with various oligomers is considerably lower than with the polymer and increases from the hexamer to the decamer. This evolution of the efficiency of interaction can be accounted for by the parallel increase of the molecular electrostatic potential in the groove involved, due to the increase of the cumulative effect produced by the increasing number of phosphate potentials. (For numerous similar examples of evolution of reactivity with increasing size of the nucleotidic substrate see (62, 90, 91).

2) As an example of the utility of the diagrams of the type of those of fig. 4 referring to the reactive properties of nucleophilic base sites we may quote a problem raised recently in conjunction with the discovery of Z-DNA. The analysis of the crystallographic data and the construction of the corresponding models have suggested to their authors (15, 41) that a number of reactive sites on the bases should be more accessible in Z-DNA than in its B-DNA counterpart. The observation concerns in particular atoms such as N7(G), O6(G) and C8(G) known to be the receptor sites for covalent bond formation with a series of carcinogenic compounds (see e.g. 90, 91). This situation led these authors to hypothesize that Z fragments interspersed within B-DNA could form particularly sensitive targets for the action of such carcinogens. They paid particular attention to C8(G) known to be the preferred site of action of the carcinogenic N-2-acetylaminofluorene (IV).

IV. N-2-acetylaminofluorene

Our calculations confirm and quantitize the increased accessibility of a number of important acceptor sites of Z-DNA, in particular of C8(G) and N2(G) with respect to the same site in the related B-DNA (see fig. 4). The comparison of the molecular electrostatic potentials

of the same sites in the same species leads, however, to the conclusion
that this potential decreases for the C8(G) receptor site in Z-DNA with
respect to B-DNA. We are therefore faced, as concerns this position,
with the dilemma of which of the two factors, accessibility or molecular
electrostatic potential, will dominate for its behaviour toward an
attacking electrophilic species.

A recent publication (92) demonstrates that the reactivity of
poly (dG-dC) . poly (dG-dC) for binding the carcinogenic N-2-acetylami-
nofluorene at the C8 position of guanine residues is substantially
smaller in conditions in which the polymer exists in the Z-form then
when it exists in the B-form. This result may be considered as an indication of the preponderance in this case of the effect of the decrease
of the electrostatic potential at C8 of guanines in Z-DNA with respect
to B-DNA, which thus seems to dominate over the greater accessibility of
that position in Z-DNA.

V. Mithramycin

On the other hand, it was shown very recently (30) that mithramycin (V), which most probably binds to the NH_2 group of poly (dG-dC) .
poly (dG-dC) (93), is relatively non discriminatory with respect to the
helix sense of this polymer. Now, unlike C8(G), the electrostatic potential of N2(G) appears to be of the same order of magnitude in the Z- and
B-forms of poly (dG-dC) . poly (dG-dC). The comparable affinity of this
antibiotic for the two forms is thus understandable and, in fact, such a
situation was predicted by us in ref. (66). Because of the significance
of the NH_2 group of guanine for the interaction of DNA with a number of
other carcinogens, in particular the active metabolic derivatives of
polycyclic aromatic hydrocarbons (for reviews see 90, 91), N-2-acetyla-
minofluorene (a minor reaction (94) but leading to an adduct excised at
a much slower rate than that occuring at C8 (95)), N-methyl-4-aminoazo-
benzene (96), 1'-hydroxyestragole (97), 1'-hydroxysafrole (98), dehy-
droretronecine (99) as well as for its interaction with antitumor antibiotics such as anthramycin and related derivatives of the pyrrole (1,
4) benzodiazepine series (100, 101) or saframycins A and C (102), this
is a particularly interesting observation.

F. A concluding comment.

It is evident from the above examples that the present theoretical analysis of such essential characteristics of the polymorphic

forms of DNA, as their molecular electrostatic potentials and the accessibilities to reactive sites, offers a tool for the understanding of the
differences in the biochemical reactivity of these forms and of the variations of this reactivity as a result of the microheterogeneity of
native DNA. A similar contribution is provided by the study of the electrostatic field in relation to hydration of DNA (70, 77, 78).

It must be underlined in this respect that the above theoretical analysis refers to the intrinsic properties of the allomorphic
forms, that is, to these properties as computed for the free, isolated
species. Now, of course, the real systems are present in what may be
referred to as screened forms, the major environmental screening factors
being water and metal cations, ubiquitously bound to the nucleic acids.
The problem of the relationship between these two types of properties,
the intrinsic and the screened, is of fundamental interest. The experi-
mentalists have, for practical reasons, a well understandable tendancy
to dismiss the former in favour of the latter. The theoreticians attribute the same importance to both, with a tendancy, also for practical
reasons, to concentrate primarily on the intrinsic properties. Intellec-
tually the understanding of such properties seems in fact a prerequisite
for the understanding of the effect of environmental factors which, by
essence, are variable. While the intrinsic properties are unique, the
properties of the screened forms are infinite in number, depending on
the precise associated environmental factors.

The important question concerns therefore the persistence of
the intrinsic properties of the nucleic acids and the extent of their
possible modifications by the environmental factors. There is, of course,
no general answer to this question as such an answer depends in each
case on the precise nature of these latter factors. We have explored this
problem in relation to nucleic acids for what appeared to us as one of
the most prominent screening effect in relation to the electrostatic
properties investigated, namely the screening due to the coordination
of metal cations to the phosphate groups. Limitations of space, again,
do not permit us to deal with this problem here and we direct therefore
the readers to the appropriate literature (65, 66, 103-106). We would
just like to state that as far as the properties considered in this paper
are concerned a general result seems to be that the impact of the environmental effect depends on the "symmetry" of the allomorphs involved. Thus
e.g. while cation screening will always strongly diminish the electrostatic potential of the nucleic acids (but not necessarily the associated
electrostatic field), the qualitative image of its overall distribution
will remain practically unchanged in B- or C-DNAs, while more drastic
transformations are produced in A- or Z-DNAs. In each case, however, the
knowledge of the intrinsic properties has an illuminative effect on the
resultant transformation. The detailed interpretation of experimental
results has to be adapted in each case to this complex situation.

Acknowledgment. This work was supported by the contract N°14 of the Institut National de la Santé et de la Recherche Médicale of France with
its ATP 77-79-109 on Chemical Carcinogenesis.

BIBLIOGRAPHY

(1) Arnott, S., Fuller, W., Hodgson, A. and Prutton, I. : Nature 220, 561 (1968).
(2) Scherer, G.E.F., Walkinshaw, M.D. and Arnott, S. : Nucl. Acid. Res. 5, 3759 (1978).
(3) Luchnik, A.N. : Mol. Biol. Rep. 6, 3 (1980).
(4) Milman, G., Langridge, R. and Chamberlain, M.J. : Proc. Natl. Acad. Sci. USA 57, 1804 (1966).
(5) Arnott, S. : Progr. Biophys. Mol. Biol. 21, 267 (1970).
(6) Zimmerman, S.B. and Pheiffer, B.H. : Proc. Natl. Acad. Sci. USA 78, 78 (1981).
(7) Selsing, E., Wells, R.D., Early, T.A. and Kearns, D.R. : Nature 275, 249 (1978).
(8) Selsing, E., Wells, D.R., Alden, Ch. J. and Arnott, S. : J. Biol. Chem. 254, 5417 (1979).
(9) Gilbert, W., Maxam, A. and Mirzabekov, A. : Control of Ribosome Synthesis eds. N.O. Kjeldgaard and O. Malvoe, Mksgaard, Copenhagen, p. 139 (1976).
(10) Selsing, E. and Arnott, S. : Nucl. Acid. Res. 3, 2443 (1976).
(11) Skakked, Z., Rabinovich, D., Cruse, W.B.T., Egert, E., Kennard, O., Sala, G., Salisbury, S.A. and Visvamitra, M.A. : Proc. Roy. Soc. B 213, 479 (1981).
(12) Conner, B.N., Takano, T., Tanaka, S., Itakura, K. and Dickerson, R.E. : Nature 295, 294 (1982).
(13) Drew, H.R. and Dickerson, R.E. : J. Mol. Biol. 151, 535 (1981).
(14) Dickerson, R.E., Drew, H.R. and Conner, B. : Biomolecular Stereodynamics, R.H. Sarma Ed., Adenine Press, N.Y., Vol. 1, 1 (1981).
(15) Rich, A., Quigley, G.J. and Wang, A.H.J. : Biomolecular Stereodynamics, R.H. Sarma Ed., Adenine Press, N.Y., Vol. 1, 35 (1981).
(16) Pilet, J. and Leng, M. : Proc. Natl. Acad. Sci. USA 79, 26 (1982).
(17) Thamann, T.J., Lord, R.C., Wang, A.H.J. and Rich, A. : Nucl. Acid. Res. 9, 5443 (1981).
(18) Ivanov, V.J. and Minyat, E.E. : Nucl. Acid. Res. 9, 4783 (1981).
(19) Sutherland, J.C., Griffin, K.P., Kerck, P.C. and Takacs, P.Z. : Proc. Natl. Acad. Sci. USA 78, 4801 (1981).
(20) Mitra, C.K., Sarma, R.H., Giessner-Prettre, C. and Pullman, B. : Int. J. Quant. Chem., Quantum Biol. Symp. 7, 39 (1980).
(21) Mitra, C.K., Sarma, M.H. and Sarma, R.H. : Biochemistry 20, 2036 (1981).
(22) Mitra, C.R., Sarma, M.H. and Sarma, R.H. : J. Am. Chem. Soc. 103, 6727 (1981).
(23) Sarma, R.H., Mitra, C.K. and Sarma, M.H. : Biomolecular Stereodynamics, R.H. Sarma Ed., Adenine Press, N.Y., Vol. 1, 53 (1981).
(24) Patel, O.J., Kozlowski, S.A., Nordheim, A. and Rich, A. : Proc. Natl. Acad. Sci. USA 79, 1413 (1982).
(25) Vorlickova, M., Kypr, J., Stokrova, S. and Sponar, J. : Nucl. Acid. Res. 10, 1071 (1982).
(26) Zimmer, Ch., Tymen, S., Marck, Ch. and Guschlbauer, W. : Nucl. Acid. Res. 10, 1081 (1982).

(27) Behe, M. and Felsenfeld, G. : Proc. Natl. Acad. Sci. USA 78, 1619 (1981).

(28) Behe, M., Zimmermann, S. and Felsenfeld, G. : Nature 293, 233 (1981).

(29) Möller, A., Nordheim, A., Nichols, S.R. and Rich, A. : Proc. Natl. Acad. Sci. USA 78, 4777 (1981).

(30) Van de Sande, J.H. and Jovin, T.M. : EMBO J. 1, 115 (1982).

(31) Sage, E. and Leng, M. : Nucl. Acid. Res. 9, 1241 (1981).

(32) Santella, R.M., Grunberger, D., Weinstein, I.B. and Rich, A. : Proc. Natl. Acad. Sci. USA 78, 1451 (1981).

(33) Santella, R.M., Grunberger, D., Broyde, S. and Hingerty, B.H. : Nucl. Acid. Res. 9, 5459 (1981).

(34) Grunberger, D. and Santella, R.M. : J. Supramol. Structure and Cellular Biol. 17, 231 (1981).

(35) Mercado, L.M. and Tomasz, M. : Biochemistry 16, 2040 (1977).

(36) Kaplan, D.J. and Hurley, L.H. : Biochemistry 20, 7572 (1981).

(37) Malfoy, B., Hartmann, B. and Leng, M. : Nucl. Acid. Res. 9, 5659 (1981).

(38) Nordheim, A., Möller, A., Lafer, E.M., Stollar, B.D. and Rich, A. : Regards sur la biochimie N°3, p. 25 (1981).

(39) Simpson, R.T. and Shindo, H. : Nucl. Acid. Res. 8, 2093 (1980).

(40) Quadrifoglio, F., Manzini, G., Vasser, M., Dinkelspiel, K. and Crea, R. : Nucl. Acid. Res. 9, 2195 (1981).

(41) Wang, A.J.H., Quigley, G.J., Kolpak, F.J., Van der Marel, G., Van Boom, J.H. and Rich, A. : Science 24, 171 (1981).

(42) Benham, C.J. : Nature 286, 637 (1980).

(43) Klysik, J., Stirdivant, S.M., Larson, J.E., Hart, P.A. and Wells, R.D. : Nature 290, 672 (1981).

(44) Wells, R.D., Klysik, J., Stirdivant, S.M., Larson, J. and Hart, P.A. : Biomolecular Stereodynamics, R.H. Sarma Ed., Adenine Press, N.Y., Vol. 1, 77 (1981).

(45) Arnott, S. and Chandrasekaran, R. : Biomolecular Stereodynamics, R.H. Sarma Ed., Adenine Press, N.Y., Vol. 1, 99 (1981).

(46) Sasisekharan, V., Bansal, M., Bramachari, S.K. and Goupta, G. : Biomolecular Stereodynamics, R.H. Sarma Ed., Adenine Press, N.Y., Vol. 1, 123 (1981).

(47) Sundaralingam, M. and Westhof, E. : Intern. J. Quantum Chem. Quantum Biol. Symp. 8, 287 (1981).

(48) Malfoy, B. and Leng, M. : FEBS Letters 132, 45 (1981).

(49) Lafer, E.M., Möller, A., Nordheim, A., Stollar, B.D. and Rich, A. : Proc. Natl. Acad. Sci. USA 78, 3546 (1981)

(50) Nordheim, A., Pardue, M.L., Lafer, E.M., Möller, A., Stollar, B.D. and Rich, A. : Nature 294, 417 (1981).

(51) Nickol, J., Behe, M. and Felsenfeld, G. : Proc. Natl. Acad. Sci. USA 79, 1771 (1982).

(52) Kolata, G. : Science 214, 1108 (1981).

(53) Hanlon, S., Johnson, R.S. and Chan, A. : Biochemistry 13, 3963 (1974).

(54) Hanlon, S., Johnson, R.S. and Chan, A. : Biochemistry 13, 3972 (1974).

(55) Lomonossoff, G.P., Butler, P.J.G. and Klug, A. : J. Mol. Biol. 149, 745 (1981).
(56) Shindo, H., Wooten, J.B., Pheiffer, B.H. and Zimmermann, S.B. : Biochemistry 19, 518 (1980).
(57) Peck, L.J. and Wang, J.C. : Nature 292, 375 (1981).
(58) Rhodes, D. and Klug, A. : Nature 292, 378 (1981).
(59) Wu, H.M., Dattagupta, N. and Crothers, D.M. : Proc. Natl. Acad. Sci. USA 78, 6808 (1981).
(60) Bubienko, E., Uniack, M.A. and Borer, Ph. N. : Biochemistry 20, 6987 (1981).
(61) Lavery, R. and Pullman, B. : Nucl. Acid. Res. 9, 4677 (1981).
(62) Pullman, A. and Pullman, B. : Quart. Rev. Biophys. 14, 289 (1981).
(63) Lavery, R., Pullman, B. and Corbin, S. : Nucl. Acid. Res. 9, 6539 (1982).
(64) Lavery, R., Corbin, S. and Pullman, B. : Theoret. Chim. Acta 60, 513 (1982).
(65) Zakrzewska, K., Lavery, R., Pullman, A. and Pullman, B. : Nucl. Acid. Res. 8, 3917 (1980).
(66) Zakrzewska, K., Lavery, R. and Pullman, B. : Biomolecular Stereodynamics, R.H. Sarma Ed., Adenine Press, N.Y., 163 (1981).
(67) Pullman, B., Lavery, R. and Pullman, A. : Europ. J. Biochem., in press.
(68) Lavery, R., Pullman, A. and Pullman, B. : Int. J. Quant. Chem. 20, 49 (1981).
(69) Pullman, A. and Pullman, B. : Chemical Applications of Atomic and Molecular Electrostatic Potentials, P. Politzer and D.G. Truhlar Eds., Plenum Press, N.Y., 381 (1981).
(70) Lavery, R., Pullman, A. and Pullman, B. : Theoret. Chim. Acta, in press.
(71) Arnott, S. and Hukins, D.W.L. : Biochem. Biophys. Res. Comm. 47, 1504 (1972).
(72) Klug, A., Jack, A., Viswamitra, M.A., Kennard, O., Shakked, Z. and Steitz, T.A. : J. Mol. Biol. 131, 669 (1979).
(73) Arnott, S. and Selsing, E. : J. Mol. Biol. 98, 265 (1975).
(74) Arnott, S., Chandrasekaran, R., Hukins, D.W.L., Smith, P.J.C. and Watts, L. : J. Mol. Biol. 88, 523 (1974).
(75) Wang, A. H-J., Quigley, G.J., Kolpak, F.J., Van der Marel, G., Van Boom, J.H. and Rich, A.: Science 211, 171 (1980).
(76) Pullman, B., Pullman, A. and Lavery, R. : in "The Living State" Ed. by R.H. Mishra, Wiley-Interscience, in press.
(77) Lavery, R., Pullman, A. and Pullman, B. : Acta Cryst., in press.
(78) Lavery, R., de Oliveira, M. and Pullman, B. : Int. J. Quant. Chem., in press.
(79) Pullman, B. and Pullman, A. : Studia Biophys. 86, 95 (1981).
(80) Zimmer, Ch. : Progr. Nucl. Acid. Res. Mol. Biol. 15, 285 (1975).
(81) Gursky, G.V., Tumanyan, V.G., Zasedatelev, A.S., Shuze, A.L., Grohovsky, S.L. and Gottikh, B.P. : Nucleic Acids-Proteins Interactions (H.J. Vogel Ed.), Academic Press, N.Y., 189 (1977).
(82) Reinert, K.E., Geller, D. and Stutter, E. : Nucl. Acid. Res. 9, 2335 (1981).

(83)   Berman, H.M., Neidle, S., Zimmer, C. and Thrum, H. : Biochem. Biophys. Acta 561, 124 (1979).

(84)   Braithwaite, A.W. and Baguley, B.C. : Biochemistry 19, 1101 (1980).

(85)   Festy, B., Sturm, J. and Daune, M. : Biochim. Biophys. Acta 407, 24 (1975).

(86)   Chidester, C.G., Krueger, W.C., Mizsak, S.A., Duchamp, D.J. and Martin, D.G. : J. Am. Chem. Soc. 103, 7629 (1981).

(87)   Mikhailov, M.V., Zasedatelev, A.S., Krylov, A.S. and Gursky, G.V. : Molecular Biol. (URSS) English Ed. 15, 541 (1981).

(88)   Ehrenpreis, S. : Georetown Med. Bull. 16, 148 (1963).

(89)   Zimmer, Ch., Luck, G. and Fric, I. : Nucl. Acid. Res. 3, 1521 (1976).

(90)   Pullman, A. and Pullman, B. : Int. J. Quant. Chem. Quantum Biol. Symp. 7, 245 (1980).

(91)   Pullman, B. and Pullman, A. : in Carcinogenesis : Fundamental Mechanisms and Environmental Effects, Eds. B. Pullman, P.O.P. Ts'o and H. Gelboin, Reidel, Holland, 55 (1980).

(92)   Santella, R.M., Grunberger, D., Weinstein, J.B. and Rich, A. : Proc. Natl. Acad. Sci. USA 78, 1451 (1981).

(93)   Waring, M.J. : Ann. Rev. Biochem. 50, 159 (1981).

(94)   Westra, J.G., Kriek, E. and Hittenhausen, H. : Chem. Biol. Interact. 15, 149 (1976).

(95)   Howard, P.C., Casciano, D.A., Beland, F.A. and Shaddock, J.G. Jr. : Carcinogenesis 2, 97 (1981).

(96)   Beland, F.A., Tullis, D.L., Kadlubar, F.F., Staub, K.M. and Evans, F.E. : Chem. Biol. Interact. 31, 1 (1980).

(97)   Phillips, D.H., Miller, J.A., Miller, E.C. and Adams, B. : Cancer Res. 41, 176 (1981).

(98)   Phillips, D.F., Miller, J.A., Miller, E.C. and Adams, B. : Cancer Res. 41, 2664 (1981).

(99)   Robertson, K.A. : Cancer Res. 42, 8 (1982).

(100)   Hurley, L.H. and Retrusek, R. : Nature 282, 529 (1979).

(101)   Maruyama, I.N. and Tanako, N. : Biochem. Biophys. Res. Comm. 98, 970 (1981).

(102)   Lown, J.W., Joshua, A.V. and Lee, J.S. : Biochemistry 21, 420 (1982).

(103)   Lavery, R., Cauchy, D., de la Luz Rojas, O. and Pullman, A. : Int. J. Quant. Chem. Quantum Biol. Symp. 7, 323 (1980).

(104)   Cauchy, D., Lavery, R. and Pullman, B. : Theoret. Chim. Acta 57, 323 (1980).

(105)   Corbin, S., Lavery, R. and Pullman, B. : Int. J. Quant. Chem. Quantum Biol. Symp. 9, in press.

(106)   Lavery, R. and Pullman, B. : FEBS Letters, in press.

Z-DNA AND CHEMICAL CARCINOGENESIS.

Marc Leng
Centre de Biophysique Moléculaire, C.N.R.S.,
1A, avenue de la Recherche Scientifique,
45045 Orléans cedex, France

It is known that ultimate carcinogens bind covalently to DNA and it is generally thought that the chemical modification is an important step in tumorogenic process (1). In general several adducts are formed and it is not yet actually known how these adducts can lead to a mutational event. Among others, the hepatocarcinogen N-hydroxy-2-acetylaminofluorene and some of its derivatives have received great attention. Some of the DNA adducts formed in rat liver after i.p. injection of the carcinogen have been clearly identified as N-(deoxyguanosin-8-yl)-N-acetyl-2-amino-fluorene, 3-(deoxyguanosin-N^2-yl)-2-acetylaminofluorene and N-(deoxy-guanosin-8-yl)-2-aminofluorene (2,3), adducts which are also found *in vitro* after reaction between DNA and metabolites of N-hydroxy-2-acetyl-aminofluorene. The percentages and the persistance of these adducts in *in vivo* modified DNA of target and not-target tissues vary greatly (2-5). Several physico-chemical studies have been devoted to the *in vitro* modi-fied DNA and possible explanations of the carcinogen effects have been proposed. For example, the covalent binding of N-acetyl-2-aminofluorene residues to the C(8) of guanine residues in native DNA induces a local denaturation of the double helix, the fluorene residue being inside the helix and stacked with the bases and the guanine residues being outside the helix (insertion-denaturation model (6-8) and base-displacement model (9-11)). On the other hand, the same adduct can induce other confor-mational changes when bound to long alternated d(GpC) sequences. Acetyl-2-aminofluorene modified poly(dG-dC).poly(dG-dC) (poly(dG-dC)AAF) is in Z-form at low ionic strength (12-15) and does not seem to be in C-form in 3.8 M LiCl (13).

In this paper, we review some physico-chemical properties of poly (dG-dC).poly(dG-dC) in Z-form and some evidences for the presence of sequences in Z-form in natural DNA. A question is to know whether the covalent binding of chemical carcinogens to natural DNA could induce the transition of some sequences from B-form to Z-form and whether this con-formational change might play a role in cancer process.

C. Hélène (ed.), Structure, Dynamics, Interactions and Evolution of Biological Macromolecules, 45–56.
Copyright © 1983 by D. Reidel Publishing Company.

INFRARED ABSORPTION OF Z-DNA

    Left-handed double stranded DNA, named Z-DNA, was first discovered
by an X-ray analysis of the alternating hexanucleotide d(CpGpCpGpCpG)
crystals (16) and then of several oligo d(C-G)$_n$ (17-19). Diffraction
patterns of poly(dG-dC).poly(dG-dC) fibers have yielded patterns consis-
tant with Z-DNA structure (20,21). Z-DNA can also be characterized by
infrared spectroscopy (22). As shown in figure 1, the infrared linear
dichroism spectra of poly(dG-dC).poly(dG-dC) hydrated oriented films
present several characteristic bands in the strong absorption regions
of phosphodiester backbone (900-1300 cm^{-1}) and of the bases (1500-1800
cm^{-1}).

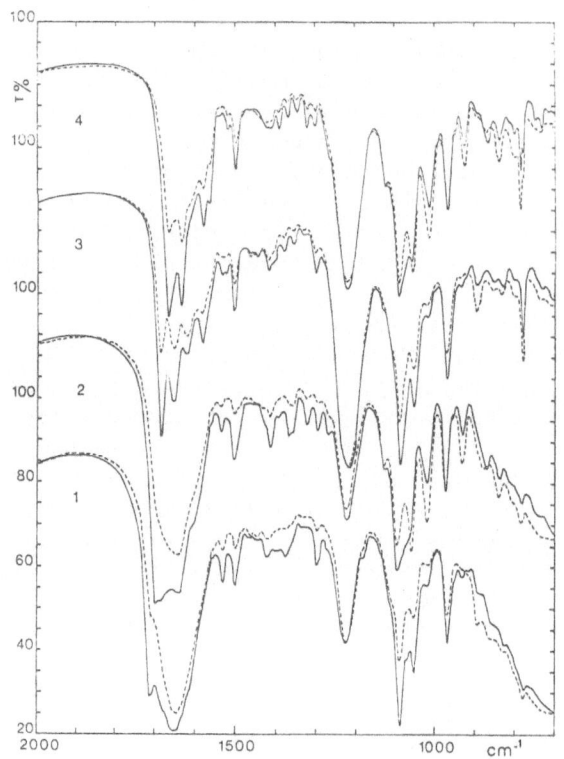

Figure 1. Infrared linear dichroism spectra of poly(dG-dC).poly(dG-dC)
hydrated film. Solid lines, transmission of polarized light with its E
vector perpendicular to the helix axis ; broken lines, transmission with
E vector parallel to the axis. Spectra : 1, film exposed to 96 % relative humidity H$_2$O vapor atmosphere, B-form ; 2, exposed to 90 % relative
humidity (H$_2$O), Z-form ; 3, to 96 % relative humidity (^{2}H$_2$O), B-form ;
4, to 90 % relative humidity (^{2}H$_2$O), Z-form, (Reprinted by permission
from Proc. Natl. Acad. Sci. US, see reference 22).

    For comparison are also shown in figure 1 the infrared linear di-
chroism spectra of B-poly(dG-dC).poly(dG-dC). The B-form - Z-form tran-

sition can be followed at several wavelengths (figure 2) as shown by the
variations of the mean angles θ of transition moments of PO_2^- stretching
modes with respect to the helix axis deduced from the dichroic ratios
as a function of the water content in the film (23,24). An interesting
result is that there is a good agreement between the calculated values
of the angles θ in the Z_I-form (18) and the experimental values (the
mean values $θ_{1220}$ corresponding approximatively to the mean direction
of the O...O vector are 52°2 (deduced from Z_I-form) and 57° (calculated
from dichroic ratio)). The midpoint of the B-form - Z-form transition
is at about 3 M Na^+.

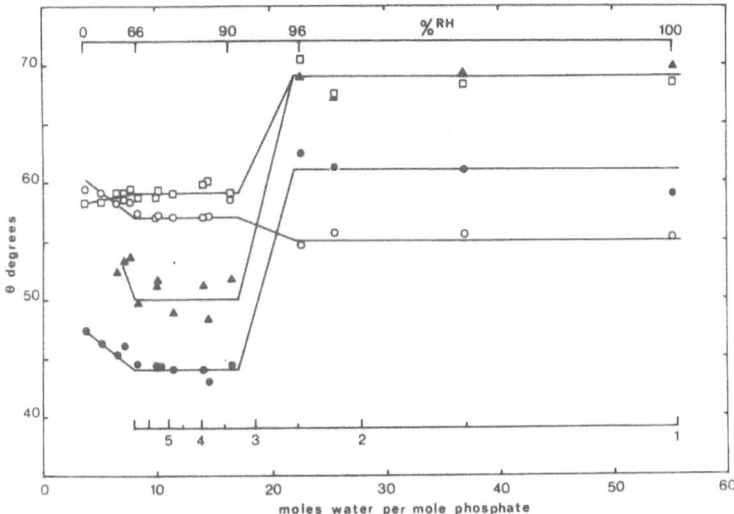

Figure 2.   Variation of mean θ angles of transition moments with respect to the helix axis deduced from the dichroic ratios at four wavelengths as a function of the water content in the films. ●, 1022 cm⁻¹ ;
▲, 1055 cm⁻¹ ; ■, 1090 cm⁻¹ ; ○, 1225 cm⁻¹ (Reprinted by permission
from Proc. Natl. Acad. Sci. US, see reference 22).

DYNAMIC STRUCTURE OF DNA

     Poly(dG-dC).poly(dG-dC) can also adopt the Z-form in high salt solu-
tion as demonstrated by Raman spectroscopy (25), nuclear magnetic resonance (26,27,28a) and dielectric dichroism (28). Circular dichroism is
very convenient to characterize the Z-form. Pohl and Jovin (29) have
shown that the circular dichroism spectra of poly(dG-dC).poly(dG-dC) in
high salt concentration is almost an inversion of the spectrum in low
salt concentration, inversion also observed in presence of some organic
solvents (30,31).

     These techniques give essentially a static picture of the conforma-
tion of the molecules. Such information although fundamental is incom-
plete because double helical nucleic acids in solution are known to be
subjected to thermal fluctuations resulting in transient conformation
with opened base pairs. A particularly useful probe of the dynamic as-

pects of nucleic acid structure is the tritium exchange technique (32).

    The exchange rates of the five protons involved in the hydrogen
bonds between guanine and cytosine residues in poly(dG-dC).poly(dG-dC)
were measured in low and high salt conditions (33).

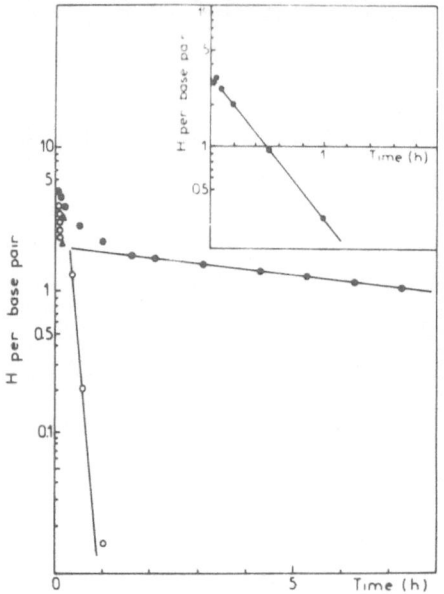

Figure 3. Hydrogen-tritium exchange curves for poly(dG-dC).poly(dG-dC)
in 3 M NaClO$_4$ (●), 1 M NaClO$_4$ (o) and 2.5 M NaClO$_4$ (▲)(Reprinted by permission from Nature, (C) 1982, Macmillan Journals Ltd, see Ref. 33).

    As shown in figure 3, there is a dramatic difference between the
two hydrogen curves, the overall exchange in high salt conditions being
much slower than in low salt conditions. In low salt, four protons are
measured and they all exchange with the same rate characterized by an
exchange half-time of 6 min. The four protons were assigned to the exocyclic amino protons of the guanine and cytosine whereas the missing proton corresponds to the guanine imino proton, whose exchange rate is too
fast to be measured by this technique. In contrast, all five protons are
measured in high salt conditions. The exchange of two of these protons is
surprisingly slow, the exchange half-time being about 7 hours. The remaining three protons have the same exchange half-time of about 20 min.

    To quantitatively describe this left-handed double helix dynamic
structure, the two slow protons which are either the amino protons of
guanine or cytosine residues had to be identified. This was solved by
the study of poly(dI-br^{5}dC).poly(dI-br^{5}dC). It was first demonstrated
by infrared absorption, ^{31}P nuclear magnetic resonance and circular dichroism that in high salt conditions, this polynucleotide belongs to the
Z-family (34).

    The proton exchange kinetics of poly(dG-dC).poly(dG-dC) and poly (dI-br^{5}dC).poly(dI-br^{5}dC) films were followed by infrared absorption (22,34). Both curves were similar and indicated the presence of slowly exchanging protons. The exchange half-times are given in table 1. From these results the slowly exchanging protons of the Z-form are identified as the cytosine amino protons.

	Poly(dI-br^{5}dC).poly(dI-br^{5}dC)		
	B	Z	
Half-time (min.)	20	90	850

	Poly(dG-dC).poly(dG-dC)		
	B	Z	
Half-time (min.)	19	67	1440

	Poly(dG-dC).poly(dG-dC)[*]		
	B	Z	
Half-time (min.)	6	20	420
Protons per base pairs	4	3	2

Table 1. Conformation dependence of the proton exchange kinetics.
[*] Tritium experiments ; the other values were deduced from infrared absorption experiments.

The kinetics were analysed according to the scheme proposed by Teitelbaum and Englander (35,36). In this scheme, a closed state is in equilibrium with an open state, the exchange process can occur only during a transient open state. From the results of Teitelbaum and Englander (35,36) and those on Z-DNA, we found that an open state is about 50 times less likely in Z-form than in B-form. Moreover, the opening rate constants are much smaller for the Z-form than for the B-form, in contrast to the closing rates which are of the same order of magnitude (34). Thus the Z conformation hinders to some extent the opening of the base pairs which might be important in biological processes that require molecular recognition.

ACETYLAMINOFLUORENE MODIFIED POLY(dG-dC).POLY(dG-dC)

The midpoint of the salt induced cooperative transition B-form Z-form is about at 2.5 M NaCl (29). Chemical modifications of cytosine or guanosine residues can shift the equilibrium.

Behe and Felsenfeld have found that the midpoint of the transition of poly(dG-m^{5}dC).poly(dG-m^{5}dC) is at 0.7 M NaCl (37). Bromine is even more efficient than methyl group since poly(dG-br^{5}dC).poly(dG-br^{5}dC) is in Z-form over 5 mM Tris-HCl, pH 7.5 - 3 M NaCl salt range (38). Poly (dG-dC).poly(dG-dC) modified on the N(7) of guanine residues by the reaction with dimethylsulfate (39) or with chlorodiethylenetriamino platinum(II) chloride (40,41) adopts the Z-form at lower NaCl concentrations than unmodified poly(dG-dC).poly(dG-dC). A similar effect is observed when mitomycin C is covalently bound to poly(dG-dC).poly(dG-dC). (42).

N-acetoxy-N-acetyl-2-aminofluorene reacts with poly(dG-dC).poly(dG-dC) and acetylaminofluorene (AAF) residues are bound to the C(8) of guanine residues (43). The covalent binding of AAF stabilizes the Z-form as mainly shown by circular dichroism (12-15) and by the reaction with the antibodies to Z-DNA (38).

In figure 4 are plotted the circular dichroism spectra of B and Z-forms of poly(dG-dC).poly(dG-dC) and of poly(dG-dC)AAF. The difference circular dichroism spectra analysis shows that the modified polynucleotide behaves as a mixture of B and Z-forms (12).

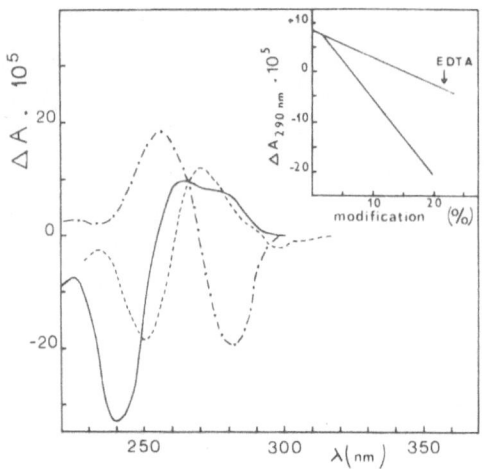

Figure 4. Circular dichroism of acetylaminofluorene modified poly(dG-dC). poly(dG-dC). Poly(dG-dC).poly(dG-dC) in 1 mM phosphate buffer (———), poly(dG-dC)AAF (6.6 % modified bases) in 1 mM phosphate buffer (---), in 40 % ethanol (—•—).

As shown in the inset of figure 4 the percentage of Z-form depends upon the level of modification. The circular dichroism signal at 290 nm is positive for the low levels of modification, decreases as the modification level increases and then becomes negative (in 1 mM phosphate buffer, poly(dG-dC)AAF is completely in Z-form when about 15-20 % of the total bases are substituted). For a given level of modification, the percentage of each form depends upon the presence of traces of multivalent cations. Addition of small amounts of EDTA decrease the percentage of Z-form (inset, figure 4). This stabilization of the Z-form by traces of multivalent cations could be very important in biological processes and has been also found on poly(dG-dC).poly(dG-dC) (40,44,45), poly(dG-m^{5}dC).poly(dG-m^{5}dC) (37) and poly(dG-dC).poly(dG-dC) modified by chlorodiethylenetriamino platinum(II) chloride (40).

An explanation for the effect of acetylaminofluorene residues can

be proposed. In the Z-form guanine residues have the syn conformation
(16). It has been demonstrated that in AAF modified guanosine and oligonucleotide the guanine residues have the syn conformation (6-11,46), the
anti conformation being prevented by the large size of AAF residues (in
fact, CPK molecular models show that the anti conformation is mainly prevented by the steric hindrance of the acetyl group in AAF residue). This
is confirmed by the study of aminofluorene modified guanosine, oligonucleotides and poly(dG-dC).poly(dG-dC) (the substitution occurs on the
C(8) of guanine residues). The aminofluorene modified guanosine has the
anti conformation (46-48) and the modified and unmodified poly(dG-dC).
poly(dG-dC) with respect to the B-form - Z-form transition behave similarly (12).

      Acetylaminofluorene residues stabilize the Z-form. It is interes-
ting to note that in poly(dG-dC)AAF, the guanine residues have the syn
conformation and are paired with the complementary cytosine residues
having the anti conformation (Z-form) and thus the fluorene residues
are outside the double helix. There is no local denaturation. This is a
major difference with the insertion-denaturation model or base displacement model (6-11) as schematically represented in figure 5.

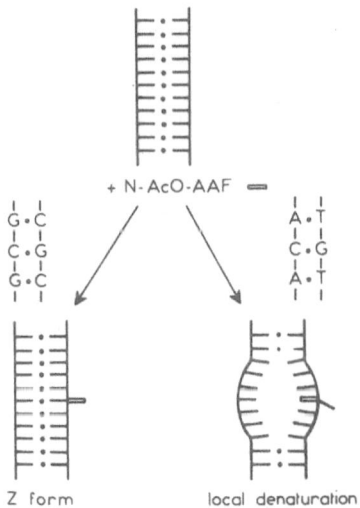

Z form   local denaturation

Figure 5.   Scheme of the reaction between N-acetoxy-N-acetyl-2-amino-
fluorene and DNA.

Thus one expects that according to the base sequences of a natural DNA,
the covalent binding of AAF residues can induce a conformational change
to the Z-form or a local denaturation. As shown further, several other
parameters can favor or hinder these conformational changes. In order
to get some more quantitative data, we have studied some oligonucleotides in solution. d(CpG)3 can undergo the B-form - Z-form transition (49).

Under various experimental conditions, we found no evidence that d(CGTACG) could adopt the Z-form. Substitution of guanine residues by acetylaminofluorene residues was not sufficient to stabilize the Z-form even at low temperature and high ionic strength (under our experimental conditions, the substituted oligonucleotide behaved as a single stranded oligonucleotide, 50).

DOES NATURAL DNA ADOPT THE Z-FORM ?

As already mentioned, it is well-established that in solution poly(dG-dC).poly(dG-dC) can adopt the Z or a Z-like form. Poly(dA-dT). poly(dA-dT) presents a dinucleotide repeat conformation in high salt conditions which is not the left-handed Z-DNA conformation (51-55). X-ray diffraction pattern of poly(dA-dC).poly(dG-dT) is similar to that of Z-poly(dG-dC).poly(dG-dC) (20) but it is not yet proven that the transition to the Z-form occurs in solution (56,57). The next question is to know whether some sequences in natural DNA can adopt a Z-like conformation. Immunological methods give a positive answer.

Z-DNA is a strong immunogen. Antibodies to Z-DNA were elicited in rabbits immunized with poly(dG-dC).poly(dG-dC) chemically modified with bromine (58) or with chlorodiethylenetriamino platinum(II) chloride (59), polymers which are in Z-form under physiological conditions. Some properties of the antibodies to Z-DNA induced by the injection of platinum modified poly(dG-dC).poly(dG-dC) can be summarized as follows (38). The antibodies react with Z-DNA but not with linear B or A-DNA, denatured DNA, RNA, guanosine or cytidine. Each antibody binding site covers about 4 nucleotide residues of poly(dG-dC).poly(dG-dC). The antibody-antigen complexes are stabilized by electrostatic interactions and there are some evidences that the exocyclic amino group of guanine residues interacts with the amino acid residues of the antibody binding sites.This is a difference in the recognition of the antigen between these antibodies and the antibodies to poly(I).poly(C) which do not interact with the base residues (60).

Form V-DNA is well-recognized by the antibodies to Z-DNA (61). Form V-DNA was prepared by the annealing of covalently closed, complementary strands of pBR322 according to the procedure of Stettler et al. (62). By radioimmunoassays, it was found that about 5 times more form V-DNA than platinum modified poly(dG-dC).poly(dG-dC) was necessary to inhibit the tracer-antibody binding in low salt concentration and 50 times more in high salt. The binding of the antibodies to form V-DNA can be visualized by electron microscopy as shown in figure 6. The antibodies to Z-DNA were first reacted with form V-DNA and the complexes were reacted with ferritin labeled goat immunoglobulins anti rabbit immunoglobulins. The experiments were performed at high ratio form V-DNA over antibodies to Z-DNA to avoid too many cross-links between the DNA molecules ant thus precipitation. Even at high ratio, some cross-links between the DNA molecules occur. On the other hand, it can be seen that the antibodies seem to be distributed all along the DNA molecules.

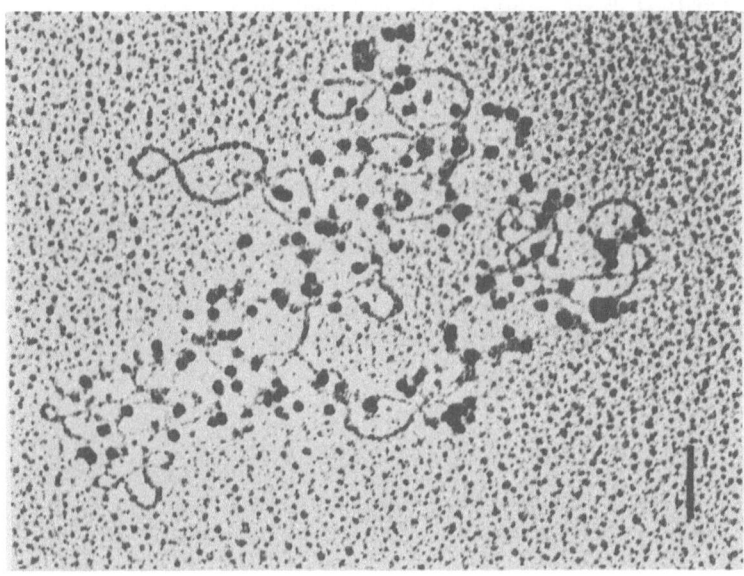

Figure 6.   Electron micrograph of ferritin-labeled goat antibodies bound
to the form V-DNA-antibodies to Z-DNA complexes. The bar represents
1000 Å

The conclusion of this study is that under topological constraints
a DNA of natural sequences can be partly in Z-form. A closer inspection
of the results strongly suggests that sequences other than alternating
(C-G) sequences can adopt a Z-like conformation (61).

Sequences of DNA in polytene chromosomes can also be in Z-form.
This was shown by indirect immunofluorescence on chromosomes of *Drosophila melanogaster* larvae (50,63,64) and on chromosomes of *Chironomus
thummi* larvae   (65). The staining was found in interband regions in
D. melanogaster (63) and in band regions in *Chironomus thummi* (65).

X-ray diffraction studies of alternating oligonucleotides $d(C-G)_n$
have demonstrated the existence of a new DNA form, named Z-DNA. In solution, poly(dG-dC).poly(dG-dC) can have the Z-conformation. Several
chemical modifications of the base residues shift the B-form - Z-form
equilibrium towards the right. Among them, the covalent binding of the
hepatocarcinogen N-acetoxy-N-acetyl-2-aminofluorene on the C(8) of guanine residues is very efficient. Sequences of DNAs in polytene chromosomes can also be in Z-form as shown by the reaction with the antibodies
to Z-DNA. This can be due to the binding of proteins which interact
specifically with the Z-form. On the other hand, topological constraints
in a naked DNA (form V-DNA) have be shown to be of major importance to
induce the Z-form. It is tempting to assume that the covalent binding
of a chemical carcinogen to some sequences of natural DNA can modify
some topological constraints and thus favor or stabilize a Z-conformation
for these sequences. This might be important for cellular events that

follow the chemical modification of DNA and more especially for the behavior of the DNA replication machinery and DNA repair enzymes.

ACKNOWLEDGEMENTS

I am pleased to acknowledge the essential role played by my collaborators   B. Hartmann, B. Malfoy, J. Pilet, J. Ramstein, P. Rio, N.
Rousseau, E. Sage and A. Soulas, on various aspects of the work reviewed
in this paper. It is also a pleasure to thank Pr. M. Daune and his collaborators. This work was supported in part by D.G.R.S.T. (contract n°
81E1213) and INSERM (contract n° 120019).

REFERENCES

1. Miller, J.A.:1970, Cancer Res. 30, 559-576.
2. Kriek, E.:1979, in Environmental Carcinogenesis, Emmelot, P. and
    Kriek, E., eds, pp. 143-164.
3. Kriek, E. and Westra, J.G.:1979, in Chemical Carcinogens and DNA,
    C.R.C. Press, Vol. 2, pp. 1-28.
4. Beland, F.A., Dooley, K.L. and Jackson, C.D.:1982, Cancer Res. 42,
    1348-1354.
5. Poirier, M.C., True, B. and Laishes, B.A.:1982, Cancer Res. 42,
    1330-1334.
6. Fuchs, R.P.P. and Daune, M.P.:1972, Biochemistry 11, 2659-2666.
7. Daune, M.P. and Fuchs, R.P.P.:1977, in Réparation du DNA, Mutagénèse,
    Cancérogénèse Chimique (C.N.R.S., Paris), pp. 83-97.
8. Lefèvre, J.P., Fuchs, R.P.P. and Daune, M.P.:1978, Biochemistry 17,
    2561-2567.
9. Nelson, J.H., Grunberger, D., Cantor, R.C. and Weinstein, I.B.:1971,
    J. Mol. Biol. 62, 331-346.
10. Grunberger, D.:1979, NCI Monograph n° 58, pp. 193-199.
11. Grunberger, D. and Weinstein, I.B.:1979, in Chemical Carcinogens
    and DNA, C.R.C. Press, vol. 2, pp. 59-93.
12. Sage, E. and Leng, M.:1980, Proc. Natl. Acad. Sci. USA 77, 4597-4601.
13. Sage, E. and Leng, M.:1981, Nucleic Acids Res. 9, 1241-1250.
14. Santella, R., Grunberger, D., Weinstein, B.I. and Rich, A.:1981,
    Proc. Natl. Acad. Sci. USA 78, 1451-1455.
15. Santella, R., Grunberger, D., Broyde, S. and Hingerty, B.:1981,
    Nucleic Acids Res. 9, 5459-5467.
16. Wang, A.H.J., Quigley, G.S., Kolpak, F.J., Crawford, J.L., van Boom,
    J.H., van der Marel, G. and Rich, A.:1979, Nature 282, 680-686.
17. Crawford, J.L., Kolpak, F.J. Wang, A.H.J., Quigley, G.J., van Boom,
    J.H., van der Marel, G. and Rich, A.:1980, Proc. Natl. Acad. Sci.
    USA 77, 4016-4020.
18. Wang, A.H.J., Quigley, G.S., Kolpak, F.J., van der Marel, G., van
    Boom, J.H. and Rich, A.:1981, Science 211, 171-176.
19. Drew, H., Takano, T., Tanaka, S., Itakura, K. and Dickerson, R.E.:
    1980, Nature 286, 567-573.
20. Arnott, S., Chandrasekaran, R., Birdsall, D.L., Leslie, A.G.W. and

Ratcliff, R.L.:1980, Nature 283, 743-745.

21. Behe, M., Zimmerman, S. and Felsenfeld, G.:1981, Nature 293, 233-235.
22. Pilet, J. and Leng, M.:1982, Proc. Natl. Acad. Sci. USA 79, 26-30.
23. Zbinden, R.:1964, in Infrared Spectroscopy of High Polymers, (Academie).
24. Nairn, J.A., Friesner, R., Frank, H.A. and Sauer, K.:1980, Biophys. J. 32, 733-754.
25. Thamann, T.J., Lord, R.C., Wang, A.H.J. and Rich, A.:1981, Nucleic Acids Res. 9, 5443-5457.
26. Patel, D.J., Canuel, L.L. and Pohl, F.M.:1979, Proc. Natl. Acad. Sci. USA 76, 2508-2511.
27. Mitra, C.K., Sarma, M.H. and Sarma, R.H.:1981, Biochemistry 20, 2036-2041.
28. Wu, H.M., Dattagupta, N. and Crothers, D.M.:1981, Proc. Natl. Acad. Sci USA 78, 6808-6811.
28a. Patel, D.J., Kozlowski, S.A., Nordheim, A. and Rich, A.:1982, Proc. Ntal. Acad. Sci. USA 79, 1413-1417.
29. Pohl, F.M. and Jovin, R.M.:1972, J. Mol. Biol. 67, 375-396.
30. Pohl, F.M.:1976, Nature 260, 365-366.
31. Ivanov, V.I. and Minyat, E.E.:1981, Nucleic Acids Res. 9, 4783-4798.
32. Englander, S.W. and Englander, J.J.:1972, in Methods in Enzymology, Hirs, C.H.W. and Timasheff, S.N., eds, XXVI, part C, pp. 406-413.
33. Ramstein, J. and Leng, M.:1980, Nature 288, 413-414.
34. Hartmann, B., Pilet, J., Ptak, M., Ramstein, J., Malfoy, B. and Leng, M.:1982, Nucleic Acids Res. 10, 3261-3277.
35. Teitelbaum, H. and Englander, S.W.:1975, J. Mol. Biol. 92, 55-78.
36. Teitelbaum, H. and Englander, S.W.:1975, J. Mol. Biol. 92, 79-92.
37. Behe, M. and Felsenfeld, G.:1981, Proc. Natl. Acad. Sci. USA 78, 1619-1623.
38. Malfoy, B., Rousseau, N. and Leng, M.:1982, Biochemistry (in press).
39. Moller, A., Nordheim, A., Nichols, S.R. and Rich, A.:1981, Proc. Natl. Acad. Sci. USA 78, 4777-4781.
40. Malfoy, B., Hartmann, B. and Leng, M.:1981, Nucleic Acids Res. 9, 5659-5669.
41. Ushay, H.M., Santella, R.M., Caradonna, J.P., Grunberger, D. and Lippard, S.J.:1982, Nucleic Acids Res. (in press).
42. Mercado, C.M. and Tomasz, M.:1977, Biochemistry 16, 2040-2046.
43. Harvan, D.J., Hass, R.J. and Lieberman, M.W.:1977, Chem. Biol. Interact. 17, 203-210.
44. Van de Sande, J.H. and Jovin, T.M.:1982, EMBO J. 1, 115-120.
45. Zacharias, W., Larson, J.E., Klysik, J., Stirdivant, S.M. and Wells, R.D.:1982, J. Biol. Chem. 257, 2775-2782.
46. Leng, M., Ptak, M. and Rio, P.:1980, Biochim. Biophys. Res. Comm. 96, 1095-1102.
47. Evans, F.E., Miller, D.W. and Beland, F.A.:1980, Carcinogenesis 1, 955-959.
48. Santella, R., Kriek, E. and Grunberger, D.:1980, Carcinogenesis 1, 897-902.
49. Quadrifoglio, R., Manzini,G., Vasser, M., Dinkelspiel, K. and Crea, R.:1981, Nucleic Acids Res. 9, 2195-2206.

56   M. LENG

50. Rio, P., Malfoy, B., Sage, E. and Leng, M.:1982, Environmental Health Perspectives (in press).
51. Volickova, M., Kypr, J., Kleinwächter, V. and Palecek, E.:1980, Nucleic Acids Res. 8, 3965-3973.
52. Kypr, J., Vorlickova, M., Budesinsky, M. and Sklenar, V.:1981, Biochem. Biophys. Res. Comm. 99, 1257-1264.
53. Patel, D.J., Kozlowski, S.A., Suggs, J.W. and Cox, S.D.:1981, Proc. Natl. Acad. Sci. USA 78, 4063-4067.
54. Klug, A., Jack, A., Viswamitra, M.A., Kennard, O., Shakked, Z. and Steitz, T.A.:1979, J. Mol. Biol. 131, 669-680.
55. Shindo, H., Simpson, R.T. and Cohe,, J.S.:1979, J. Biol. Chem. 254, 8125-8128.
56. Zimmer, C., Tymer, S., Marck, C. and Guschlbauer, W.:1982, Nucleic Acids Res. 10, 1081-1091.
57. Vorlickova, M., Kypr, J., Stokrova, S. and Sponar, J.:1982, Nucleic Acids Res. 10, 1071-1080.
58. Lafer, E.M., Möller, A., Nordheim, A., Stollar, B.D. and Rich, A.: 1981, Proc. Natl. Acad. Sci. USA 78, 3546-3550.
59. Malfoy, B. and Leng, M.:1981, FEBS Letters 132, 45-48.
60. Leng, M., Guigues, M. and Genest, D.:1978, Biochemistry 17, 3215-3220.
61. Lang, M.C., Malfoy, B., Freund, A.M., Daune, M.P. and Leng, M.: 1982, EMBO Journal (in press).
62. Stettler, U.H., Weber, H., Koller, Th. and Weissmann, C.:1979, J. Mol. Biol. 131, 21-40.
63. Nordheim, A., Pardue, M.L., Lafer, E.M., Möller, A., Stollar, B.D. and Rich, A.:1981, Nature 294, 417-422.
64. Leng, M., Harmann, B., Malfoy, B., Pilet, J., Ramstein, J. and Sage, E.:1982, 47th Symposium on Quantitative Biology, Cold Spring Harbor (in press).
65. Lemeunier, F., Derbin, C., Taillandier, E., Malfoy, B. and Leng, M.: 1982 (submitted).

INITIATION OF CARCINOGENESIS : FROM THE LOCAL STRUCTURAL CHANGE OF DNA
TO THE SETTING UP OF A MUTATION

Michel DAUNE
Laboratoire de Biophysique, Institut de Biologie Moléculaire
et Cellulaire du CNRS, 67084 Strasbourg cedex

We present a survey of the results obtained in the last few years in
our laboratory to trace the way from the structural defects of DNA
introduced by covalently bound carcinogenic amides to the corresponding
mutational event set up in bacteria. The main role of excision repair,
the unequal ability of carcinogens to block the replication machinery
and some aspects of the recA-controlled SOS functions are successively
analysed. By using directed mutagenesis inside a specified fragment of
a plasmid DNA, frameshift mutation with characteristic hotspots is for
the first time detected directly at the genomic level. Several implications are discussed in terms of problems to be studied in the future.

INTRODUCTION

The recent advances in the field of carcinogenesis give evidence of
the existence in the eukaryotic genomes of cancer genes or oncogenes
(1-2). Their corresponding coded protein is unknown except in the case
of the src gene which codes for a phosphokinase able to phosphorylate
tyrosine in several target proteins (3-4). In the present state of the
cancer problem it appears therefore that a cancerous cell could be
induced in any individual provided the repression of the regulation of
the transcription of cancer genes be suppressed or strongly perturbed.
If the transcription of oncogenes like that of other genes, is on the
dependence of the strength of their promoter, any modification of the
DNA in the vicinity of (or inside) the promoter sequence can be considered as a triggering event of a carcinogenic process. Such an effect
can be conceived, in a speculative way, as structural changes at the
level of the promoter sequences affecting their specific recognition
process by the RNA polymerase. Therefore any molecular event like long
terminal repeat sequences of viruses (5), transposable elements (6-7),
chemical or physical modification of DNA, could be involved in such an
alteration of the transcription of oncogenes. On the other hand, the
large evidence accumulated in testing carcinogens as mutagens in bacteria (8) shows convincingly that carcinogens are able to induce a
variety of mutational damages. When these mutations are occuring in
regions of the DNA critical for the control of the transcription, the

57

C. Hélène (ed.), Structure, Dynamics, Interactions and Evolution of Biological Macromolecules, 57–68.
Copyright © 1983 by D. Reidel Publishing Company.

correlation between mutagenesis and carcinogenesis is obvious. We have been therefore interested in the problem of tracing the way of mutagenesis from a known mutational event back to the physical defect introduced in a given place of the DNA. We have tried to unravel at least some parts of the "black box" which is generally taken as granted in any experimental approach of mutagenesis and in which repair processes and replication machinery are mainly involved. Such an approach is only possible in procaryotic systems which are the only ones to offer large possibilities of mutant strains and can also be considered as the simplest systems to be analysed.
This paper will be divided accordingly in three parts :
- Perturbation of the DNA structure at the level of the adducts.
- Study of the efficiency of some repair mechanisms.
- Analysis of induced mutations.
The mutagenic (and carcinogenic) agents which will be used all along this study are the acetylaminofluorene AAF and two parent compounds : iodoacetylaminofluorene IAAF and aminofluorene AF.

1. PERTURBATION OF THE DNA STRUCTURE

Covalent binding of an electrophilic ultimate carcinogen like N-acetoxy N-2-acetylaminofluorene to guanine takes place in several successive steps.
1) The major binding site is the C-8 which however is generally not (or at least poorly) accessible when the DNA is in the B-form. Some breathing process was already proposed more than ten years ago in order to explain the reactivity of the C-8 atom (9). Recent advances in the structural and dynamical properties of the helix permit us to explain the accessibility of C-8 on the basis of a simple dynamic equilibrium (10). Thermal fluctuations are supposed to affect mainly the sugar ring puckering and their effect can be interpreted within the conceptual frame of pseudo rotation (11). The conformational change from a C 2' endo to C 1' exo or O1' endo structure proceeds through a small potential barrier. Because this pseudorotational mobility the C-8 moves in such a way as to become accessible and able to react with the electrophilic carcinogen. As soon as the covalent binding is made, the local structure is modified according to the nature of the carcinogen (size of the aromatic ring, nature of the substitution, etc.) and of the resultant steric conflicts between the carcinogenic moiety and the phosphate group. In many cases, however, one cannot speak of a permanent local structure, which differs of the native one, but of a dynamic state in which two or more local conformations are in equilibrium.
a) In the case of AAF (acetylaminofluorene) a model of "insertion - denaturation" was proposed by us ten years ago (12-13), in which the fluorene ring takes the place of the guanine moiety and pushes outside of the helix both guanine and cytosine, disrupting the base pairing and decreasing locally the stability of the double helix.
A less rigid and more realistic picture would be to consider the binding as a sort of "crankshaft" motion, which was proposed recently (10), in which the sugar and not the guanine is rotated gradually in a syn

conformation relative to the base. Concerted rotations about the backbone bonds are such as to switch the original g^-g^- conformation of the P-O bonds in the B-DNA, to a g^+g^+ conformation similar to that observed in Z-DNA. Actually, AAF binding to poly(dG-dC) was shown to allow the change of B to Z-form at a much lower salt concentration than that necessary in the case of the unmodified polynucleotide (14-15). However, this g^+g^+ conformation could be in equilibrium with a partially disrupted structure, in which a few base pairs are opened giving rise to loop formation which could be enzymatically or chemically detected.

b) In the case of the 7-iodo derivative IAAF, the g^+g^+ conformation would not be accompanied by an anti to syn transconformation. The iodo-fluorene ring remains outside, partially stabilized by interaction between the iodine atom and neighbour phosphate. Such a binding process was called "outside binding" (16) in which the guanine remains inside, and the cytosine pairing is only but little affected. However, in both cases a) and b) there is a concerted rotation around the C4'-C5' bond from the g^+ state to the transstate.

c) In the case of the deacetylated carcinogen AF (which represents in vivo the majority of the adducts), the structural perturbation is much smaller in view of a possible interaction between the NH of the fluorene ring and the phosphate group, which favors an outside position of the aromatic ring. As soon as the covalent binding is achieved, the torsional angle ψ remains in a g^+ conformation and the backbone chain in a g^-g^- conformation. The only structural change would be the presence of a C1'-exo or O1'-endo sugar linked to the modified guanine. This general scheme is able to take into account the structural data gained so far in the field of carcinogenesis by aromatic amides.

1) This is the local g^+g^+ conformation which induces the unwinding of the double helix as detected in the case of modified covalently closed DNA by the progressive decrease of the density of supercoiling in fonction of the number of adducts (17-18). Actually AAF and IAAF are equally efficient to unwind the DNA of about 20° but this is not the case with AF (unpublished results). It is worth noting that the binding of benzo(a)pyrene 7-8-dihydrodiol-9-epoxide to the N2 of guanine induces the same unwinding as AAF or IAAF (17). Instead to assume a conformation of the adduct in which the aromatic ring is lying in the small groove (19), we would rather propose a partial intercalation of the pyrene ring into base pairs leading to an unwinding similar to that induced by intercalating dyes. Recent physical measurements (20 and Vigny, personal communication) seems to favor such a model.

2) The presence of looped structure around each adduct which was largely documented (21) was explained in terms of insertion of the fluorene ring in place of the guanine residue.

In the present model, such a local disordered structure was supposed to be in dynamic equilibrium with a Z-type structure in which the fluorene ring is laying outside. Actually endonuclease S1 does not react with Z-structure, and only poorly with IAAF-modified DNA (22). In the latter case, specific antibodies directed against Guo-AAF (23) as well as the tripeptide KWK (24) are able however to detect the adduct, indicating the presence of a small percentage of inserted fluorene and disorganized regions, which could be increased by the specific interaction with the

probe.
In vitro studies of the covalent binding of carcinogens to DNA lead to
a final number of modified guanines which is many orders of magnitude
higher than that observed in vivo. This amplification of the process
allows us to determine its structural aspect and the perturbations
induced in DNA structure ; but the local geometry around each adduct
remains the same in vitro and in vivo.
We have now to understand the relationship between these structural
modifications of the DNA and the perturbations which are finally introduced in the genome as somatic mutations. Actually, a series of enzymatic systems are involved in the repair of the genome and we have first
to examine their behavior when they have to cope with locally modified
DNA molecules.

2. REPAIR MECHANISMS

In the bacteria, several enzymatic mechanisms are able to repair UV
damages, i.e. presence of pyrimidine dimers (25). In the first one,
dimers are monomerized in presence of visible light by a photoreactivating enzyme. The second one, which takes place before replication and
termed excision repair, proceeds with the excision of modified regions
by specific endonucleases followed by a polymerisation using the intact
opposite strand as a template and ligation. A series of uvr genes are
involved in this process (26) and a few of the corresponding proteins
have been isolated. Excision repair which reform the original sequence
of the DNA molecule is said to be error-free and cannot give rise to
mutagenesis. When all of the pyrimidine dimers are not completely removed
DNA replication is blocked at each time the polymerase encounters one
remaining defect. However the replication can be resumed beyond the
dimer, and as a consequence a single-stranded gap is introduced. There
exist two mechanisms of by-pass of pyrimidine dimers, both under the
control of the recA protein. Either the gaps are removed by insertion
of homologous DNA derived from sister chromosome (recombinational repair)
or they can be filled by a polymerizing process which is unable to gua-
rantee the fidelity of the copy and then introduces errors in the genome.
While the first mechanism is error-free, and was recently elucidated (27)
the second one is error-prone and is still at least partially unknown
and described more generally as SOS function (28). It was necessary to
study the role of these different repair mechanisms when the local
defects are no more induced by pyrimidine dimers but by chemical adducts.
To this purpose several experimental approaches have been designed.
1) In vivo experiments were made by following the survival of a plasmid
(pBR322) which was modified with different carcinogens, inside different
E. coli repair strains. Actually the plasmid is able or not to give
ampicillin resistance to the bacteria. From the curves of fig. 1 several
observations can be made.

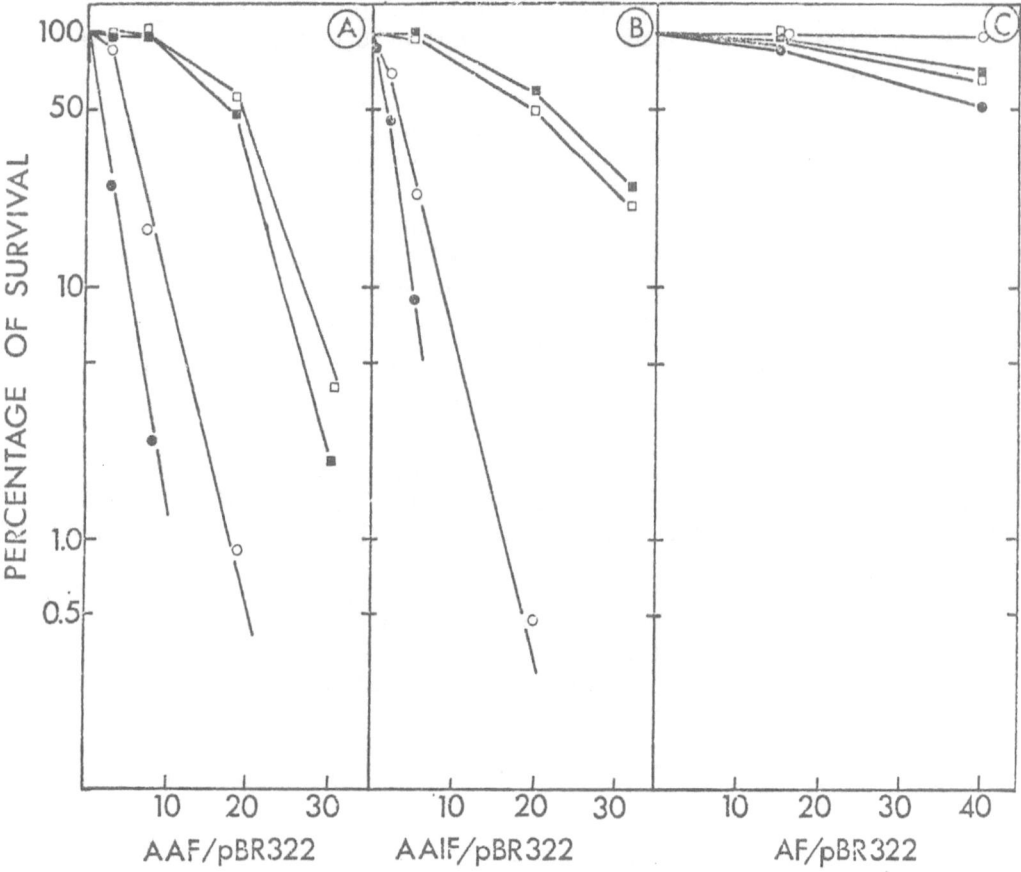

Fig. 1 - Survival of plasmid DNA damaged with the reactive fluorene
derivatives in different repair deficient mutants of E. coli.
□   Wild-type strain   ○   uvrA mutant
■   recA mutant   ●   uvrA recA mutant
(from Fuchs R.P. and Seeberg E., to be published).

a) The number of AAF and AAIF adducts per lethal hit (37% survival)
ranges between 20-25 adducts per plasmid, indicating a repair of at
least 95% of the adducts inside the wild type strain. This result has
to be compared with UV damage. In the same wild type strain, the number
of dimers per lethal hit is significantly lower (5-7) indicating a more
efficient repair of fluorene adducts than of pyrimidine dimers.
b) With the double recAuvrA mutant there is still 2 to 2.5 adducts per
lethal hit as opposed to 1 dimer in the case of UV lesions.
An additional repair pathway is thus operative and this assumption could
be fully supported by data obtained with AF adducts. In the later case
more than 98% of the defects are repaired via a pathway independent of
both recA and uvrA, since the survival curve is about the same for the
three repair mutants. However, another explanation could be given : -AF

adducts are simply unable to block the replication machinery since in
the case of the double mutant uvrArecA, up to 50 adducts are tolerated
before to give a lethal hit.
2) In vitro experiments were realized with the same modified plasmid
considered now as a substrate for uvr gene products. Their endonuclease
activity transforms the supercoiled DNA into the relaxed form. In these
experiments 0.53, 0.28 and $\sim$ 0.25 breaks were introduced per -AAF, -AAIF
and -AF adducts respectively. On the other hand, the level of repair is
comparable when either an AAF adduct or a cyclobutane ring between two
adjacent thymines is present. It was shown recently (29) that thymine
dimers were able to unwind circular DNA of about the same value as AAF
and a recent conformational study of the DNA around the cyclobutane
ring concludes to conformational changes similar to that introduces by
the insertion of a fluorene ring (30). Notwithstanding that the local
geometry of the helix around the -AF adduct differs strikingly from
that of -AAF adduct both are recognized during the excision process
which apparently does not depend too much on the nature of the adduct
and on the type of local structural change of the DNA (Fuchs and Seeberg,
to be published).
3) The level of induced recA protein, when the bacterial DNA was modified with AAF and AF, was measured by using an immunochemical tool.
With AAF the effect is similar to that obtained with UV dimers, i.e. a
30 fold increase of the background level, but with AF the increase is
one order of magnitude lower (Salles and Lang, to be published).
4) Although repair and mutagenesis have been studied for genomic and
phage DNA in E. coli, little is known about the effects of the different repair systems of the host on plasmid DNA.

We have thus initiated the study of the repair of a plasmid DNA which
was previously modified in vivo by a carcinogen (31). This approach
offers two main advantages : the number of damaged sites per DNA molecule can be quantitatively assessed prior to exposure to in vivo systems. Any other toxic or damaging effect of the carcinogen on other
cell structures or processes is excluded. The plasmid used (pKO482)
contains both an ampicillin gene and a gal K gene coding for galactokinase. The recipient bacterial strains are ampicillin sensitive and
unable to synthetize galactokinase. Selection is made on agar containing ampicillin and galactose. With an intact plasmid, colonies are red
but any mutation in the plasmid gal K gene results in a visually distinguishable white colony. Host cells with wild type DNA repair capa-
cities were compared to uvrA, recA and double mutant uvrArecA strains.
In addition, the effect of UV induction of SOS function was also studied.

Table 1

AAF adducts per lethal hit and mutation frequency calculated
at 1% survival for plasmid pKO482 transformed into E. coli
strains with differing DNA-repair capabilities.

Transformed strain	AAF adducts per plasmid molecule	AAF adducts per lethal hit	Mutation frequency
AB 1157 uvr[+] rec[+]	37.2	8.1	0.033
AB 2463 uvr[+] recA	32.4	7.0	0.010
AB 1886 uvrA rec[+]	11.6	2.5	0.030
AB 2480 uvrA recA	8.4	1.8	0.006

a) The survival curves are similar to those observed with pBR322.
The uvrA dependent excision repair is found again to be responsible for
the majority of repair of AAF adducts ($\sim$ 80%) and recA functions only
make a small but measurable contribution ($\sim$ 15%).
These results are in contrast with those obtained with genomic DNA in
which the repair process appears to be equally dependent on both uvr
and recA genes. With plasmid DNA and in agreement with other recent
experiments the repair depends almost completely on excision processes.

b) At 1% survival level, the mutation frequency for AAF adducts is
about three times higher than that measured in presence of UV dimers.
Moreover, in the case of the uvrArecA double mutant, the mutation frequency was not higher than the spontaneous one in the case of pyrimidine
dimers, but about 50 times higher in the case of AAF adducts. It is thus
clear that an additional non-recA mediated mechanism of mutagenesis is
operating when AAF adducts are present. There are two possible explanations :

- During DNA replication a direct base mispairing occurs past the
site of a particular AAF adduct. According to our discussion on the
structure of the adduct such a process could be present only in the case
of AAF binding to N2 of guanine, for which a model of transversion
mutation could be imagined.

- One has to assume the existence of a new recA-independent mutagenesis pathway.

c) When plasmid DNA is modified with AF instead of AAF, the transformation of the uvrArecA double mutant yields 37% survival (one lethal
hit) with an average of 45 adducts per molecule, a figure comparable
to that found with pBR322. It is thus possible that AF adduct does not
hinder the DNA replication. Such a result could be interpreted in terms
of a different local structure. As shown above, -AF adducts do not perturb greatly the B-structure but without blocking the DNA polymerase,
it would be able to create miscoding and then point mutations. It was
found indeed a low but dose-dependent level of AF-induced mutation
frequency regardless the genotype of the strain (Schmid, Fuchs and
Daune, to be published).

3. MUTAGENESIS

The covalent binding of the ultimate carcinogen to guanine residue
creates what can be called a premutational lesion. In vivo it is processed by the repair, replication and recombination enzymes and we have
studied some aspects of these processes. Finally the lesion may be
converted into a stable heritable mutation. We have now to investigate
the relationship between the premutational lesion and this final mutation, i.e. to relate directly the two events through a "black box"
which, in this type of approach, has not to be explained.
The strategy is to analyse forward mutations induced in the tetracycline-resistance gene of the plasmid pBR322 by directing the chemical
reaction of the carcinogen to a small restriction fragment of 275 bp
inside the tetracycline-resistance gene (32). After reinserting this
small piece of modified DNA into the non-reacted large restriction

fragment the reconstituted "hybrid" plasmid was used to transform recipient bacteria cells. After extraction of the plasmid DNA of tetracy-
cline-sensitive mutants the sequence of the small piece (275 bp) is
analysed and compared to that of the wild type. In this so-called
"directed" mutagenesis we have only observed frameshift mutations and
in each mutant a deletion of at least one GC pair. One challenging
result is the finding of four mutants (out of nine) at the same site,
i.e. of a hot spot of mutagenesis. Similar results were obtained with
about the same material but using reversed mutations (33). In order to
discuss the issue more carefully, it seems worthwhile to summarize
briefly the present state of frameshift mutation theory.

A first explanation was proposed in 1966 (34) : after a single-strand
break, local denaturation could take place leading to a slipped mispairing in a region with repetitive sequence. When the slippage occurs in
the parent strand a deletion appears, and conversely an addition corresponds to the slippage of the daughter strand. Intercalating agents were
supposed to stabilize the slipped-mispairing intermediate state during
a time long enough to let the error copied by the replication machinery.
However the model can be strongly opposed by observing that correlation
between intercalation and frameshift is only poor - many other stabilizing agents are not frameshift mutagens - covalently bound carcinogens
are generally destabilizing agents.
We have ourselves proposed, only in the case of non covalently bound
intercalating agents, a model in which it is assumed that the intercalating drug acts at the level of the replicating fork and that the
replication process is dissymetric (35). Frameshift mutation would thus
correspond either to the deletion of a GC pair or to the addition of
base pairs by a kind of "stuttering" of the polymerase.
Finally it was proposed a modification of the first model in which
frameshift mutagens would stabilize the loops which result from the
slippage of one strand in a repetitive sequence (36) and some experimental data are in favor of this model (37-38). Some recent experiments
(39) with alkylated acridines (ICR 191) and lac I gene have emphasized
the role of sequence of non alternating GC pairs but at the present time
there is no satisfactory model of frameshift mutation.
If now we examine our limited number of results the existence of hot
spots could be a clue to the understanding of frameshift mutagenesis.
At first one can argue that such hot spots reflect some specificity at
the level of the initial covalent binding of carcinogens. In other
words, the adduct with G, would be formed only in regions of DNA presen-
ting special structural features :
a) a local Z-form favored by a sequence of alternating GC. However we
have found that in the Z-form, the C-8 of G is no more reactive than in
the B-form (unpublished results).
b) A partial destabilization under the influence of a torsional strain.
It was shown that the existence of thymine dimers favor the denaturation
of a given sequence but the reverse was never proven.
It is thus better to assume that hot spots of mutagenesis are created
during the processing of the premutational lesion. When analysis is
made of sequence properties around the hot spots, a short hairpin secon-

dary structure can only be built after removal of a GC pair (fig. 2).
As stated in the insertion - denaturation model, the local change of
structure at the level of the modified G, might create a hinge point
and favor the hairpin structure which could also be stabilized by
interaction with an enzyme able to excise the "outside" G and C.
Actually, due to the multicopy state and to the recessivity of the
mutations which are scored in our system, the conversion of the premu-
tagenic lesion into a stable mutation most likely occurs simultaneously
in both strands prior to replication.

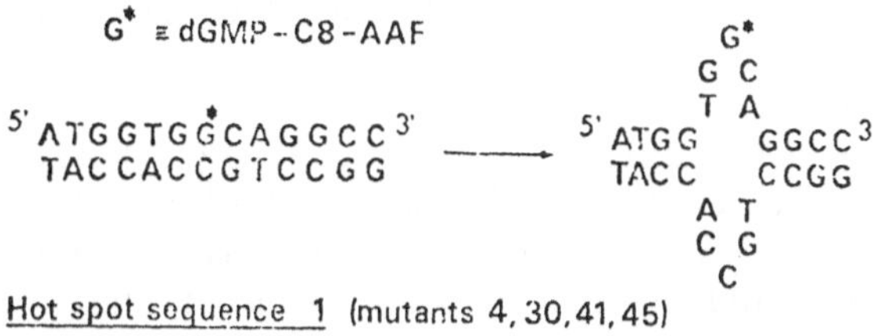

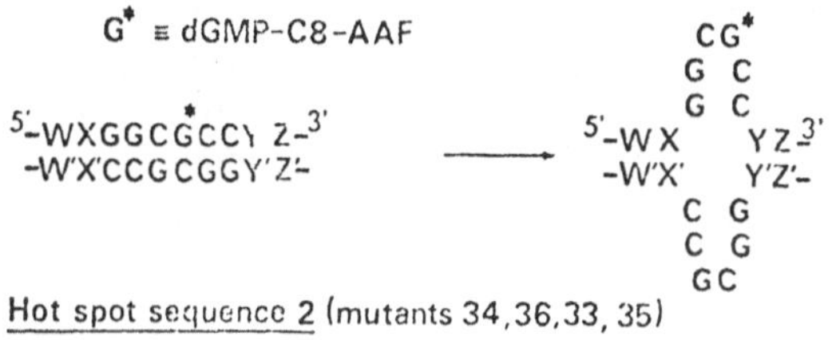

Fig. 2 - Hypothetical hair pin structure at hot spot sequences 1 and 2.
According to the insertion-denaturation model proposed by
Fuchs and coworkers (12-13) there is a local denaturation of
the helix around the guanine-AAF adduct that might favor the
hair pin structure shown in this figure.

CONCLUSION

From structural modifications at the DNA level to mutations directed
in a bacteria, our results bring some information on the processing of
a premutational lesion and comparison between fluorene derivatives was
often rewarding. Actually many problems remain unsolved but way is now
opened to new investigations.
1) The recognition process between DNA and excision repair enzymes is
not yet understood. We have observed a lack of specificity since different structural defects are excised with about the same efficiency.
However the recognition pattern has to be fine enough to distinguish

between a B-DNA and a slightly perturbed structure. One could assume
for example that a bulky substituent either is recognized when standing
out of the DNA molecule or creates a local hinge point (12 and 30). When
large amounts of pure isolated enzymes will be available for physical
studies, the knowledge of the recognition pattern between a modified
DNA and the enzyme will be of the utmost importance not only in the
problem of carcinogenesis but also in that of ageing.
2) The methodology of site-directed forward mutagenesis we have introduced to relate the nature of the mutagens with the final stable mutation was so far used in the case of AAF and the corresponding frameshift
mutation. Obviously the work has to be extended
      a) by increasing the number of examples of frameshift mutagenesis
by AAF in order to obtain a definite model of this process,
      b) by comparing AAF with parent compounds AAIF and AF to elucidate
the type of correlation which exists between a given structural defect
of the double helix and the corresponding mutational event,
      c) by using other types of carcinogens (like alkylating agents)
able to induce mispairing and then transversions or transitions instead
of frameshift mutagenesis.
3) The precise location and characterization of a mutation cannot generally tell us what are its consequence at the protein level. The deletion of a single base pair leads to a totally different polypeptide
chain after the site of mutation. In our case, some of the mutants create a non-sense codon which stops early the translation. We have already
stressed the importance of a mutation inside a promoter or close to a
promoter, to perturb the transcription of an oncogene. The analysis must
therefore go further by predicting what could be the modification of
promoter-polymerase interaction which results from the induced mutation.
Work is now in progress in our laboratory
      a) to find a new methodology in the approach of the dynamical
aspect of a promoter-RNA polymerase complex,
      b) to direct a mutation no more inside a structural gene (as that
controlling tetracycline resistance) but inside a regulation gene as
the origin of replication of E. coli.
4) Finally the use of prokaryotes to study carcinogenesis is justified
as a first approach which deals with living system offering not only the
highest simplicity of the process but also the possibility to obtain
many mutants as tools to decipher the enzymatic mechanisms. Nevertheless
mutagenesis in prokaryotes is an oversimplified view of carcinogenesis
in eukaryotes and such a work has obviously to be extended to eukaryotic
systems and especially to cells in culture.

REFERENCES

1. Bishop, J.M. : 1982, Scientific American 246, pp. 69-78.
2. Weinberg, R.A. : 1982, Trends Biochem. Sci., 7, pp. 135-136.
3. Brugge, J.S. and Erikson, R.L. : 1977, Nature 269, pp. 346-348.
4. Hunter, T. : 1980, Cell 22, pp. 647-648.
5. Temin, H.M. : 1982, Cell 28, pp. 3-5.
6. Klein, G. : 1981, Nature 294, 313-318.

7. Cairns, J. : 1981, Nature 289, pp. 353-357.
8. McCann, J. and Ames, B.N. : 1976, Proc. Natl. Acad. Sci. USA 73, pp. 950-954.
9. Kapuler, A.M. and Michelson, A.M. : 1971, Biochim. Biophys. Acta 232, pp. 436-450.
10. Sundaralingam, M. and Westhof, E. : 1981, Intern. J. Quant. Chem., 8, pp. 287-306.
11. Altona, C. and Sundaralingam, M. : 1972, J. Amer. Chem. Soc., 94, pp. 8205.
12. Fuchs, R. and Daune, M. : 1972, Biochemistry 11, pp. 2659-2666.
13. Fuchs, R. and Daune, M. : 1973, FEBS Lett., 34, pp. 295-298.
14. Sage, E. and Leng, M. : 1980, Proc. Natl. Acad. Sci. USA, 77, pp. 4597-4601.
15. Santella, R.M., Grunberger, D., Weinstein, I.B. and Rich, A. : 1981, Proc. Natl. Acad. Sci. USA, 78, pp. 1451-1455.
16. Fuchs, R.P.P., Lefèvre, J.F., Pouyet, J. and Daune, M.P. : 1976, Biochemistry 15, pp. 3347-3351.
17. Drinkwater, N.R., Miller, J.A., Miller, E.C. and Yang, N.C. : 1978, Cancer Res., 38, pp. 3247-3255.
18. Lang, M.C., Freund, A.M., de Murcia, G., Fuchs, R.P.P. and Daune, M.P. : 1979, Chem. Biol. Interactions 28, pp. 171-180.
19. Geacintov, N.E., Gagliano, A., Ivanoviv, V. and Weinstein, I.B. : 1978, Biochemistry, 17, pp. 5256-5262.
20. Hogan, M.E., Dattagupta, N. and Whitlock, J.P. : 1981, J. Biol. Chem., 256, pp. 4504-4513.
21. Daune, M.P., Fuchs, R.P.P. and Leng, M. : 1981 in "Carcinogenic and Mutagenic N-substituted aryl comounds" NCI Monograph 58, pp. 201-210.
22. Fuchs, R.P.P. : 1975, Nature, 257, pp. 151-152.
23. Sage, E., Spodheim-Maurizot, M., Rio, P., Leng, M. and Fuchs, R.P.P. 1979, FEBS Lett., 108, pp. 66-68.
24. Toulme, F., Helene, C., Fuchs, R.P.P. and Daune, M.P. : 1980, Biochemistry 19, pp. 870-875.
25. Hall, J.D. and Mount, D.W. : 1981, in "Progress in Nucleic Acids Research and Molecular Biology", Cohn W.E. ed. (Academic Press, New York) vol. 25, pp. 53-126.
26. Seeberg, E., Rupp, W.D. and Strike, P. : 1980, J. Bacteriol., 144, pp. 97-104.
27. Livneh, Z. and Lehman, I.R. : 1982, Proc. Natl. Acad. Sci. USA, 79, pp. 3171-3175.
28. Little, J.W. and Mount, D.W. : 1982, Cell, 29, pp. 11-22.
29. Ciarrocchi, G. and Pedrini, A.M. : 1982, J. Mol. Biol. 155, pp. 177-183.
30. Broyde, S., Stellman, S. and Hingerty, B. : 1980, Biopolymers, 19, pp. 1695-1701.
31. Schmid, S.E., Daune, M.P. and Fuchs, R.P.P. : 1982, Proc. Natl. Acad. Sci. USA, in press.
32. Fuchs, R.P.P., Schwartz, N. and Daune, M.P. : 1981, Nature, 294, pp. 657-659.
33. Yoon, K., Shelegedin, V.N. and Kallenbach, N.R. : 1982, Mutation Res.,
34. Streisinger, G., Okada, Y., Emrich, J., Newton, J., Tsugita, A.,

Terzaghi, E. and Inouye, M. : 1966, Cold Spring Harbor Symposium on Quantum Biology 31, pp. 77-84.

35. Schreiber, J.P. and Daune, M. : 1974, J. Mol. Biol. 83, pp. 487-501.
36. Drake, J.W. and Baltz, R.H. : 1976, Ann. Rev. Biochem. 45, pp. 11-37.
37. Lee, C.H. and Tinoco, I. : 1978 : Nature, 274, pp. 609-610.
38. Helfgott, D.C., and Kallenbach, N.R. : 1979, Nucl. Acids Res., 7, pp. 1011-1017.
39. Calos, M.P. and Miller, J.I. : 1981, J. Mol. Biol. 153, pp. 39-66.

BREATHING REACTIONS IN NUCLEIC ACIDS AND PROTEINS

Neville R. KALLENBACH
Department of Biology – Leidy Laboratories
University of Pennsylvania
Philadelphia, Pennsylvania 19104

ABSTRACT

A class of conformational fluctuations in biopolymers is revealed by hydrogen exchange (HX) rate measurements on molecules in their native state. In model double helices [poly (rA)· poly (rU)and poly (rI) · poly (rC)], two rate constants are measured in ultraviolet detected ^2H –^1H exchange experiments.   In both systems the faster rate is independent of pH, salt and general base catalysis, and is attributed to a conformational opening reaction in which the hydrogen bond involving U or I N_3–H severs and is exposed to solvent or base catalysts. The slower rate corresponds to exchange of the A or C NH_2 groups, which are slower chemically and permit estimation of the equilibrium constant for opening. A simple model for this open state is described, which relates the opening to torsional fluctuations of the double helix. The enthalpy of these fluctuations represents a relatively minor fraction (< 25%) of the total energy involved in pre-melting transitions, measured by IR or CD spectroscopy. The contrasting situation in the highly α -helical globular protein, myoglobin, has been investigated by ^1H NMR saturation transfer and line broadening experiments.

The organic base piperidine is found to be capable of catalysing exchange of the proximal histidine (F8) side chain located in the heme pocket of myoglobin. The effect of piperidine does not conform with predictions from a local unfolding model. Instead, a progressive mode of fluctuational "opening" of the molecule is indicated rather than one dominant partially unfolded intermediate. The exchange event does not occur in conditions approximating bulk solvent.

C. Hélène (ed.), Structure, Dynamics, Interactions and Evolution of Biological Macromolecules, 69–79.

1. HYDROGEN EXCHANGE AND CONFORMATIONAL FLUCTUATIONS IN MACROMOLECULES

Despite the high degree of internal order apparent in crystals of proteins and nucleic acids, it is now understood that the interior regions of both molecules possess considerable flexibility in solution as well as in the solid state. Individual residues as well as larger segments of structure can be mobile [1,2]; the time scale of these internal motions may extend from picoseconds to perhaps seconds or slower [2]. The existence of conformational flexibility within ordered systems has potentially significant functional as well as structural and energetic implications. Given the range of dynamic processes accessible to a biological macromolecule, it is difficult to sort out interesting motions from those lacking functional significance. Macromolecular dynamics have to be investigated by as many different methods as possible to obtain more data. For some thirty years, hydrogen exchange (HX) rate measurements have served as a probe of fluctuations in the native structure of proteins [4,5]. The underlying idea is simple: participation of peptide N-H groups in α-helical structures, for example, should retard the rate of exchange with solvent. This could be monitored by intensity or frequency changes in appropriate infrared bands on substituting ^2H for ^1H or _vice versa_. Practical refinements in methodology involve use of ^3H isotopic labelling and gel filtration [6]. ultraviolet absorbance detection of $^2H-^1H$ exchange [7], and extensive $^2H-^1H$ or $^1H - ^1H$ NMR exchange studies [8,9]. More recently still, neutron diffraction $^1H - ^2H$ exchange studies on crystalline molecules have been reported [10,11].

Nucleic acid HX measurements were first reported in 1965, with the advent of the ^3H -Sephadex method [6]. The conclusion from HX studies on both proteins and nucleic acids is clear: N-H (and O-H or S-H) protons in native molecules can exchange with solvent under conditions remote from denaturation, even in crystals, at rates slower than or equal to those of the same groups in the free state. Since the mechanism of proton transfer reactions requires hydrogen bonded transition states between donor and acceptor species [12], exposure of internal exchangeable residues to solvent or catalysts must occur. A class of conformational fluctuations ("breathing" reactions) is postulated to account for this. The nature of these fluctuations is uncertain at present: in one view, local cooperative unfolding of structure is proposed to lead to exposure of internal groups [13], while a second class of model postulates penetration of solvent molecules to the site of each exchangeable residue via more localized motions [9. 14]. It is shown here that while opening of the nucleic acid double helix involves a major open state in which all H of the base pair are accessible to solvent or ion catalysts, in the case of the heme pocket in myoglobin the exchange event does not occur in a medium resembling the bulk solvent. The degree of unfolding is more localized than foreseen originally.

2. THE OPEN-STATE IN A DOUBLE HELIX: POLY (rA) POLY (rU).

Analysis of a breathing reaction in the synthetic duplex, poly (rA)
• poly (rU), has provided the clearest evidence for a local opening
or partial unfolding of the duplex in which both hydrogen bonds of the
A·U base pair sever in concert [15, 16].   The evidence is summarized
in Fig. 1 and Table 1. Two kinetic processes are observed in poly (rA)
• poly (rU) as a D_2O solution is diluted rapidly into H_2O.
Exchange of ^2H for ^1H in the bases is accompanied by a difference
in UV abosrbance near 290 nm; analysis of the traces in a stopped-flow
spectrophotometer reveals a fast rate (near $1s^{-1}$ at 20°C, pH7), and a
slow rate, which is sensitive to general base catalysis as shown in
Fig. 1.

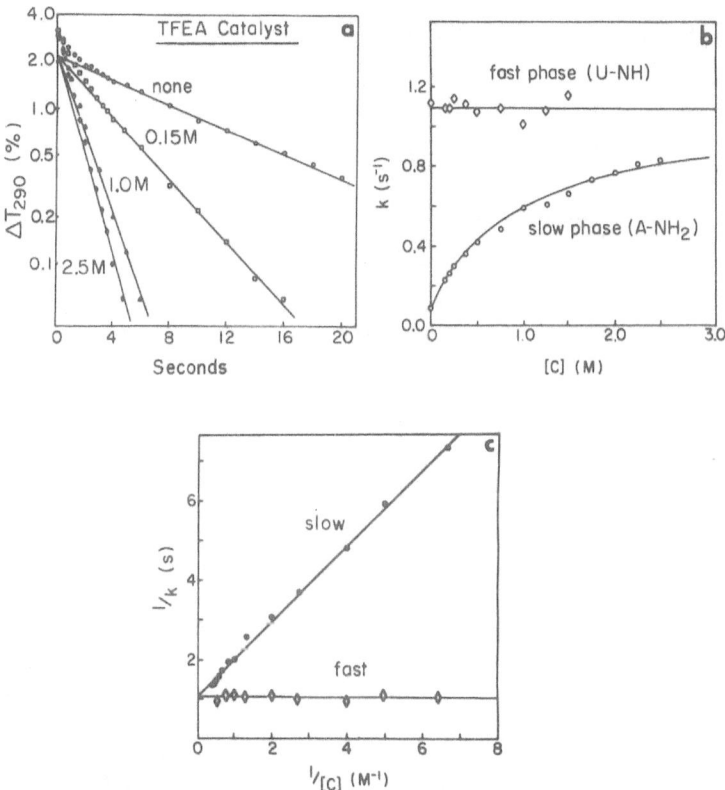

Fig 1a. Catalysis of hydrogen exchange of poly (rA) • poly (rU) by a
general base, trifluoroethylamine (TFEA). In the absence of TFEA, the
faster rate is that of U N_3-H, the slower that of A NH_2.   Only the
latter responds to buffer base. b. Dependence of the rates on
concentration of TFEA. c. Reciprocal plot of the data in panel b,
according to eqn (2) of text.

Table 1. Hydrogen Exchange Characteristics of Uridine, Amp and the two
          rates in poly(rA) · poly (rU).*

			Poly (rA) · poly (rU)	
	U N_3-H	Ap-NH_2	Fast (UN_3-H)	Slow (A-HN_2)
Rate constant pH 7 20 C, s^{-1}	1.6×10^3	8.5	1.1	.14
Activation energy (kcal/mol)	14	11.5	15	18
Catalytic response	OH⁻,B	pH independent, HB	None	pH independent HB
k_{cat} values, s^{-1}		410 [HIm] 630 [TFEA]		2.5 [HIm] 1.2 [TFEA]

*B denotes a buffer base, HB the conjugate acid.   Data from
references [7,15]. HIm is acid form of imidazole, TFEA is trifluoro
ethyl amine.

Comparison with the HX behavior of free bases shows that the latter
process is due to the A NH_2 groups; the key point is that the rate of
catalysis by a buffer, B, is proportional to the product [H⁺] [B] or
[HB], the concentration of buffer acid (Table 1). This is a
consequence of the fact that the dominant pathway for exchange of A
NH_2 is _via_ protonation of the ring. which overcomes the unfavorable
pK of the amino group [16, 17].

     The arguments that exchange of the U N_3-H and A NH_2 proton
proceeds via a concerted open state are the following:

1. The U N_3H and one A NH_2 are hydrogen bonded in the intact
    duplex, surrounded by stacked base pairs above and below. Hence
    neither is available for exchange.
2. In contrast to free uridine, the exchange rate of U N_3-H is
    independent of both pH and buffer base; it thus has the
    characteristics of a conformational reaction, not a chemical one.
3. Despite involvement of one A-NH_2 hydrogen in the H-bond and
    apparent free availability of the other, both exchange at the same
    rate (from an open state).
4. Maximal catalysis of the HX rate of A-NH_2 increases its rate
    precisely to that of the faster (conformational) step, with a
    quantitative dependence predicted from the simple opening limited
    kinetic scheme (see 15):

$$[H]_{Native} \; \underset{k_{21}}{\overset{k_{12}}{\rightleftarrows}} \; [H]_{Intermediate} \; \overset{k_{chem}}{\longrightarrow} \; [H]_{Exchanged}$$

which gives the rate:

$$k_{ex} = \frac{k_{12}\, k_{chem}}{k_{12}+k_{21}+k_{chem}} \qquad (1)$$

Under catalysis by a buffer. B, $k_{chem} = k_{cat}\,[B]$ and

$$\frac{1}{k_{ex}} = \frac{1}{k_{12}} + \frac{1}{k_{cat}\, K_{op}\,[B]} \qquad (2)$$

As shown in Fig. 1. equation (2) describes the data with respect to trifluoro ethylamine extremely well. Eqn. (1) also predicts the behavior of the faster U N_3-H because if. as in this case $k_{chem} \gg k_{12}$ or k_{21}, then eqn (1) reduces to $k_{ex} = k_{12}$, and exchange proceeds at the rate limited by the conformational process.

3.   NATURE OF HX OPEN STATES IN NUCLEIC ACIDS

What are the properties of the open state in poly (rA) · poly (rU), or in the similar poly (rI) · poly (rC) duplex?   Comparison of the equilibrium constants and thermodynamic values for the open state defined from the exchange data (17) shows first that the thermodynamics seem inconsistent with a complete "denaturational" or melting type of intermediate. The conformational opening rate is independent of salt concentration; melting is highly sensitive, on the other hand. The ΔG° and ΔH° values imply no major loss of stacking free energy in either duplex. Comparison with detailed IR and CD spectral data on these systems at temperatures below the denaturation point shows that the ΔH° for HX opening is substantially lower than that for equilibrium "pre-melting" in both duplexes.

A reasonable model for the opening is that it represents a torsional untwisting of the duplex which breaks the H-bonds and permits solvent or base to reach the N-H sites. Since the catalytic constants are always lower than those observed for the free bases, the actual exchange event may occur in a region that is within the duplex. While torsional oscillations flex the duplex, it can be imagined that at the same time solvent molecules or catalytic ions are brought closer to the interior. In terms of model double helical structures, a pure torsional movement without some change in the distance between bases seems difficult to carry out. It should be noted that exchange from a polymer-polymer duplex is slower than from a corresponding oligomer-polymer structure [17]. This could reflect two factors: 1) faster torsional motion of the short strand regions or 2) facilitated translation of mobile open states along the duplex in the latter case [18].

Why are the opening and closing reactions both so slow when torsional oscillations in principle could be very rapid? Part of the answer lies in the fact that the rates in both directions have high activation energies, suggesting that the deformation involved is more complex than simple sliding of bases over each other. One intriguing suggestion is that in the open state improper hydrogen bonding reforms, so as to freeze in the deformation. Theoretical calculations by Kollman indicate that such states with single hydrogen bonds have energies about 4-6 kcal above the native structure; they arise spontaneously in simulations where a dodeceamer is allowed to relax following a strong deformation of an internal base pair.

Finally, a note concerning the relative HX behavior of B and Z DNA duplexes. Workers at Orleans have reported that two of the five hydrogens per base pair in the high salt form of poly (dGC) · poly (dCG) are remarkably slowed in exchange relative to their rate in the low salt B form [19]. This is an important observation in connection with the nature of open states because if only the G N_1 ring proton is open limited as in poly (rA) · poly (rU), then one has to conclude that no single open-state exists in the Z structure. The argument follows from eqn (1) in its preequilibrium limit, i.e.

$$k_{ex} = K_{op} \, k_{chem} \qquad\qquad\qquad (3)$$

If in both B and Z duplexes exchange occurs from a concerted open state then the rates of exchange of the four non-open limited hydrogens should be related by a single constant:

$$k_{ex}(Z)/k_{ex}(B) = K_{op}(Z)/K_{op}(B)$$

since the chemistry is the same. The fact that this is not seen implies one of two possibilities. First, the conformational rate(s) in the Z structure are intrinsically slower than in B, and lead to a situation in which three (or more) protons are open-limited in Z but not in B. Second, the character of torsional oscillations in Z could differ fundamentally from those in B. Since the backbone in the Z structure runs alternately from one side of the helix to the other twisting or untwisting might well distinguish HX of certain of the protons in the dinucleotide repeat from that of the others in a manner not possible in the more symmetric B duplex. Clearly this issue needs to be investigated in more detail, because it sheds light on the nature of opening reactions in a situation where the chemical steps are relatively unaffected but the duplex structure changes profoundly.

4. BREATHING REACTIONS IN A GLOBULAR PROTEIN: CYANMET MYOBLOBIN

If the DNA double helix can open predominantly via the simple local torsional unwinding mode proposed, one can ask if any similar mechanism operates in the context of the much more complex structures in globular proteins. The exchange behavior of proteins is a field in itself, and has been intensively studied for years [6,8,9]. In the case of peptide N-H groups, the pK for base catalysed exchange is so unfavorable that general catalysis by base other than OH⁻ itself is unlikely, at least without gross perturbation of the native structure. However, the exchange of certain side-chain groups does present the possibility for introducing buffer bases as a probe of the unfolding reactions as we have done in the nucleic acids.

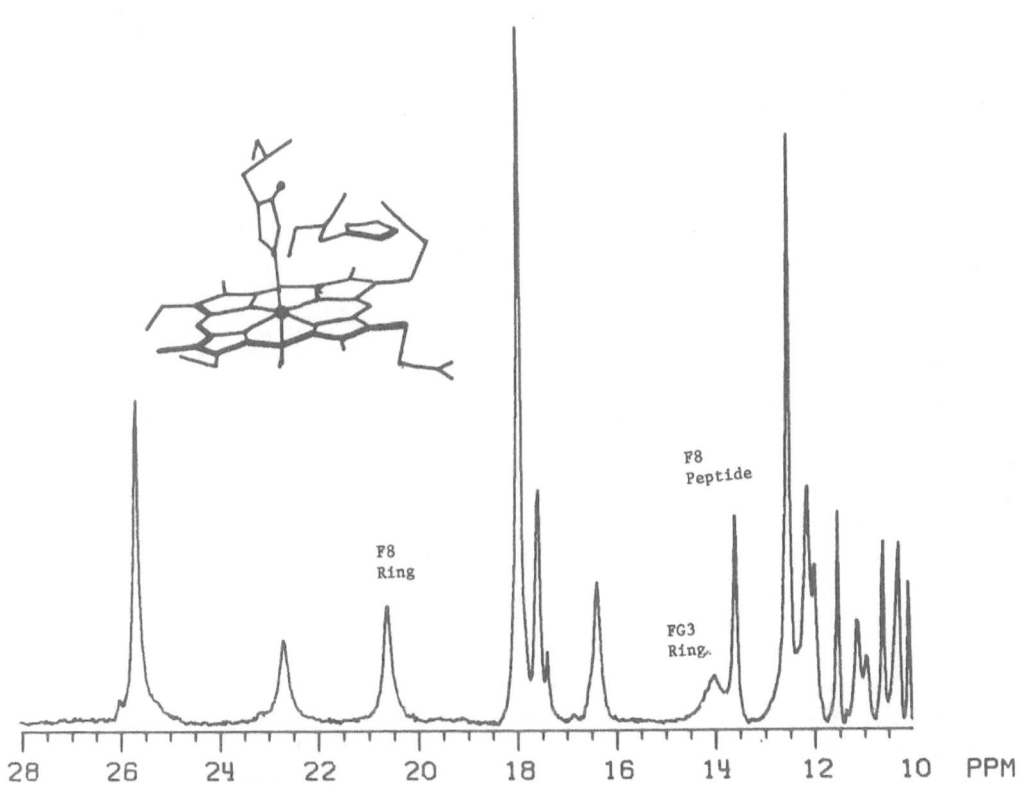

Fig. 2. The low field [1] NMR spectrum of MbCN, in .3M KCl. 10^{-4}M EDTA, 10% D_2O, showing the resonances assigned to the F8 histidine ring and peptide protons oriented in the heme pocket as indicated by the insert.

Fig 2 outlines a favorable structural situation; the F -helical segment in myoglobin clearly is one of the components responsible for burying the insoluble or hydrophobic heme group from the aqueous environment. Recent work by LaMar and his colleagues [19,20] has established the assignment of four exchangeable N-H in the 'H NMR spectrum of cyanmet-myoglobin (MbCN) in aqueous solutions - the distal histidine side chain N-H (E7), the proximal histidine side chain (F8) and its peptide in the F helix, and a third histidine side chain, that of His FG3 in the interface between the F and G helices. They have shown that both the side-chain and peptide of His F8 exchange via a OH⁻

dependent pathway.   The pH dependence of the exchange rate of the F8
sidechain and peptide (Fig 3) indicates that no significant deviation

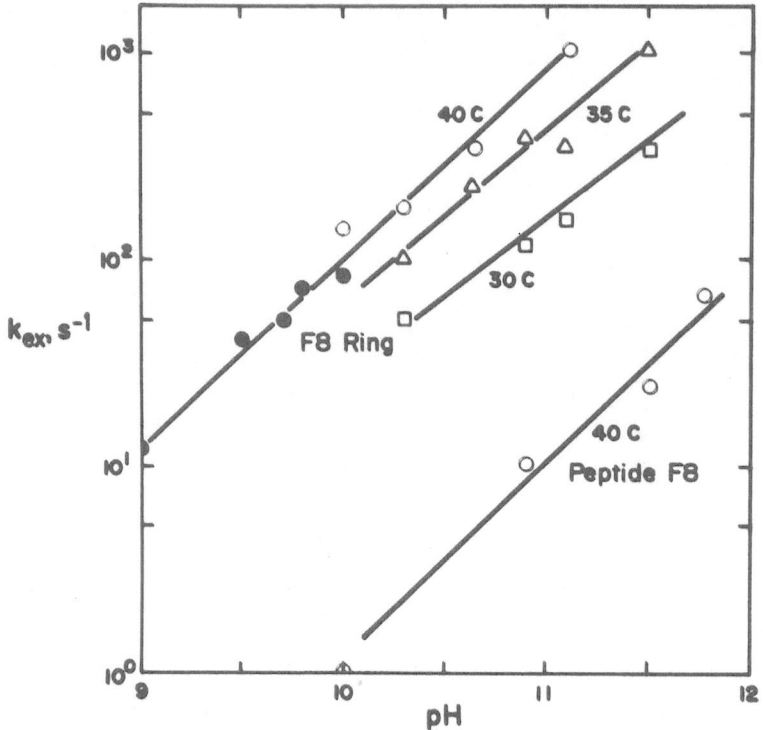

Fig. 3.   pH dependence of the exchange rate of the F8 histidine ring
and peptide N-H in MbCN as a function of pH and temperature.   The
filled circles are from measurements by Cutnell et al (1981).
Linearity of the rates of both protons indicates a preequilibrium
pathway for exchang as in eqn (3).

from a preequilibrium exchange pathway occurs from pH 8 to pH 11
(20,22).   Using a soluble imidazole Vitamin B_{12a} model for the open
state in myoglobin, the value of the equilibrium constant in eqn (3)
can be calculated; it is 10^{-6} at 40°C, and independent of pH even up
to pH 11 or higher.

    The organic base piperidine catalyzes the exchange of the F8
histidine side-chain, but not the peptide.   (Fig 4) The behavior of
piperidine as a catalyst does not conform to the predictions of a
minimal local unfolding model.   To be specific, suppose the local
unfolding event is the cooperative loss of the F α-helix unit, which
exposes the heme pocket to solvent.   Eqn. (3) then predicts that the
catalysis by a buffer with pK above that of the F8 side chain will be
independent of pH below pH 11; experimentally this is not observed [see
Fig 4].   Second, the local unfolding model predicts only a weak

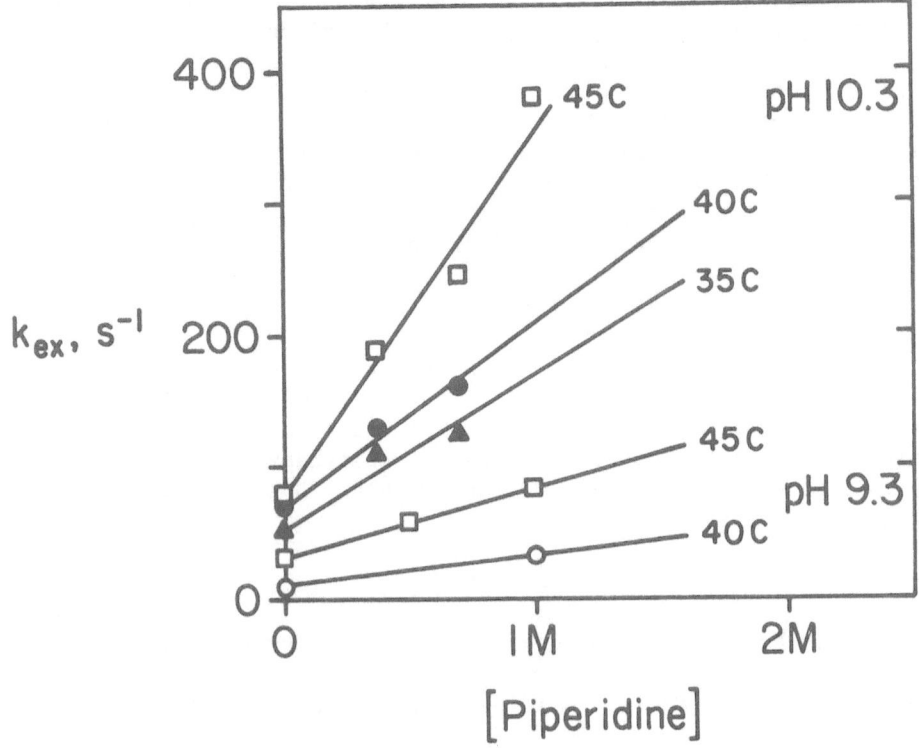

Fig 4. Catalysis of the exchange rate of the His F8 side chain N-H
group of myoglobin by piperidine at two pH values and different
temperatures. AT 30°C, pH 9.3 piperidine is ineffective. Neither the
of pH nor temperature is consistent with a minimal local unfolding
model.

temperature dependence of the piperidine catalysic rate; again this is
not seen. Since any increase in unfolding of the protein due to added
piperidine, which is itself a denaturant, should be detectable as an
increase in the exchange rate of the F8 peptide group the explanation
does not seem to be that increased unfolding by piperidine is
involved. Instead, the fluctuations that permit access of (hydrated)
OH^- ions to the heme pocket differ from those responsive to
piperidine. In any case the data strongly suggest that the exchange
event does not occur in a medium corresponding to external bulk
solvent. Models which postulate a progression of fluctuations of
increasing efficiency (amplitude?) below the denaturation region can
explain the piperidine data (Kallenbach and Kim, 1982).

                              CONCLUSION

     To summarize this discussion, breathing fluctuations in nucleic
acid double helices can be associated fairly firmly with torsional
unwinding of the duplex, which also may increase exposed solvent
accessibility of the interior of the double helix. A dominant pathway

in which coordinate opening of both hydrogen bonds of the A·U base
pair in the poly (rA) · poly (rU) model takes place is indicated.

     In the contrasting case of the much less facile exposure of side
chains in the heme pocket of myoglobin to exchange, no comparably
simple mechanism is operative. General base catalysis by piperidine
suggests that this base functions _via_ fluctuations in the protein
distinct from those that permit access of OH⁻.   Exchange does not
take place in a medium resembling bulk solvent.

ACKNOWLEDGEMENT

     This research was supported by grant PCM 77 26740 from the US
National Science Foundation and a fellowship F33GM08466 from the United
States Public Health Service. It is a pleasure to dedicate this
article to Prof Charles Sadron on his 80th birthday. The contributions
of my colleagues R. Preisler, C. Mandal. S.W. Englander and P.S. Kim
are gratefully acknowledged.

REFERENCES

1. Gurd, F.R.N. and Rothgeb, M.: 1979, Adv. in Protein Chemistry 33,
   pp 73-110.
2. Sarma, R.H. ed. Nucleic Acid Geometry + Dynamics. Pergamon Press
   1980.
3. Careri. G. (1974) in Quantum Statistical Mechanics in the Natural
   Sciences, B. Kursunoglu S.L. Mintz and S.M. Widmayer Eds. Plenum
   Press N.Y. pp 15-35.
4. Linderstrom-Lang. K.U. and Schellman, J.A.: 1959. in the Enzymes,
   P. Boyer, H. Lardy and K. Myrback Eds. Vol. 1, 2nd ed. Academic
   Press, pp 443-510.
5. Hvidt, A.: 1973. In Dynamic Aspects of Conformational Changes in
   Macromolecules, C. Sadron Ed. Reidel Holland pp 103-115.
6. Englander, S.W. and Englander, J.J.: 1978. Methods Enzym. 49, pp
   24-39.
7. Cross, D.G.: 1975. Biochemistry 14. pp 357-362.
8. Wagner, G. and Wuthrich, K.: 1979. J. Mol. Biol. 130, pp 31-37
   34 pp 75-94.
9. Woodward, C. and Hilton, B.D.: 1979. Ann. Rev. Biophys. Bioeng.
   8. pp 99-127.
10. Kossiakoff, A.A.: 1982. Nature 296, pp 713-721.
11. Wlodawer, A. and Sjolin, L.: 1982. Proc Nat Acad Sci USA 79. pp
    1418-1422.
12. Eigen, M.: 1964. Angew. Chemie Int. Ed. 3, pp 1-19.
13. Englander, S.W. et al 1980. Biophysics J. 32, pp 577-587.
14. Woodward, C.K. and Hilton, B.D.: 1980. Biophys. J. 32. pp 561-
    572. 15. Mandal, C., Kallenbach, N.R. and Englander, S.W.: 1979.
    J. Mol. Biol. 135, pp 391-411.
16. Teitelbaum, H. and Englander, S.W.: 1975. J. Mol. Biol. 92, pp 55-
    78.
17. Preisler, R. _et al_ 1981. in Biomolecular Stereodynamics Vol 1.

R.H. Sarma Ed. Adenine Press. pp 405–415.

18. Englander, S.W. et al 1980. Proc Natl. Acad. Sci US 77, pp 7222–7226.
19. Ramstein, J. and Leng, M.: 1980. Nature 288, pp 413–414.
20. Cutnell, J.D., LaMar, G.N., and Kong, S.B.: 1981. J. Am. Chem. Soc. 103, pp 3567–72.
21. LaMar, G.N., Cutnell, J.D., and Kong, S.B.: 1981. Biophys J. 34, pp 217–225.
22. Kallenbach, N.R. and Kim, P.S.: 1982 Proc Natl Acad Sci USA (submitted).

18.
19.
20.
21.
22.

CONFORMATION OF SYNTHETIC AND NATURAL POLYELECTROLYTES

G. WEILL
Centre de Recherches sur les Macromolécules - CNRS -
and Université L. Pasteur - Strasbourg - France

Coulombic interactions introduce many features in the properties of solutions of charged particles. One of the most obvious is the ionic strength dependence of the molecular conformation of highly charged flexible macromolecules such as synthetic polyacids, polynucleotides or charged polysaccharides. It reflects for example in the viscosity dependence of solutions of polyelectrolytes with the concentration and the nature of added salt. This rather old problem is not totally solved. Its quantitative understanding is important in such domain as the viscous properties and stability of polyelectrolytes solutions used in secondary oil recovery, the gel forming properties of natural polysaccharides or the evaluation of intrinsic flexibility and folding possibilities of DNA molecules. New theoretical and experimental methods have been recently developed. It is the purpose of this paper to critically review these methods and present some of their still somewhat discording results, focussing onto the ionic strength dependence of the persistence length of DNA.

1. ELECTROSTATIC CONTRIBUTION TO THE CHAIN STIFFENING AND SWELLING

Early theoretical approaches to the ionic strength dependence of polyelectrolytes conformation were based on an extension of the excluded volume theory of uncharged polymers. The electrostatic contribution to the excluded volume integral was calculated using a simple Debye potential with a screening length κ^{-1} related to the concentration of added salt[1]. These theories failed to reproduce the very general Strauss-Fuoss law[2] relating the intrinsic viscosity at finite concentration of added salt $[c_s]$ to that at infinite ionic strength.

$$[\eta] = [\eta]_{\mu = \infty} + A[c_s]^{-1/2} \tag{1}$$

They predicted a c_s^{-1} dependence unless some very unrealistic assumptions were made on the counterion distribution or on the degree of hydrodynamic interaction[3].

It has been recently remarked that electrostatic repulsions contri-

C. Hélène (ed.), Structure, Dynamics, Interactions and Evolution of Biological Macromolecules, 81–88.
Copyright © 1983 by D. Reidel Publishing Company.

bute both to a local stiffening and a long range swelling. Using the
worm like chain model, the electrostatic contribution q_e to the persistence length can be calculated from the free energy of bending of a
cylinder. Such a calculation has been performed independently by Odijk[4]
and by Skolnick and Fixman[5]. They obtain a very simple expression :

$$q_e = \frac{1_B}{4\kappa^2 b^2} \qquad (2)$$

where b is the mean distance between unit charges along the polymer
backbone and 1_B the Bjerrum length at which their electrostatic interaction in water equal the thermal energy kT. The total persistence length
q_t is then given by :

$$q_t = q_p + q_e \qquad (3)$$

when q_p is the bare persistence length of the polymer, ie the limit of
q_t at very high ionic strength.

Considering the very inhomogeneous distribution of counterions
around highly charged polyelectrolytes a simple approximation for the
evaluation of κ is given by Oosawa[6] and Manning's[7] condensation model
which predict in a satisfactory way the colligative properties of the
solution. Due to the confinement of part of the counterions in a very
thin layer around the locally stiff polyion, segments interact as if
they would have a reduced charge corresponding to a mean distance $Z1_B$,
where Z is the charge of the counterion. Therefore, for $b < Z1_B$[8] :

$$q_e = \frac{1}{4\kappa^2 Z^2 1_B} \qquad (4)$$

and κ must now be calculated form the concentration of added salt c_s
plus the fraction $b/Z1_B$ of uncondensed counterion coming from the concentration c_p of polymer. In the limit $c_s \gg c_p$ this leads for $Z = 1$ to

$$q_e = 0,32/[c_s] \qquad \overset{\circ}{A} \qquad (5)$$

In order to release the assumption of counterion condensation, Le
Bret[8,9] has calculated the electrostatic free energy of bending from a
numerical solution of the non linearized Poisson-Boltzmann equation for
a rod and a torus. The results are function of b and κa, where a is the
radius of the molecule taken as a cylinder. The main differences with
that of relation (1) and (2) in the case of DNA ($a = 10$ Å, $b = 1,7$ Å) is
that q_e remains large and a function of c_s above $10^{-2}M$

An interesting remark, easily derived from relation (2) is that due
to the preferential condensation of divalent counterions, q_e is very
sensitive to the presence of small concentrations of divalent counterions
of the order of c_p. If one uses for added salt, a divalent monovalent
salt such as $MgCl_2$ relation (3) becomes :

$$q_e = 0,046/[c_s] \qquad \overset{\circ}{A} \qquad (6)$$

Taking in account the ionic strength dependence of q_t one can calculate an electrostatic excluded volume parameter Z_{e1} and use the two
parameter theory of polymer solutions to calculate the swelling factor
α of the system[10]. In a simple approximation, the excluded volume integral β can be calculated as if the segments were cylinders with an apparent hard core diameter proportional to κ^{-1}. One predict then for
$L \gg q_t$ a variation of the radius gyration :

$$R^2 \ \alpha \ L^{6/5} \qquad q_t^{2/5} \qquad \kappa^{-2/5} \tag{7}$$

Alternatively, the use of the two parameter theory implies that
Z_{e1} can be extracted form experimental values of the second virial coefficient and used, together with a self consistent value of q_t, to correct the radius of gyration for excluded volume effects.

2. HYDRODYNAMIC METHODS FOR THE MEASUREMENT OF CHAIN DIMENSIONS

Intrinsic viscosity, sedimentation, translational and rotational diffusion coefficients measurements by quasi elastic light scattering and
electric or flow birefringence and dichroïsm are well documented techniques for the characterization of polymer conformation. The interpretation of their results for polyelectrolytes as a function of c_p and c_s
may be rather difficult. This is especially true when the overall length
L of the molecule is such that L/q_t changes from high to low values so
that the change from a coil to a flexible rod has to be taken in account
using a worm like chain model with both excluded volume and variable hydrodynamic interaction[11]. Moreover, electroviscous effects related to the
distorsion of the counterion layer may affect these dynamic experiments

It is probably for that reason that recent experiments of DNA have
been performed either on high molecular weight T_2 and T_7 phage DNA where
$L \gg q_t$ in the whole range of c_s, or on restriction fragments in the
flexible rod limit.

Some results for the NaCl concentration dependence of the persistence length derived from flow birefringence[11,12](F.B.) and dichroïsm[13]
(F.D.) on high molecular weight samples are given in Table I together
with those derived from rotational diffusion constants measured from
electric birefringence (E.B.) relaxation on low molecular weight samples.
(14).

The discrepancy between the T_7 and T_2 results may reflect the difficulty to correct for excluded volume and partial draining effects. It
must be noted that the intrinsic viscosity is well predicted by the
electrostatic theory for q_t and α with no adjustable parameter down to
$2 \cdot 10^{-3}$M NaCl, assuming complete hydrodynamic interaction[15]. There is essentially no overlap with the E.B. results which give in the low ionic
strength range q_e values about 3 times smaller as predicted by relation
(5). These experiments have also been performed with $MgCl_2$ as added salt.
Here q_t is found constant down to $5 \cdot 10^{-4}$M, increasing to 800 A in 10^{-4}M.

$[c_s]$	FBT$_2$ [11]	FBT$_7$ [12]	FDT$_7$ [13]	EB Fragments [14]
	460	400	463	
10^{-1}M	740	450	478	
$2\,10^{-2}$M	1130		510	
10^{-2}M		600	564	
$5\,10^{-3}$M	1540	650	669	500
$2\,10^{-3}$M		750		580
10^{-3}M				590
$5\,10^{-4}$M				700
$2\,10^{-4}$M				1130
10^{-4}M				1800

Table 1. DNA persistence length q_t from hydrodynamic measurements

3. SCATTERING METHODS

The well established scattering techniques (light, X ray or neutron
scattering) suffer severe limitations due to the multicomponent character of the system and to the fact that, at low ionic strength, strong
interactions preclude the description of the departure from ideality by
a virial expansion. An example is provided by the neutron scattering of
solutions of rigid spherical micelles or proteins[16,17]. The q = 0 zero
angle scattered intensity S(q) is nearly zero due to the osmotic incom-
pressibility resulting from long range electrostatic repulsion. S(q) has
a peak which is not due to any ordering but can be seen as resulting
from the exclusion of other particles in a volume of radius $\sim \kappa^{-1}$ much
larger than the geometrical radius R[18]. While no molecular weight can be
directly measured from S(0) a full analysis of S(q) provides the size
and charge of the rigid particle. No such model is available for defor-
mable flexible coils. Neutron scattering brings however new possibilities
to separate intramolecular (S_s) and intermolecular (S_I) contributions to
S(q) at low ionic strengths. For a mixture of identical protonated and
deuterated identical molecules in a contrast matching solvent of one of
the species :

$$S(q) = x\, S_s(q) + x^2\, S_I(q) \tag{8}$$

where x is the proportion of scattering molecules. Working at variable
x permits to extract $S_s(q)$[19]. Its low q variation gives the radius of
gyration and its large q behaviour can give direct access to the persistence length by curve fitting to a worm like chain model with excluded
volume. This approach is however presently limited to synthetic polyacids of medium molecular weight and requires extremely long experiments
since the scattering power is low and the substraction of the incoherent
scattering crucial in the high q range.

At higher ionic strength where S(q) is a monotonously decreasing

function of q, ordinary neutron or light scattering can in principle
be used. The interactions may however still be large enough for the
approximation involved in the Zimm procedure of extrapolation not to be
any more valid. One should always control before studying the change in
the initial slope of $S^{-1}(q)$ $\left[P^{-1}(\theta)\right.$ in light scattering language] that
the q = 0 extrapolation gives a constant molecular weight, using the
proper contrast factor $(dn/dc)_\mu$ measured between the solution and the
solvent in dialysis equilibrium.

    Such measurements have been carried out between approximately 10^{-3}M
and 4 M NaCl on a well characterized Col E_1 plasmid DNA[20]. The values of
q_t uncorrected and corrected for excluded volume effects with $Z_{e\ell}$ deduced from the second virial coefficient[20] or from a direct calculation[21]
are given in Table II. The table illustrates the difficulty in correc-
ting excluded volume effects at low ionic strength. The main new result
is the large decrease of q_t between 0,2M and 1 M NaCl and the small value of q_p as compared to previous estimates. There is however a conflicting report on a sonicated polydisperse sample of much smaller weight
average size L = 1670 Å where the uncorrected mean radius of gyration
has been found to vary from 459 ∓ 5 Å in .2 M NaCl to 479 ∓ 5 Å in
10^{-3}M NaCl[22] corresponding to a small change of q_t from 550 to 610 Å.

$[c_s]$	q_t uncorrected	q_t corrected	
		$Z_{e\ell}$ from A_2	$Z_{e\ell}$calc
0,007	914	530	820
0,012	780	539	700
0,05	648	461	600
0,1	549	398	510
0,2	498	366	465
1	360	286	330
2	341	269	320
3	331	269	310
4	320	264	295

Table 2. DNA persistence length q_t from light scattering experiments[20,21]

4. MAGNETIC BIREFRINGENCE

Considering the difficulties arising from the use of either hydrodynamic or scattering experiments in the whole range of ionic strengths,
another static measurement of the dimensions would be of much help.
Magnetic birefringence has been proposed a few years ago[23] and a correct theory recently developed[24] and applied to polystyrene sulfonate[25]
and DNA. In contrast with electrooptic methods where the orientation
at a given field is ionic strength dependent because of the change in
ionic polarisability, the magnetic anisotropy is independent of c_s.
Therefore, the equilibrium value of the magnetic birefringence is directly related to the chain conformation. In the simple model of a chain

made of segments with axial tensors of magnetic and optical anisotropy, with differences in principal values ΔX and $\Delta \alpha$, directed along the chain axis, the magnetic birefringence of a solution at concentration c is given by :

$$\Delta n = \frac{2\pi}{15} \frac{1}{n} \frac{Nc}{M} \quad \Delta\alpha \; \Delta X \; \Sigma\Sigma \; < \frac{3\cos^2\theta_{ij} - 1}{2} > \quad \frac{H^2}{kT} \qquad (9)$$

n is the solvent index of refraction, θ_{ij} the angle between segments i and j.

In the absence of excluded volume effects the double sum can be calculated for a worm like chain model and the final result writes :

$$\frac{\Delta n}{CH^2} = \frac{4\pi}{45} \frac{N}{nkT} \quad \frac{\Delta\alpha \; \Delta X}{m_o \; \ell_o} \quad q \left[1 - \frac{q}{3L} \left(1 - \exp - \frac{3L}{q} \right) \right] \qquad (10)$$

where m_o and ℓ_o are the segment mass and length. The variation of Δn with the ionic strength provides therefore a direct measurement of the variation of q. Absolute values depend upon the determination of $\Delta\alpha$ and ΔX.

We have found experiments on DNA at very low ionic strength diffi-cult because of possible denaturation, and all the experiments have been carried out at 4°C with a check on the melting profile before and after measurement. Denaturation resulted in all cases in a reduced va-lue of q. Working with low c_p values of the order of c_s we have checked that the results are c_p independent provided one uses the isodilution technique, ie maintains the total equivalent ionic strength $[c_s]$ + 0,12 $[c_p]$ constant.

From the results shown in Fig 1. three points should be noticed : i) the persistence length seems not to vary whithin experimental error ($\mp$ 5%) for $c_s > 10^{-2}M$
   ii) the high $[c_s]$ limit of q, has been calculated using $\Delta\alpha$ values known from combined measurements of electrical birefringence and dich-roïsm, and ΔX from magnetic anisotropy measurements on oriented films which have to be corrected for many systematic errors. The value presen-tly obtained q_p = 600 Å has a large uncertainty.
   iii) for that reason the low ionic strength variation of Δn has been fitted to relation (5) using different values of q_p from 400 Å to 700 Å. The agreement is good.

We have also compared measurements in NaCl and $MgCl_2$. In the lat-ter case, the change in magnetic birefringence in $10^{-4}M$ $MgCl_2$ is only 20%.

One question arises wether agreement is somewhat fortuitous since we have neglected the effect of excluded volume. Experiments with the much more flexibles PSS reveal a large effect of the chain length. This seems however mostly due to the fact that the PSS experiments have been there performed at concentrations where chain considerably overlap. This is much less the case with our DNA (M $\sim$ 4 10^6,c_p $\sim$ .1mgr/mℓ).

A simple argument, based on the comparison of the double summation over $<3\cos^2\theta_{ij} - 1>$ in relation (7) as compared to a double summation over $<\cos\theta_{ij}>$ for the chain dimension would indicate a reduced influence of the excluded volume effects. This point deserves however further studies.

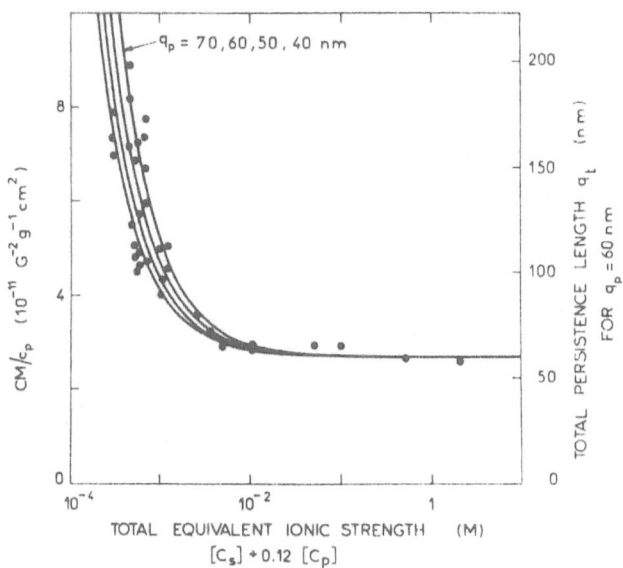

Fig 1. Magnetic birefringence of DNA solutions

4. CONCLUSION :

There is a need for concerted experiments on well defined polyelectrolyte systems. The possibility to cover a wide range of molecular weight
with essentially monodisperse samples makes DNA a ideal system to study
the relative importance of local stiffening and overall swelling. The
possibility to prepare protonated and deuterated PSS allows to use the
full capability of neutron experiments. The preliminary results obtained
with divalent counterions would suggest that they should systematically
be used to reduce the importance of local stiffening versus long range
excluded volume effects.

REFERENCES

1. Noda,I., Tsuge,T., and Nagasawa,M.,:1970,J.Phys.Chem. 74,710
2. Fuoss,R.M., and Strauss,U.P.,:1949,Ann.N.Y.Acad.Sc. 51,836
3. Imai,N.,:1980,Rep.Progr.Pol.Phys.Japon,23,95
4. Odjik,T.,:1977,J.Pol.Sc.Pol.Phys. 15,477
5. Skolnick,J., and Fixman,M.,:1977,Macromolécules 10,944
6. Oosawa,F.,:1971,Polyelectrolytes,Dekker,M.,N.Y.
7. Manning,G.S.,:1969,J.Chem.Phys. 51,924
8. Le Bret,M.,: 1981, C.R.A.C.Sc. 292,II,291
9. Le Bret,M., and Zimm,B.H.,:1981,Pullmann,B.,Ed.,"Intermolecular
    Forces".,Reidel,D.,Pub.Co.,p.257
10. Odjik,T., and Houwart,A.C.:1978,J.Pol.Sc.Pol.Phys. 16,627
11. Harrington,R.E.,:1978,Biopolymers 17,919
12. Cairney,K.L., and Harrington,R.E.,:1982,Biopolymers 21,923
13. Rizzo,V., and Schellmann,J.,:1981,J.Biopolymers 20,2143
14. Hagerman,P.J.,:1981,Biopolymers 20,1503
15. Odjik,T.,:1981,Biopolymers 18,3111
16. Hayter,J.B., and Penfold,J.:1981,Mol.Phys. 42,109
17. Hayter,J.B., and Penfold,J.:1981,J.Chem.Soc.Far.Trans.I 77,1851
18. Benmouna,M., Weill,G., Benoit,H., and Akcassu,Z.:1981, to appear in
    J.Phys.
19. Williams,C.E. and al.:1979,J.Pol.Sc.Pol.Letters 17,379
20. Borochov,N., Eisenberg,H., and Kam,Z.:1981,Biopolymers 20,231
21. Manning,G.S.:1981,Biopolymers 20,1751
22. Mandel,M. and Schouten,J.:1980,Macromolecules 13,1247
23. Maret,G. and Dransfeld,K.,:1977,Physica 86/88B 1077
24. Weill,G.:1981,Molecular Electrooptics,Krause,S.,Ed.,Plenum.Press,
    pp.473
25. Weill,G., and Maret,G.,submitted to Polymer

DNA CONFORMATION AND FOLDING: FROM SOLUTION TO THE HIGHER ORDER
STRUCTURE OF CHROMATIN

Juan Ausio, Nina Borochov, Zvi Kam, Michael Reich,
Dalia Seger and Henryk Eisenberg
Department of Polymer Research, The Weizmann Institute of
Science, Rehovot 76100, Israel

We describe the folding of DNA in solution, in the nucleosome and
in chromatin. From measurements of plasmid $ColE_1$-III linear LiDNA in
LiCl solutions we find a limiting value for the persistence length a
of 29 nm at high concentrations of salt. No condensation of DNA occurs
up to 5 M LiCl. DNA in chromatin, at low concentrations of salt, is
compacted because of the folding of the DNA in the nucleosome. To DNA
of chromatin comprising about 45 nucleosomes, would correspond a radius
of gyration R_g of 253 nm, if free in solution. The measured value for
the "10 nm" fiber is 74 nm. Considering the folding due to the linker
DNA we calculate about 83 nm for R_g. Upon addition of up to 75 mM NaCl
or 0.3 mM $MgCl_2$ chromatin folds into the "30 nm" higher order fiber
and R_g for this sample is reduced to about 28 nm. We calculate 5.9
nucleosomes in the helical repeat, if a regular solenoidal structure
is assumed.

INTRODUCTION

Ever since the structure of the high humidity B form double helix
DNA was described in 1953 (1), it was believed to represent the para-
mount DNA form encountered in all natural systems. Only recently has
there been a growing realization that, in addition to the already
earlier defined right-handed A and C helical forms, additional families
of left-handed, zig-zag forms and mixed structures are feasible (2). A
crucial point in these studies is the question relating to the biological significance of these newly established structures (3). Though
the flurry of contributions which started with the pathbreaking work
of Pohl and Jovin (4), and is now concentrating mostly on the high resolution analysis of model crystal structures (5,6) is continuing unabated,
it is fair to say that the B structure of DNA will continue to represent the structure of the vast majority of DNA extant, and any other
structure may, at best, fulfill a specialized, and maybe temporal role,
adapted to particular, even though vital situations.

C. Hélène (ed.), Structure, Dynamics, Interactions and Evolution of Biological Macromolecules, 89–100.
Copyright © 1983 by D. Reidel Publishing Company.

The shape and conformation of the B DNA structure has been a favorite research topic for many years. Basically a straight linear structure (though note some very interesting arguments postulating the existence and rationale for naturally bent DNA (7)), DNA bends in solution as a result of random thermal motion balanced by a natural elastic response. Thus, whereas an intact unfolded DNA molecule of a human chromosome may have a molar mass as high as 1.6×10^{12} g/mole, contain 2.4×10^{8} base pairs (b.p.), a contour length L_c of 8.2 cm and a thread diameter of about 2.5 nm, it would fold in solution, if intact into a spherical coil of radius of gyration R_g about 0.035 mm only. Such large DNA molecules are inherently unstable and are easily broken under the slightest shear. A rough practical limit for pure specimens of unbroken DNA in solution is bacteriophage T2 virus DNA, $M \cong 1.15 \times 10^{8}$ g/mole.   Under special circumstances DNA preparations freshly isolated from $E.\,coli$ bacteria, containing unbroken DNA ($M \cong 3.9 \times 10^{9}$ g/mole), have been investigated in imaginative ways (8,9).

It is an essential characteristic of biological systems that huge stretches of unbroken DNA can be organized in compact, neat packages, capable of undergoing reversible processes of folding and unfolding to cope with functional demands. Morphological units, such as chromosomes, go through a variety of resting and active stages when their DNA becomes available for transcription and replication. Another particular striking example is the release of the total amount of DNA packed in the head of a phage, upon attachment of the latter to a bacterial cell wall and its penetration, or upon osmotic shock (10). The DNA molecule is thus capable of considerable folding and compaction. It is known that drastic shape changes and condensation of DNA can be achieved by the addition of non-aqueous solvents and/or polyvalent ions (11), without apparent loss of the classical B form. In nature, folding of DNA into well organized structures is usually achieved by interaction with well specified proteins, histones in the case of chromosomal structures. The interaction is finely balanced and involves a minimum of stabilizing energy. Thus, "at the flip of a coin", packaged DNA can unfold "for action", or vice-versa. The general principle is clear, yet very little is known about the actual steps involved in these conformational transitions.

The conformation of DNA in solution is best interpreted in terms of the descriptive concept of the wormlike coil (12). A natural connection exists between the persistence length a the statistical parameter of the theory, and the energetics of the bending of the stiff rod (13). Whereas in solution, radii of curvature of the smooth DNA coil are of the order of a few hundred Å, in the nucleosome core particle (14), for instance, superhelical DNA, still in the B form, curves over a radius of about 50 Å. Earlier it was believed (15,16) that, in order to achieve this degree of coiling the DNA must kink and base pairs unstack and rearrange in alternate ways. This ghost has happily   been laid to rest (17,18) and it is now well accepted that DNA can easily and smoothly bend in the nucleosome type structure.

In the following we will review some quantitative aspects of the folding of DNA in solution, in the nucleosome and unfolded chromatin, concluding our present quest at the stage of the so-called higher order chromatin structure. Further stages of considerable DNA compaction are observed in the chromosome.

The Persistence Length of DNA in Solution

We have recently investigated the conformational properties of a homogeneous sample of linear $ColE_1$ DNA over a wide range of NaCl concentrations (19,20). This DNA has a molar mass of 4.35×10^6 g/mole (in the Na form), is composed of 6594 base pairs, and has a contour length L_c of 2230 nm in the B form (0.34 nm repeat per b.p.). We have determined the following experimental quantities: velocity sedimentation coefficients $s_{20,w}^0$, molar mass M and radii of gyration R_g by elastic total intensity light scattering and translational diffusion coefficient $D_{20,w}$, by quasielastic light scattering. The molar masses calculated from elastic light scattering and from the Svedberg equation agree closely with this quantity derived from DNA sequencing study. The inner consistency of the experimental data is thus established.

From R_g and L_c it is possible to calculate a by an analytical expression due to Benoit and Doty (21). This expression is correct in the absence of excluded volume corrections. We have therefore calculated reduced R_g values corrected for intramolecular excluded volume from measured values of the intermolecular excluded volume (obtained from the second virial coefficient A_2, determined from the concentration dependence of the scattered light, extrapolated to zero scattering angle) (19). This procedure may lead to overcorrection, in particular at very low values of the ionic strength, but is believed to be quite trustworthy at the higher values (above 0.1-0.2 M NaCl) of the ionic strength. A better estimate results (Table II, Ref. 20) from an application of a theory (22) of scattering curves of wormlike chains with excluded volume. Rather similar results have been obtained in an application by Manning (23) of the theory of Yamakawa and Stockmayer (24) of the excluded volume of finite chains and more recently in a Monte Carlo analysis applied to DNA (25). The final answer can only be given by extensive investigation of additional DNA samples, of different contour length, to establish whether a chain length dependence indeed exists.

In the present work we have proceeded in a different way. Whereas, in NaCl solutions, the A_2's of DNA solutions do not vanish (or become negative) even at the highest concentrations of the salt, LiDNA in LiCl is expected to behave in a different way. Thus it is believed, that at very high LiCl concentrations (above 5-6 M), DNA condenses and collapses into a compact form (26,27). It is to be expected that, prior to molecular collapse repulsive interactions, characteristics of a positive excluded volume, transform, with increasing salt concentrations, into attractive interactions. When these latter interactions compensate volume exclusion due to geometric exclusion, akin to the van der Waals

Table 1. Molecular properties of linear ColE$_1$ form III DNA
in solutions of LiCl[*]

	M, LiCl				
	0.2	2	3	4	5
s_{20w}^O, S	15.4	16.2	17.2	17.3	17.0
$D_{20,w}^O \times 10^8$, cm^2/sec	2.07	2.36	2.50	2.60	2.60
R_g, nm	176	146	143	143	142
$A_2 \times 10^4$, ml mol/g^2	3.3	0.6	0.3	0.4	0.01
a, nm	44.3	30.0	29.1	29.1	29.1

[*] Solutions also contain phosphate buffer (4.8 meq Na$^+$) 2 mM
NaEDTA, pH 7. 5 M LiCl solution contains only 4 mM NaEDTA,
pH 7.

real gas situation, the virial coefficients vanish (28). We have therefore examined the behavior of LiDNA solutions in LiCl, and the experimental results are summarized in Table 1. A detailed analysis of these
data will be presented elsewhere. Here we would only like to point out
that A$_2$ indeed becomes very small at high LiCl concentration and vanishes
(within experimental error and as compared to the much higher values previously reported for NaCl solutions (19)) between 4 and 5 M LiCl. The
variation of R$_g$ and of the hydrodynamic data s$_{20,w}$ and D$_{20,w}$ with LiCl
concentration is smooth and no incipient condensation or molecular collapse
is noticed. This is in agreement with the findings from X-ray diffraction of fibers immersed in various media (29) that the B form of LiDNA
is maintained in LiCl concentrations as high as 8 M. Extension of our
experiments to higher LiCl concentrations, towards negative A$_2$'s and
second order phase transitions (intramolecular condensation prior to
intermolecular aggregation and precipitation) are in progress.

    For the LiDNA-LiCl solutions for which A$_2$ vanishes we can calculate
a without applying any further excluded volume correction. We find
$a \cong$ 29 nm, which is rather close to the value of NaDNA-NaCl solutions,
derived at high concentration (2-4 M) of NaCl. The corresponding values
in NaCl were $a \cong$ 33 nm without and $a^O \cong$ 27 nm after applying an excluded
volume correction . From the present, and the previously reported data
we thus conclude, that the persistence length of DNA tends to a limiting
value of about 29 nm, at high concentrations of salt. This is somewhat
lower than a lower limit of about 45-50 nm determined by other workers
from an evaluation of electro-optical and hydrodynamic data (30,31).
Our conclusion is that DNA is somewhat more flexible than previously
believed which facilitates smooth folding into organized structures. A
detailed discussion will be presented elsewhere.

    In the analysis of quasielastic light scattering the apparent diffu-
sion constant D$_{app}$ (obtained from 1/2 τ q^2, where τ is the decay time
of a single exponential fit to the autocorrelation function of the

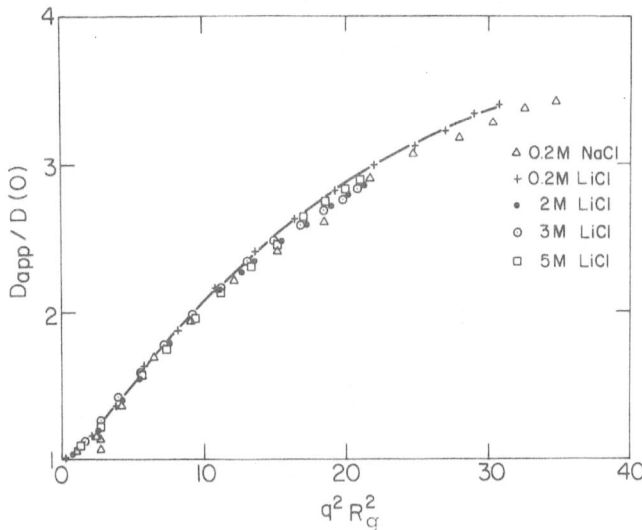

Figure 1. Dependence of the normalized diffusion coefficient
$D_{app}(q)/D(o)$ on $q^2 R_g^2$ for the plasmid $ColE_1$-III Li DNA at
various concentrations of LiCl. The behavior of the Na form
of the plasmid in NaCl (20) is shown for comparison.

scattered light fluctuations and q is the scattering vector
$(4\pi n/\lambda)\sin(\theta/2)$, has been identified with the translational diffusion at
low values of q. We have previously shown that $D_{app}(q)/D(o)$ scales with
$q^2 R_g^2$ for all values of the ionic strength (20). This relation now extends to the LiCl data (Fig. 1). The striking dependence of $D_{app}(q)/D(o)$
on $q^2 R_g^2$ has not been fully explained. Note below that in the more compact, or perhaps less mobile, chromatin samples the dependence of D_{app} on
scattering angle is much reduced.

Folding of DNA into Chromatin Structures

    Considerable compaction of DNA occurs when it complexes with histones
and non-histone proteins to form chromatin-like structures. Intact chromatin fragments were obtained from mature chicken erythrocytes nuclei by
mild micrococcal digestion, dialysis into hypotonic solution(0.25 mM EDTA),
and separation on sucrose gradients. The distribution of DNA length in
the chromatin fragments was carefully determined by sedimentation analysis
of the purified DNA. Here we present initial results on two chromatin
samples. The number N of nucleosomes in each sample has been determined
by dividing the number of DNA base-pairs by 210, the average repeat
of base-pairs per nucleosome, in the chicken erythrocyte genome. The
correct number, n, weight w, or higher average of N has been used, consistent with the specific experimental quantities derived (32).

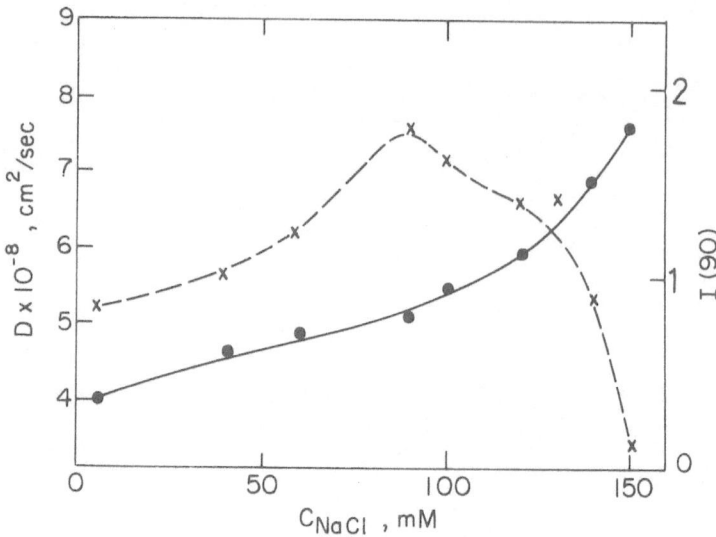

Figure 2. Scattering behavior of chromatin as a function of
NaCl concentration (x) D(90) and (●) I(90), arbitrary units.
Solutions also contain 1 mM Tris, 0.1 mM EDTA, pH 8.A_{260}^{1cm} = 0.7.

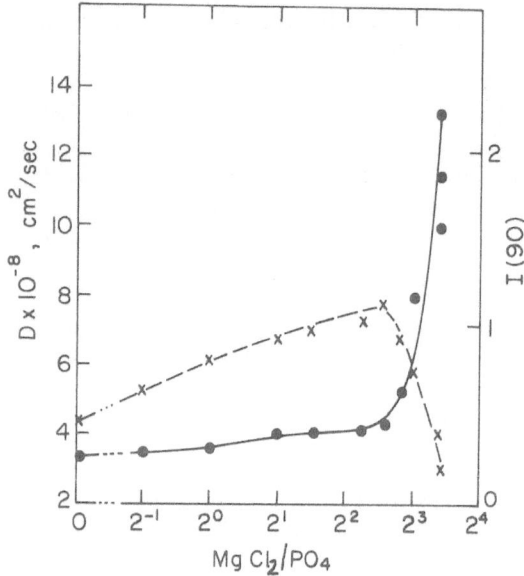

Figure 3. Scattering behavior of chromatin as a function of
$MgCl_2/PO_4$ ratio. A_{260}^{1cm} = 0.7 (x) D(90) and (●) I(90), arbitrary
units. Solutions also contain 5 mM NaCl, 1 mM Tris, pH 8.

At very low ionic strength chromatin, containing a full complement of core histones as well as histone H1 (and H5 in the case of chicken erythrocytes) and some non-histone proteins, is in the form of the "10 nm" coil. With increase in NaCl or $MgCl_2$ concentration, this chromatin undergoes a further compaction to the more rigid higher order structure, the "30 nm solenoid" (33). Data presented below refer to these two structures. An earlier light scattering study on this system has been reported by Cotter *et al.* (34).

For a preliminary understanding of the effect of either NaCl or $MgCl_2$ on chromatin systems, and to establish limits of solubility, we have undertaken a study of D_{app} and of the scattering intensity I at 90° scattering only, in the 50 µl transparent test tube of the Beckman Airfuge (35). A sampling of data is shown in Figs. 2 and 3. Upon increase of salt concentration in both instances (yet at considerably lower values in the case of $MgCl_2$, much lower than corresponding to equal ionic strength) D_{app} increases, qualitatively commensurate with increasing compaction of the chromatin. At well defined critical points D_{app} decreases dramatically indicating strong aggregation and incipient precipitation. In the critical region I(90) increases considerably as well. It is thus found (Fig. 2) that precise measurements in the NaCl system should not be undertaken above 80 mM NaCl; for the $MgCl_2$ system (Fig. 3, and data not shown) we found that, independent of chromatin concentration, over a very wide range of concentrations, there exists a critical concentration of 0.5-0.6 mM $MgCl_2$ for the solubility of average length (20-50 nucleosomes) chromatin. The chromatin length dependence is under study.

The behavior of the sedimentation coefficients for the two fractions is shown in Fig. 4. These data (for NaCl) are in good agreement with recently published values for chicken erythrocyte chromatin (36). The behavior of the diffusion coefficients is shown in Fig. 5. Notice the relatively small dependence of D_{app} on scattering angle Θ, when compared to the DNA data, in Fig. 1. Results from elastic scattering experiments are not shown but are summarized, together with the other data, in Table 2.

We see in Table 2 that the weight average molar masses M_w from elastic light scattering are in good agreement with M_w from the Svedberg equation and with M_w from the independently derived DNA distribution (assuming a molar mass of 268,000 for the repeating unit comprising 210 base-pairs). This establishes the consistency in the analysis. Both $s_{20,w}$ and $D_{20,w}(0)$ decrease with increase in salt concentration (the values at 75 mM NaCl and 0.3 mM $MgCl_2$ + 5 mM NaCl are very similar) whereas R_g decreases in both cases. We can now make some preliminary quantitative estimates.

We consider the measurements in 5 mM NaCl to correspond to the unfolded "10 nm" coil. (Additional measurements in 1 mM NaCl are now in progress.) For a random coil the measured R_g corresponds to the z-average, M_z, of the molar mass. Consider the longer chromatin sample,

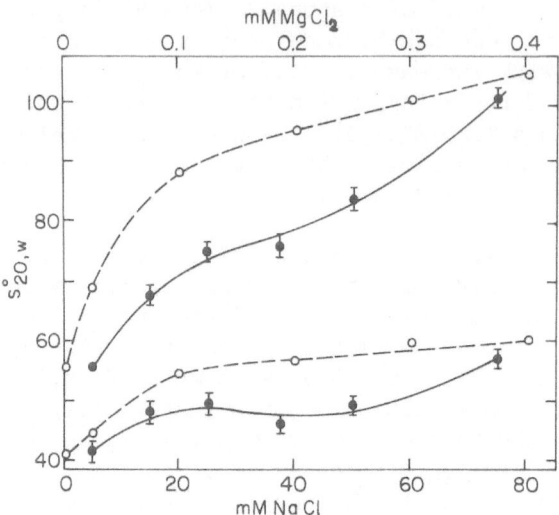

Figure 4. Sedimentation coefficients $s_{20,w}$ of chromatin samples fraction A (lower two curves) and fraction B (upper two curves) cf. Table 2, as a function of NaCl (●) and MgCl$_2$ (o) concentration. NaCl solutions also contain 1 mM Tris, 0.1 mM EDTA, pH 8 and MgCl$_2$ solutions also contain 5 mM NaCl, 1 mM Tris, pH 8.

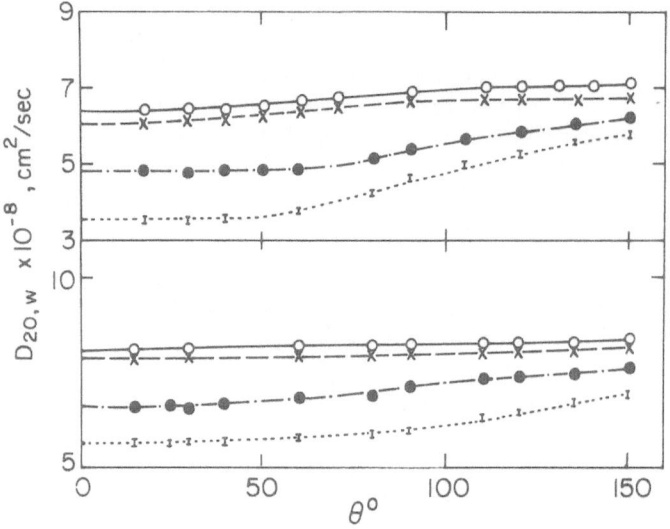

Figure 5. Diffusion coefficients $D_{20,w}$ of chromatin fraction A (lower panel) and fraction B (upper panel) as a function of scattering angle Θ,(I) 5 mM NaCl; (●) 37.5, mM NaCl; (o) 75 mM NaCl; (x) 0.30 mM MgCl$_2$. For exact composition of solutions *cf.* Fig. 4.

Table 2. Molecular Properties of chicken erythrocyte chromatin fractions

Fraction A

$M_n=2.4 \times 10^6$ $M_w=4.5 \times 10^6$ $M_z=5.85 \times 10^6$ $M_{z+1}=7.20 \times 10^6$

salt	$s_{20,w}$	$D_{20,w} \times 10^8$	R_g	$M_w(s,D) \times 10^6$	$M_w(LS) \times 10^6$
mM	S	cm^2/sec	nm	g/mol	g/mol
5, NaCl	41.5	5.9	42.6	4.89	4.43
37.5	48.5	7.1	26.4	4.75	4.51
75	58	8.7	16.6	4.64	4.32
0.3, $MgCl_2$	59	8.3	17.7	4.94	4.58

Fraction B

$M_n=4.4 \times 10^6$ $M_w=9.4 \times 10^6$ $M_z=12.1 \times 10^6$ $M_{z+1}=13.7 \times 10^6$

salt	$s_{20,w}$	$D_{20,w} \times 10^8$	R_g	$M_w(s,D) \times 10^6$	$M_w(LS) \times 10^6$
mM	S	cm^2/sec	nm	g/mol	g/mol
5, NaCl	55	3.8	74.3	10.0	10.0
37.5	77.5	5.2	45.4	10.4	9.7
75	100	6.8	27.8	10.2	9.8
0.3, $MgCl_2$	100	6.3	31.6	11.0	8.6

Values for the density $(\partial \rho / \partial c)_\mu = 0.350$ and refractive index increments $(\partial n / \partial c)_\mu = 0.180$ were used in the calculation of the molecular weights and in the reduction of s and D to standard conditions.

fraction B ($M_z = 12.1 \times 10^6$, $N_z = 45$). If the DNA would be free in solution we calculate $R_g = 253$ nm (cf. Fig. 1 of Ref. 20, assuming $a = 60$ nm) which is to be compared with the experimental value $R_g = 74.3$ nm at this concentration of salt (Table 2). This compaction is the result of the organization of 165 base-pairs of DNA in two superhelical turns per nucleosome. Assuming that the entry and exit positions of the DNA from the nucleosomes roughly coincide, we now calculate whether the assumption that the overall "10 nm" coil conformation corresponds to 45 x N_z base-pairs of DNA (45 base-pairs is the linker length) freely coiling in solution ($a = 60$ nm), is realistic. We find $R_g = 105$ nm, which is larger than, but much closer to, the experimental value $R_g = 74.3$ nm. In another model we assume (E. Trifonov, private communication) that the DNA entry and exit trajectories form an angle of 125°-130°, that the 45 base-pairs long linkers are roughly linear, and random rotation for the addition of successive nucleosomes. From standard treatment of bond rotations in polymers (37) we calculate $R_g = 83$ nm, again to be compared with the experimental value $R_g = 74.3$ nm.

The second model thus conveys better the physical situation, that the "10 nm" coil is considerably compacted as compared to free DNA, and that the DNA sequestered in nucleosome formation accounts for a large part of the compaction achieved. The models presented express the fact that the structure is not rigid but possesses residual degrees of freedom of coiling, leading to some additional compaction. This may be due to the constraints imposed by histones H1 and H5, which have not been included in the rough calculation.

We now consider the structure formed at 75 mM NaCl, or 0.3 $MgCl_2$, corresponding to the "30 nm solenoid". Let us again examine the larger fraction B (Table 2). Assuming this to be a cyclindrical rodlike struc-ture the experimental value R_g = 27.8 nm at 75 mM NaCl (Table 2) corres-ponds to the z (z+1) average of N (32). This value is $N_{z,z+1}$ = 48. From the value of the cross-sectional radius of gyration R_c = 9.3 nm, determined by small angle X-ray scattering, we calculate R = 14.7 nm for the radius of the cylinder. Using

$$R_g^2 = \frac{R^2}{2} + \frac{L^2}{12}$$

we calculate a length $L_{z,z+1}$ = 89.3 nm for the cylinder. If we assume that the nucleosomes are organized helically in the higher order struc-ture and that the pitch P is equal to 11 nm, then we calculate a value

$$n = \frac{N_{z,z+1}\, P}{L_{z,z+1}}$$

equal to 5.9 for n. A similar result is obtained for the shorter chroma-tin fragment. This is very close to the value of 6 postulated as the nucleosomal repeat per helical turn in the "30 nm solenoid" structure (33). A detailed evaluation of this and the feasibility of alternate models, as well as calculation of the frictional parameters, will be given elsewhere.

In summary, we have indicated in this report, how the process of DNA compaction proceeds from free DNA in solution to the "10 nm" chromatin coil and further to the "30 nm solenoid". Both further experimental work and theoretical evaluations are in progress and we would like, at this stage, to consider our presentation in the narrower sense of con-veying a semiquantitative feeling for the compaction process, rather than as a definitive discussion of the precise structural aspects involved.

Acknowledgement

This work was supported by a grant from the United States-Israel Binational Science Foundation, Jerusalem, Israel. J.A. is supported by an EMBO long term fellowship.

References

1. Watson, J.D. and Crick, F.H.C.: 1953, Nature 171, pp.737-738.
2. Wang,A.H.-J., Quigley, G.J., Kolpak, F.J., van der Marel, G.,
       van Boom, J.H. and Rich, A.: 1981, Science, 211, pp.171-176.
3. Rich, A.: This volume.
4. Pohl, F.M. and Jovin, T.M.: 1972, J.Mol.Biol. 67, pp.375-396.
5. Drew, H.R., Wing, R.M., Takano, T., Broka, C., Tanaka, S.,
       Itakura, K. and Dickerson, R.E.: 1981, Proc.Natl.Acad.Sci.U.S.A.
       78, pp.2179-2183.
6. Shakked, Z., Rabinovich, D., Cruse, W.B.T., Egbert, E., Kennard, O.,
       Sala, G., Salisbury, S.A. and Vismamitra, M.A.: 1981, Proc.R.
       Soc.London Ser.B, 213, pp.479-487.
7. Trifonov, E.N.: 1980, Nucl.Acids Res. 9, pp.4041-4053.
8. Weissman, M., Schindler, H. and Feher, G.: 1976, Proc.Natl.Acad.Sci.
       U.S.A. 73, pp.2776-2780.
9. Klotz, L.C. and Zimm, B.H.: 1972, J.Mol.Biol. 72, pp.779-800
10. Kleinschmidt, A.K., Lang, D., Jacherts, D. and Zahn, R.K.: 1962,
       Biochim.Biophys.Acta 61, pp.857-864.
11. Widom, J. and Baldwin, R.L.: 1980, J.Mol.Biol. 144, pp.431-453.
12. Kratky, O. and Porod, G.: 1949, Rec.Trav.Chim.Pays-Bas 68, pp.1106-
       1122.
13. Landau, L. and Lifshitz, E.: 1958. "Statistical Physics",
       Pergamon Press, Oxford, pp.478-482.
14. Finch, J.T., Lutter, L.C., Rhodes, D., Brown, R.S., Levitt, M.
       and Klug, A.: 1977, Nature 269, pp.2936.
15. Crick, F.H.C. and Klug, A.: 1975, Nature 255, pp.530-533.
16. Sobell, H.M., Tsai, C.C., Gilbert, S.G., Jain, S.C. and
       Sakore, T.D.: 1976, Proc.Natl.Acad.Sci.U.S.A. 73, pp.3068-3072.
17. Sussman, J.L. and Trifonov, E.N.: 1978, Proc.Natl.Acad.Sci.U.S.A.
       75, pp.103-107
18. Levitt, M.: 1978, Proc.Natl.Acad.Sci.U.S.A. 75, pp.640-644.
19. Borochov, N., Eisenberg, H. and Kam, Z.: 1981, Biopolymers 20
       pp.231-235.
20. Kam, Z., Borochov, N. and Eisenberg, H.: 1981, Biopolymers
       20, pp.2671-2690.
21. Benoit, H. and Doty, P.: 1953, J.Phys.Chem. 57, pp.958-963.
22. Sharp, P. and Bloomfield, V.A.: 1968, Biopolymers 6, pp.1201-1211.
23. Manning, G.S.: 1981, Biopolymers 20, pp.1751-1755.
24. Yamakawa, H. and Stockmayer, W.H.: 1972, J.Chem.Phys. 57,
       pp.2843-2854.
25. Post, C.B.: 1982, Biopolymers, submitted.
26. Wolf, B., Berman, S. and Hanlon, S.: 1977, Biochemistry 16,
       pp.3655-3662.
27. Parthasarathy, N. and Schmitz, K.S.: 1980, Biopolymers 19,
       pp.1137-1151.

28. Flory, P.J.: 1953, "Principles of Polymer Chemistry", Cornell
    University Press, Ithaca, New York.
29. Zimmerman, S.B. and Pheiffer, B.H.: 1980, J. Mol.Biol. 142,
    pp.315-330.
30. Rizzo, V. and Schellman, J.: 1981, Biopolymers 20, pp.2143-2163.
31. Hagerman, P.J.: 1981, Biopolymers 20, pp.1503-1535.
32. Eisenberg, H.: 1971, in "Procedures in Nucleic Acids Research",
    Vol. 2 (G.L. Cantoni and D.R. Davies, Eds.), Harper and Row,
    New York, pp.137-175.
33. Finch, J.T. and Klug, A.: 1976, Proc.Natl.Acad.Sci.U.S.A. 73,
    pp.1897-1901.
34. Campbell, A.M., Cotter, R.I. and Pardon, J.F.: 1978, Nucl.Acids
    Res. 5, pp.1571-1580.
35. Shaikevitch, A. and Kam, Z.: 1981, J.Biochem.Biophys.Methods,
    5, pp.287-292.
36. Bates, D.L., Butler, P.J.G., Pearson, E.C. and Thomas, J.O:
    1981, Eur.J.Biochem. 119, pp.469-476.
37. Flory, P.J.: 1969, "Statistical Mechanics of Chain Molecules",
    Interscience Publishers, New York.

HIGHER ORDERS OF CHROMATIN STRUCTURE

Gary Felsenfeld, James D. McGhee and Donald C. Rau
Laboratories of Molecular Biology and Chemical
Physics, National Institute of Arthritis,
Diabetes, and Digestive and Kidney Diseases
National Institutes of Health, Bethesda, MD 20205

We describe recent studies of the 30 nm fiber, in which we have used
electric dichroism to determine the orientation of chromatosomes within
the chromatin fiber. We show that data obtained with chromatin from a
variety of sources, and with varying spacer length, suggest a similar
packing of the chromatosomes in all 30 nm fibers. We also discuss
factors that might affect the stability of the fiber, particularly in
such a way as to render the DNA it contains more accessible for transcription.

INTRODUCTION

The chromatin subunit called the nucleosome core particle consists
of 145 bp of DNA wrapped in a superhelical path around an octamer of
histones (Finch et al., 1977; Pardon et al., 1977; Suau et al., 1977;
McGhee and Felsenfeld, 1980). In chromatin containing long DNA, somewhat more DNA (166 bp) can be tightly bound to the core histones; the
corresponding particle, with two full turns of DNA wrapped around the
histones, is called the chromatosome (Simpson, 1978). The chromatosomes
in chromatin are connected to each other by 0-80 base pairs of spacer
DNA (Kornberg, 1977); the spacer length depends upon the source of the
chromatin.

Although we now have a rather good understanding of the architecture
of nucleosomes, we are only beginning to learn how they are assembled
into higher order structures. Physical and electron microscopic studies
have shown that at low ionic strength and in the absence of divalent
cations, chromatin assumes the form of an extended filament, 10 nm in
diameter, composed of alternating chromatosomes and spacers (Ris and
Korenberg, 1979). If small amounts of divalent cations or larger
amounts of monovalent cations are added, the 10 nm filament is condensed
into a fiber with a diameter of about 30 nm. The 30 nm fiber is a
structural component commonly observed in electron microscopic studies
of chromatin within the nucleus (Davies, Murray and Walmsley, 1975;
Rattner and Hamkalo, 1978; Marsden and Laemmli, 1979; Olins and Olins,
1979; Ris and Korenberg, 1979; Adolph, 1980). Electron microscopy

C. Hélène (ed.), Structure, Dynamics, Interactions and Evolution of Biological Macromolecules, 101–112.
Copyright © 1983 by D. Reidel Publishing Company.

(Finch and Klug, 1976; Thoma et al., 1979) and neutron scattering
(Baldwin et al., 1978; Suau et al., 1979) suggest that the conversion
from 10 nm filament to 30 nm fiber is accomplished by winding the 10 nm
filament in a helical coil; the resulting solenoid has a pitch of 11 nm,
and contains about six chromatosomes per turn.

   In this paper, we use sedimentation velocity and electric dichroism
to study the transition between 10 nm filament and 30 nm fiber both for
bulk chromatin and for defined regions of the eukaryotic genome. The
electric dichroism results further permit us to place stringent limits
on the orientation of chromatosomes within the 30 nm fiber; our recent
results show that this orientation is nearly invariant with the source
of the chromatin, or the average length of spacer DNA.

ORIENTATION OF CHROMATOSOMES IN THE 30 NM FIBER

   Electric dichroism is a useful technique for studying the conformation in solution of DNA and its complexes with other molecules. The
theory of dichroism is well understood (for references see McGhee et
al., 1980). Furthermore, both the orientation and the optical properties
of the bases in the DNA duplex are well known. It is therefore possible
to predict, for an arbitrary path of DNA relative to an orienting
field, the expected dichroism. As we will show, the experimental data
obtained for chromatin limit the possible paths which the DNA can
assume.

   In the case of electric dichroism, the orienting field is provided
by a pulse of direct current. Although the applied voltage is not
large enough to orient the particle perfectly, extrapolations to infinite
voltage (and thus perfect alignment) can be made. The definition of
the reduced dichroism is shown in Fig. 1.

$$\text{Reduced Dichroism, } \rho = \frac{A_\| - A_\perp}{A}$$

$$\rho_\infty = \lim_{1/E \to 0} (\rho)$$

Figure 1. The definition of dichroism in terms of absorbance parallel
and perpendicular to the field E.

In order to apply such methods to the 30 nm fiber, we first prepare a sample partly fractionated with respect to fiber length. In our first experiments, nuclei purified from adult chicken erythrocytes are gently digested with micrococcal nuclease. The chromatin fragments released from the nucleus are polydisperse and can then be fractionated on a sucrose gradient. The fractions used in our studies are in the range 20-100 nucleosomes in length.

The results of a typical dichroism measurement on such fractions is shown in Fig. 2. The values obtained for the reduced dichroism are essentially independent of the size of solenoid, as would be expected for a rigid rodlike structure of length sufficient to make end effects negligible. Furthermore, the relaxation behavior (aligned to random) measured when the field is removed reveals a single relaxation time with a value consistent with the expected solenoid dimensions (McGhee et al., 1980).

In order to interpret these data, we must construct a series of models for the 30 nm fiber, and calculate the dichroism expected for each. As a first approximation, we consider the possible simple arrangements of chromatosomes, neglecting the spacer. Three such arrangements are shown in Fig. 3. It is easy to show (McGhee et al., 1980) that Model C cannot be consistent with the data, but that variants of Models A or B might be approximately correct. Electric dichroism cannot distinguish A from B: They are optically equivalent in solution because they are both free to rotate about the long solenoidal axis. However, Suau et al. (1979) have used neutron scattering to eliminate models related to B; such models have too large a predicted radius of gyration to be consistent with the scattering data.

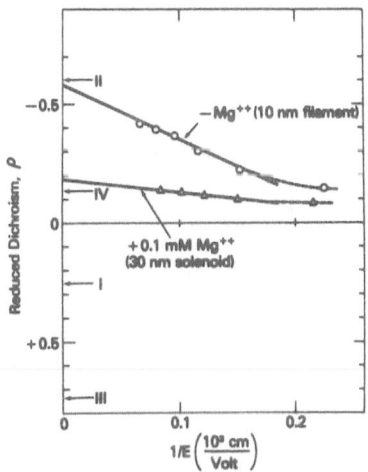

Figure 2. Dependence of the reduced dichroism ρ on (field strength)$^{-1}$, for a chromatin sample 45 nucleosomes long (O---O) 0.2 mM Na$^+$, 0.03 mM EDTA (pH 7); (Δ---Δ) the same sample in the presence of 0.1 mM MgCl$_2$. (From McGhee et al., 1980.)

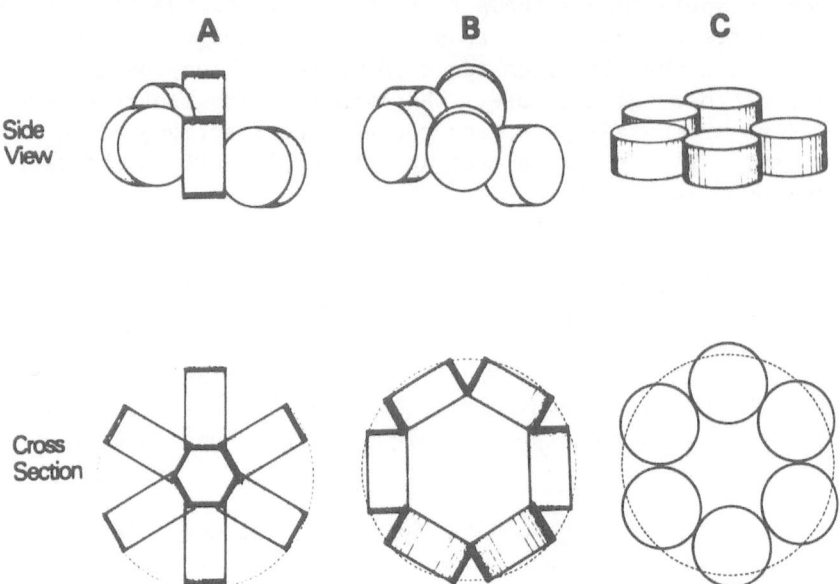

Figure 3. Drawings of three possible orthogonal positions of chromatosomes within the solenoid. The chromatosome faces are arranged
(A) radially, (B) tangentially, and (C) perpendicular to the solenoid
axis. Drawings are made approximately to scale, with a solenoid diameter
of 30 nm, chromatosome dimensions of 11 nm diameter and 5.5 nm thick,
and with six chromatosomes per turn of the left-handed solenoid.
(From McGhee et al., 1980.)

     Model A is the starting point from which we construct a detailed
model of the 30 nm fiber. To do so, we now introduce the spacer elements, which are about 44 bp in length in chicken erythrocyte chromatin.
We assume that this spacer DNA follows a simple curved path between
chromatosomes; the path traces out a part of a supercoil with pitch for
the moment unspecified (Fig. 4). In order to fit the measured value of
the dichroism, we must now tilt the chromatosomes away from their
perfectly vertical orientation relative to the solenoid axis. For each
tilt angle γ, there is a corresponding spacer pitch angle ϕ which fits
the data. The range of allowed pairs of tilt and pitch angles is shown
in Figure 5. Even in the absence of detailed calculation, it is evident
that the chromatosome tilt angle cannot exceed 25 degrees. Further
examination places even more severe limits on this value. In order to
understand why this is so, it is only necessary to note that the length
of each spacer is fixed, and that it must span the gap between adjacent
chromatosomes. Since the fiber is about 30 nm in diameter, the size of
the gap is known, and therefore the pitch angle of the spacer is determined. (The sine of the pitch angle is roughly the ratio of the gap
length to the spacer length.) Such a calculation yields a spacer pitch
angle of 8 to 18 degrees for solenoid diameters between 25 and 35 nm.
The chromatosome tilt angle deduced from Fig. 5 would therefore lie .
between about 23 and 25 degrees.

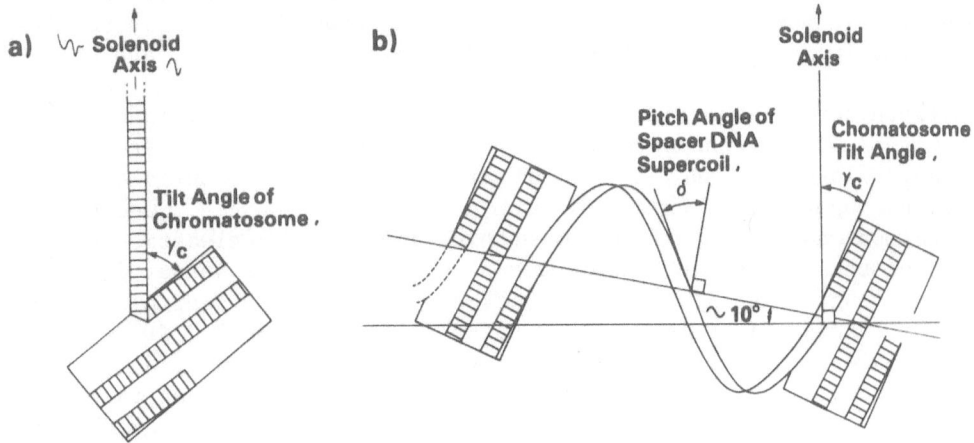

Figure 4.   Structural parameters of the 30 nm fiber.

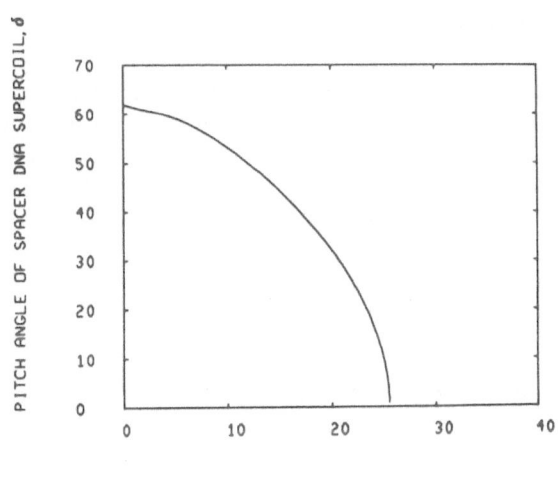

Figure 5.   Allowed values for the chromatosome tilt angle and spacer DNA
pitch angle consistent with the dichroism data for chicken erythrocyte
chromatin.   (From McGhee et al., 1980.)

    Although the assumption of a supercoiled path for the spacer DNA
may seem somewhat arbitrary, the conclusions reached in the preceding
discussion are rather general. Even if the path of the spacer is more
complicated, it must still make an average angle relative to the field
that is related to the pitch angle of the simpler model. It should be
noted that although the simple model described above yields a unique
value of the pitch angle, it does not specify the pitch (or, equivalently,
the diameter) of the putative spacer supercoil. The dimensions of the
supercoil are not determined by the data, but reasonable limits can be
estimated from geometrical arguments (McGhee et al., 1980).

The analysis described above applies to data obtained with chicken erythrocyte chromatin. It is important to know whether similar results can be obtained with chromatin isolated from other sources, particularly those in which the spacer length differs. (It should also be noted that erythrocyte chromatin carries histone H5, rather than histone H1; this might also have an effect on the structure.) In recent studies (McGhee, Felsenfeld, and Rau, in preparation) we have measured the dichroic properties of chromatin from a variety of sources. One such preparation was obtained from sea urchin sperm. The nucleosome repeat length of this chromatin is 245 base pairs, so that the spacer is about twice as long as that present in chicken erythrocyte chromatin. We find that the reduced dichroism of this material is about -0.2. If an analysis similar to that described above, and taking into account the greater spacer length, is applied to these data, we find that the range of allowed chromatosome tilt angles is roughly 20° ± 5°. Thus, the tilt angle varies very little in two 30 nm fibers with quite different spacer lengths. Similar results have been obtained with chromatins isolated from several other sources possessing a variety of spacer sizes.

STABILITY OF THE 30 NM FIBER

As noted earlier, the 30 nm fiber is stabilized by cations. In the preceding experiments, Mg^{++} was used to stabilize the solenoid (the use of electric fields requires solvents with low conductivity) but NaCl at higher concentrations will also confer stability. At low ionic strength, the solenoid unfolds into the extended 10 nm filament. Addition of either $MgCl_2$ or NaCl will then result in reformation of the 30 nm fiber. The process can be followed using either electric dichroism (Mg^{++} only) or sedimentation velocity. Typical results are shown in Figs. 6 and 7. The conversion from 10 nm filament to 30 nm fiber is accompanied by a decrease in the negative value of the dichroism, a decrease in the relaxation time, and an increase in the sedimentation coefficient. The change in sedimentation accompanying the folding process has been well characterized by Butler and Thomas (1980).

The change in physical properties is particularly useful in studying the effects of various changes in chromatin composition or added components on the folding reaction. For example, in a recent study we examined the effects of the high mobility group proteins on chromatin folding (McGhee, Rau and Felsenfeld, 1982). The high mobility group proteins HMG 14 and 17 are known to be associated with transcriptionally active chromatin, and to be essential for conferring on such chromatin sensitivity to pancreatic DNase (Weisbrod and Weintraub, 1979). It seemed possible that these proteins might act in part by altering the stability of the 30 nm fiber. To test this, we measured the salt dependence of both dichroism and sedimentation in the presence of saturating amounts of a mixture of the two HMG proteins. The sedimentation results are shown in Fig. 7. These data, and the dichroism data as well, show that the HMG proteins have no measurable effect on the folding of bulk chromatin.

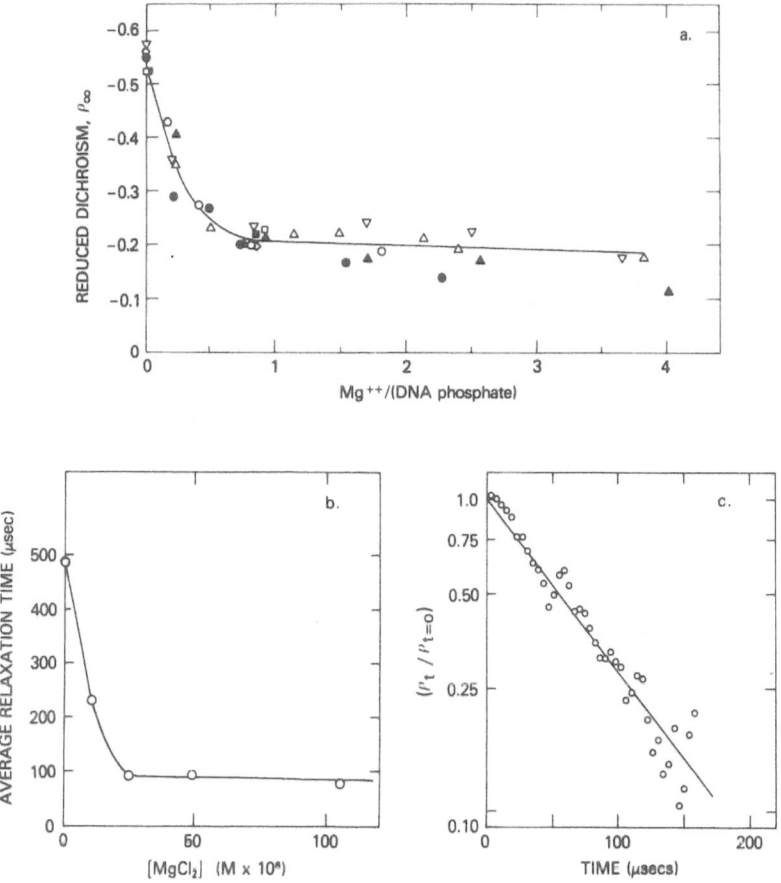

Figure 6.   Chromatin condensation induced by divalent ion.   (a) Limiting reduced dichroism as a function of the ratio of Mg^{++} added per DNA phosphate.   The different symbols refer to different chromatin fractions and preparations.   (b) Average relaxation times (μsec) for the chromatin sample used in Figure 3, as a function of MgCl$_2$ concentration (chromatin concentration = 70 μM in DNA phosphate).   (c) Semilogarithmic plot of the decay of the dichroism signal for the sample used in (b) at 50 μM MgCl$_2$. (From McGhee et al., 1980.)

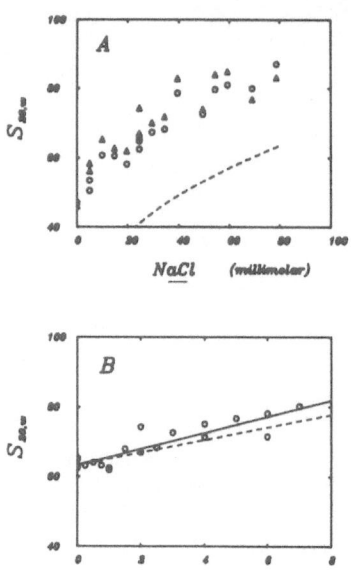

Figure 7A. $^{s}20$,w, (the median sedimentation coefficient corrected to standard conditions) of chromatin as a function of added NaCl concentration. Unfractionated chromatin was obtained from 14-day erythrocytes and had a weight average DNA size of 26 nucleosomes, estimated from the sedimentation coefficient of the purified DNA. All solutions contained 5 mM Tris Hcl, pH 8.0, 0.1 mM EDTA, 1 mM sodium butyrate. 0 = no added HMG 14/17; Δ = 2.0 ± 0.1 HMG 14/17 added per nucleosome;----expected sedimentation behavior if histones H1 and H5 were removed.

B. $^{s}20$,w plotted as a function of HMG 14/17 added per nucleosome, for the same chromatin preparation used in A. Solvent was 25 mM NaCl, 5 mM Tris HCl, 0.1 mM EDTA, 1 mM Na butyrate, pH 8. Solid line = best linear fit to data; dashed line = expected sedimentation behavior if all added HMG 14/17 were bound but caused no changes in the frictional properties of the chromatin particles. (From McGhee, Rau and Felsenfeld, 1982.)

Are these properties shared by the very small chromatin component that is capable of being expressed in the chicken erythrocyte nucleus? The adult beta globin genes in these nuclei are sensitive to pancreatic DNase and might be expected to have a structure in some way different from that of bulk chromatin. To answer this question, we carry out a sucrose gradient sedimentation experiment in 25 mM NaCl, a salt concentration at which bulk polynucleosomes are partly folded, and assay each fraction for globin gene abundance (Fig. 8). The assay reveals that the HMG proteins have an effect on the sedimentation properties of the globin polynucleosomes that is identical to their effect on the bulk material. It can also be seen from the control experiment in Fig. 7 that the globin fraction behaves identically to bulk chromatin in the absence of added HMG 14/17.

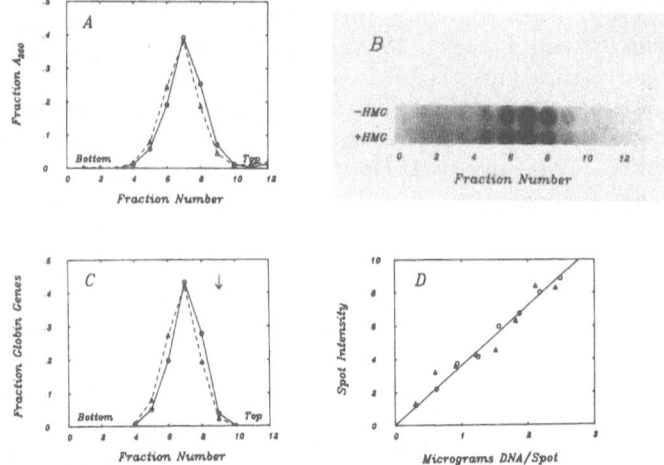

Figure 8. Erythrocyte chromatin from 14-day embryos was stripped of
endogenous HMG 14/17 and a size fraction containing ~20 nucleosomes
sedimented on an isokinetic sucrose gradient containing 25 mM NaCl with
or without added HMG 14/17. Direction of sedimentation is from right
to left.

        A. Normalized A_{260} profile of gradients; 0---0, no added HMG
14/17; Δ---Δ = +2 HMG 14/17 per nucleosome.

        B. Typical dot blot hybridization assay measuring the adult
β-globin gene content of each fraction of A.

        C. Normalized globin gene content of gradients shown in Fig. 4A.
0---0 = no added HMG 14/17; Δ---Δ = +2 HMG 14/17 per nucleosome. Arrow
marks expected sedimentation position if this chromatin were stripped
of histones H1 and H5.

        D. Standard curve for dot blot assay, relating spot intensity
(i.e. globin gene content in arbitrary units) to amount of chicken DNA
applied per spot, either as chromatin (0) or as protein-free DNA (Δ).
(From McGhee, Rau and Felsenfeld, 1982.)

DISCUSSION

     Our electric dichroism studies show that chromatosomes are packed
radially in the 30 nm fiber, with an average inclination with respect to
the solenoid axis of 20-30 degrees. The value of this angle appears to
be nearly invariant in a number of chromatin preparations with varying
lengths of spacer DNA. This suggests that the interactions stabilizing
the 30 nm fiber are also invariant. We do not yet know the nature of
these interactions. It is possible that direct interactions are involved
between the core histones of neighboring chromatosomes. However, this
would require that the chromatosomes be locked into a very limited

number of fixed positions relative to one another. We have suggested
(McGhee et al., 1980) that another form of interaction may be involved
in stabilizing the 30 nm fiber: The core histones (in particular the
positively charged amino terminal histone tails) of one chromatosome may
interact with the DNA of a neighbor. Since the 80 base pairs of DNA in
one turn on the surface of a chromatosome appear in roughly the same
external orientation once every 10 base pairs, rotation about the cylindrical axis of the chromatosome by 360/8 degrees will place the DNA in
an equivalent orientation relative to an external binding element. Such
a model satisfies the requirement for regular packing of the chromatosomes, while allowing sufficient flexibility to accommodate varying
spacer lengths without exerting any extra strain on the spacer DNA.
Some recent results from other laboratories are at least consistent with
such a model. It has been reported by Allan et al. (1981) that if
polynucleosomes are treated with trypsin to remove the amino terminal
tails of the core histones, the 30 nm fiber can no longer be reconstituted by addition of histone H5. Furthermore, Mirzabekov et al. (1982)
have shown that within a chromatin fiber, some part of the histone
component of a given chromatosome can be crosslinked to the DNA of
another chromatosome.

The data presented here concerning the chromatin structure of the
globin genes in chicken erythrocyte chromatin suggest that in 14-day-old
cells there is no distinguishable difference between most of the globin
region and bulk chromatin, and that interactions with the high mobility
group proteins HMG 14 and 17 are also similar for these two fractions.
We have obtained similar results (data not shown) with respect to the
structure of chromatin from 10-day-old chicken erythrocyte nuclei, which
are at the peak of production of adult beta globin message. It is
possible that some portion of the globin region is unfolded, but it
seems unlikely that a large fraction of this region (which retains its
sensitivity to pancreatic DNase after fractionation of the 30 nm fiber
preparations) can be significantly unfolded.

How can such results be reconciled with reports that the chromatin
structure of other transcribed genes (such as the heat shock genes of
Drosophila (Wu et al., 1979)) is disrupted? It is first necessary to
note that the globin gene region we have studied is probably transcribed
at a considerably lower rate than are the fully expressed heat shock
genes. It seems likely that a very active region of the genome will be
disrupted in structure by the very frequent passage of polymerase molecules. In any case, the primary distinction that must be made is between
genes in cells that are determined for the expression of that gene, and
the same gene in cells that will never express that gene. Our results
show that the step of determination does not involve the alteration of
an extended region of chromatin structure in the neighborhood of the
gene (though discrete segments, near the promoter, may have a distinctive
structure (McGhee, Wood and Felsenfeld, 1981)). Thus, the Drosophila
heat shock genes at low temperature, where they are in principle active
but in fact are not expressed, are also covered with a regular array of
nucleosomes (Wu et al., 1979).

Although disruption of higher order chromatin structure may not play a major role in determination of gene expression, it is clear that this structure is disrupted at some point during the transcription process itself. We know little about the lifetime of the disrupted state, or about its structure. The methods presented here, which combine classical physicochemical techniques with the use of probes for specific DNA sequences, should help us to attack these problems.

REFERENCES

Adolph, K.W.: 1980, Exp. Cell Res. 125, pp. 95-103.

Allan, J., Harborne, N., Rau, D.C., and Gould, H.: 1982, J. Cell Biol. 93, pp. 285-287.

Baldwin, J.P., Carpenter, B.G., Crespi, H., Hancock, R., Stephens, R.M., Simpson, J.K., Bradbury, E.M., and Ibel, K.: 1978, J. Appl. Crystall. 11, pp. 484-486.

Butler, P.J.G., and Thomas, J.O.: 1980, J. Mol. Biol. 140, pp. 505-524.

Davies, H.G., Murray, A.B., and Walmsley, M.E.: 1974, J. Cell Sci. 16, pp. 261-299.

Finch, J.T., and Klug, A.: 1976, Proc. Nat. Acad. Sci. USA 73, pp. 1897-1901.

Finch, J.T., Lutter, L.D., Rhodes, D., Brown, R.S., Rushton, B., Levitt, M., and Klug, A.: 1977, Nature 269, pp. 29-36.

Kornberg, R.D.: 1977, Ann. Rev. Biochem. 46, pp. 931-954.

Marsden, M.P.F., and Laemmli, U.K.: 1979, Cell 17, pp. 849-858.

McGhee, J.D., Rau, D.C., and Felsenfeld, G.: 1982, Nucl. Acids Res. 10, pp. 2007-2016.

McGhee, J.D., Rau, D.C., Charney, E., and Felsenfeld, G.: 1980, Cell 22, pp. 87-96.

McGhee, J.D., Wood, W.I., Dolan, M., Engel, J.D., and Felsenfeld, G.: 1981, Cell 27, pp. 45-55.

Mirzabekov, A.D., Karpov, V.L., Preobrazhenskaya, O.V., Bavykin, S.G., Ebralidze, K.K., Tuneev, V.M., Melnikova, A.F., Goguadze, E.G., Chenchick, A.A., and Beabealashvili, R.S.: 1982, Cold Spring Harbor Symposium in Quantitative Biology, 47, in press.

Olins, A.L., and Olins, D.E.: 1979, J. Cell Biol. 81, pp. 260-265.

Pardon, J.F., Worcester, D.L., Wooley, J.C., Cotter, R.I., Lilley, D.M.J., and Richards, B.M.: 1977, Nucl. Acids Res. 4, pp. 3199-3214.

Rattner, J.B., and Hamkalo, B.A.: 1978, Chromosoma 69, pp. 363-379.

Ris, H., and Korenberg, J.: 1979, Cell Biology, Prescott, D.M., and Goldstein, L. (Eds.) New York: Academic Press, 2, pp. 267-361.

Simpson, R.T.: 1978, Biochemistry 17, pp. 5524-5531.

Suau, P., Bradbury, E.M., and Baldwin, J.P.: 1979, Eur. J. Biochem. 97, pp. 593-602.

Suau, P., Kneale, G.G., Braddock, G.W., Baldwin, J.P., and Bradbury, E.M.: 1977, Nucl. Acids Res. 4, pp. 3769-3786.

Thoma, F., Koller, T., and Klug, A.: 1979, J. Cell Biol. 83, pp. 403-427.

Weisbrod, S., and Weintraub, H.: 1979, Proc. Nat. Acad. Sci. USA 76, pp. 631-635.

Wu, C., Wong, Y.-C., and Elgin, S.C.R.: 1979, Cell 16, pp. 807-814.

The submitted manuscript has been authored by an employee of the U.S. Government. Accordingly, the U.S. Government retains a nonexclusive, royalty-free licence to publish or reproduce the published form of this contribution, or allow others to do so, for U.S. Government purposes.

STRUCTURE AND DYNAMICS OF PEPTIDE-NUCLEIC ACID COMPLEXES.

Thérèse MONTENAY-GARESTIER[†], Francine TOULME[×], Judit FIDY[†§],
Jean-Jacques TOULME[†], Trung LE DOAN[†] and Claude HELENE[†×]
†Laboratoire de Biophysique, INSERM U.201, ERA CNRS 951,
Muséum National d'Histoire Naturelle, 61, rue Buffon,
75005 Paris, France
×Centre de Biophysique Moléculaire, CNRS,
45045 Orléans cedex, France

INTRODUCTION

Understanding the molecular mechanisms which govern the selective association of proteins with nucleic acids requires a detailed knowledge of the interactions involving the functional groups of both molecules (1). These interactions can be better characterized at the level of simple systems such as oligopeptide-oligonucleotide complexes even though extrapolation of these results to real biological systems may raise difficulties. However it is hoped that the physico-chemical characteristics of defined interactions between nucleic acid bases and amino acid side chains will help identify them in protein-nucleic acid complexes. A number of studies have been devoted to oligopeptide-nucleic acid interactions during the past ten years (for a recent review see reference 2). In this paper we further characterize electrostatic and stacking interactions involved in the binding of oligopeptides containing basic and aromatic residues to nucleic acids.

Previous work from our laboratory had mainly been focused on tripeptides whose general sequence was Lys-X-Lys where X is an aromatic residue. In the present paper, tetra and pentapeptides containing either one or two aromatic residues are investigated. Structural features of these complexes are deduced from fluorescence, fluorescence decay and NMR experiments. Dynamic parameters of the peptides in their complexes with nucleic acids are inferred from fluorescence anisotropy and quenching measurements.

MATERIALS AND METHODS

The tetrapeptides Lys-Gly-Trp-Lys(OtBu) and Lys-Trp-Gly-Lys(OtBu) were purchased from Bachem. The pentapeptides Lys-Trp-Ala-Tyr-Lys(NHEt) and Ac-Lys-Trp-Ala-Tyr-Lys(NHEt) were synthesized in Orléans by R. Mayer.

§Present address : Institute of Biophysics, Semmelweis Medical University, Budapest, Hungary.

C. Hélène (ed.), Structure, Dynamics, Interactions and Evolution of Biological Macromolecules, 113–128.

Fluorescence experiments were carried out with a Fica 55000 or an Aminco SPF 500 spectrofluorometers. Fluorescence decays were measured with an Edinburgh Instrument apparatus (199 M).

Fluorescence measurements were carried out in a buffer containing 1 mM Na cacodylate, 1 mM Na chloride and 0.2 mM EDTA (referred to as standard buffer). The pH was adjusted to 6 or 7 as indicated.

RESULTS

1. Fluorescence quenching

All the peptides containing tryptophan that we have investigated here (see materials) exhibit a quenching of their fluorescence when they are bound to nucleic acids. This quenching is not complete even under conditions where all peptide molecules are bound. The simplest scheme which accounts for the fluorescence data assumes that only one complex is formed whose fluorescence quantum yield (ϕ_c) is lower than that of the free peptide (ϕ_F) (see equation 1) :

$$\text{(1)} \qquad \text{Peptide + Nucleic Acid} \xrightleftharpoons{\quad K \quad} \text{Complex}$$

However in many cases the fluorescence decay of a peptide complete-ly bound to a nucleic acid is indistinguishable from that of the free peptide (3). This result means that at least two complexes are formed : one with the same fluorescence quantum yield and lifetime as the free peptide (Complex I) ; a second one which does not emit fluorescence at all (Complex II).

$$\text{(2)} \qquad \text{Peptide + Nucleic Acid} \xrightleftharpoons{\quad K_I \quad} \text{Complex I} \xrightleftharpoons{\quad K_{II} \quad} \text{Complex II}$$

$$(\phi_{II} = 0)$$

Previous experiments had shown that stacking interactions of tryp-tophan with nucleic acid bases were characterized by a complete quenching of the fluorescence of tryptophan (4). Energy transfer is not expected to play an important part in the quenching of tryptophan fluorescence by nucleic acid bases (5). Monoanionic phosphates −such as those found in the phosphodiester backbone of nucleic acids− have no effect on the fluo-rescence of tryptophan (6). Only dianionic phosphates which may be pre-sent at the end of oligonucleotides are good quenchers of tryptophan fluorescence. For a long polynucleotide chain it is thus likely that stacking of the tryptophyl ring with nucleic acid bases in complex II provides the main contribution to the fluorescence quenching observed in peptide-nucleic acid complexes. This is confirmed by NMR experiments which show that the resonance lines of Trp are shifted upfield in the presence of poly(A) or denatured DNA (2).

Analysis of the fluorescence data was therefore carried out accor-ding to the two-step model of scheme 2. The method used to calculate K_I and K_{II} is identical to that previously described (7) (see equation 5 below). It should be noted that the two schemes ($\underline{1}$) and ($\underline{2}$) are formally equivalent with $K = K_I (1 + K_{II})$ and

$$\phi_c = \frac{\phi_F}{1 + K_{II}}$$

Figure 1 and tables 1 and 2 present some of the data obtained with tetra or pentapeptides and polynucleotides, native DNA or heat-denatured DNA. A comparison with previous results obtained with the tripeptide Lys-Trp-Lys- are also included in tables 1 and 2.

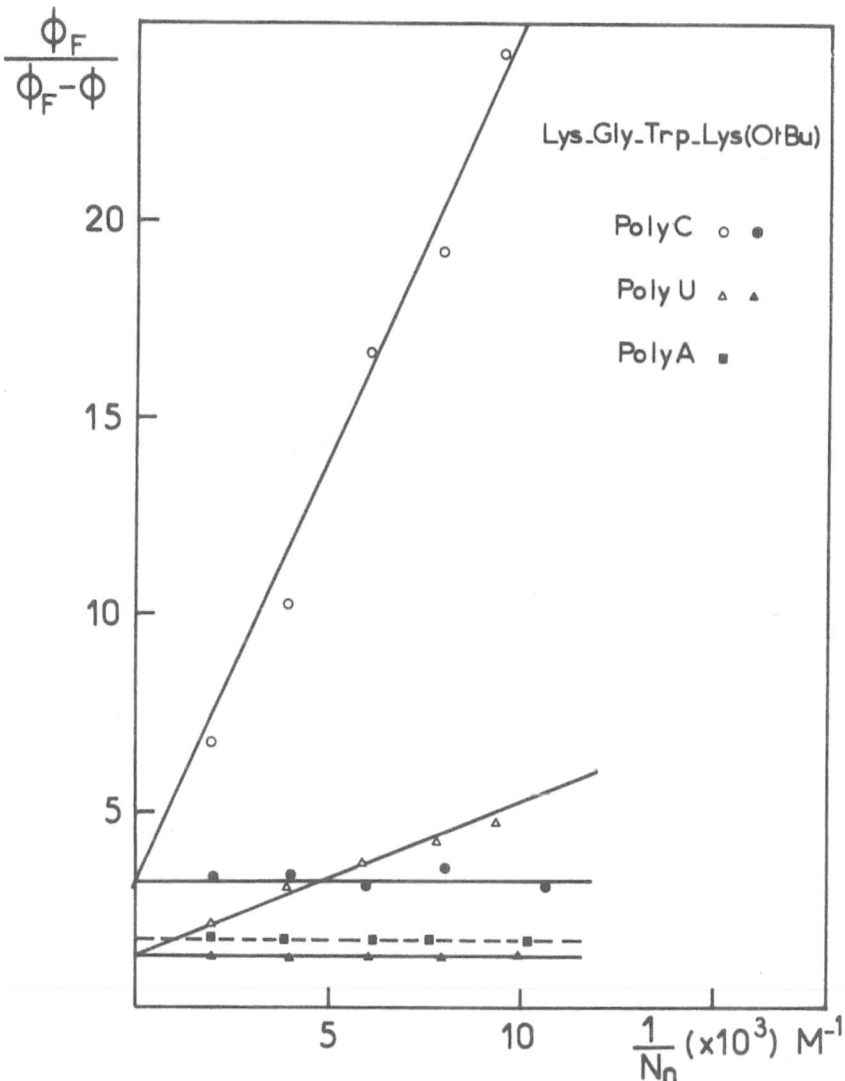

Figure 1 - Plot of $\phi_F/(\phi_F-\phi)$ *versus* $1/N_0$ according to equation ($\underline{5}$) for the binding of Lys-Gly-Trp-Lys(OtBu) to poly(C), poly(U) and poly(A). Filled symbols refer to the pH 7 standard buffer (1mM Na cacodylate, 1mM NaCl, 0.2 mM EDTA). Open symbols correspond to a 10 times more concentrated buffer.

	poly(A)	poly(U)	poly(C)
Lys-Gly-Trp-Lys(OtBu)	1.45	3.35	0.45
Lys-Trp-Lys	2.3	4	1.05

Table 1 - K_{II} values for the binding of oligopeptides to polynucleotides according to scheme (2) and equation (5). All values refer to a pH 7 standard buffer (1 mM Na cacodylate, 1 mM NaCl, 0.2 mM EDTA) at 14°C.

	KWAYK(NHEt)	(Ac)KWAYK(NHEt)	KGWK(OtBu)	KWGK(OtBu)	KWK
Denatured DNA	8.3	5.0	6.7	5.7	4.3
Native DNA	0.83	0.28	0.1	0.45	0.36

Table 2 - K_{II} values calculated according to scheme (2) and equation (5) for the binding of oligopeptides containing tryptophan to denatured and native *E.coli* DNA in a pH 6 standard buffer (1 mM Na cacodylate, 1 mM NaCl and 0.2 mM EDTA). The one-letter symbols are used for amino-acids (K = Lys, W = Trp, A = Ala, Y = Tyr).

As shown in figure 1, K_{II} values (scheme 2) or ϕ_c values (scheme 1) do not appear to depend on the ionic strength. The ionic strength dependence of the K_I value was therefore determined from the dependence of fluorescence quenching at a low peptide-to-nucleic acid ratio ($\approx$ 0.01). According to polyelectrolyte theory the binding of a positively charged ligand to a nucleic acid in a solution containing monovalent cations is accompanied by the release of $m\psi$ positive ions where m is the number of electrostatic bonds involved in the complex and ψ is the average number of monovalent cations thermodynamically bound per phosphate group of the nucleic acid.

(3) $\log K = \log K_0 - m\psi \log[Na^+]$

Using the values of ψ tabulated by Record *et al.* (8) the number of electrostatic bonds in an oligopeptide-nucleic acid complex can be calculated from the variation of log K with $\log[Na^+]$. Within experimental error, this value corresponds to the number of positive charges borne by the oligopeptide at pH 6 (2 and 3 for the acetylated and non-acetylated peptides, respectively ; in all peptides investigated here the carboxylic group is substituted by a ter-butyl (tBu) or an ethylamide (NHEt) group and therefore does not bear any negative charge).

2. Fluorescence decay measurements

As previously described (3), the fluorescence decay of the tryptophan-containing peptides that we have investigated is the superimposition
of two exponential decays. The third component which was previously reported (9) was probably due to the presence of an impurity in the former
sample of the tetrapeptide Lys-Gly-Trp-Lys(OtBu). As shown in figure 2,
the analysis with two exponentials gave a quite good fit with the experimental data. Introducing a third exponential did not improve the fit.

The two components in the fluorescence decay of Lys-Trp-Lys, Lys-
Gly-Trp-Lys(OtBu) and Lys-Trp-Gly-Lys(OtBu) might originate from two
conformers of the peptides in which the tryptophyl ring has different
environments and thus different fluorescence quantum yields and lifetimes. If this interpretation is correct the interconversion between the
two conformers must be slow with respect to the fluorescence lifetimes.

We previously reported that the fluorescence decay of the tripeptide
Lys-Trp-Lys was not markedly affected upon binding to native and denatured DNA (3). The same behavior was observed for Lys-Trp-Gly-Lys(OtBu)
complexes with denatured DNA at pH 6 ; the average lifetime of the bound
peptide was not affected as compared to that of the free peptide (free
peptide : 1.88 ns ; bound to denatured DNA : 1.90 ns ; at 10°C in the
pH 6 standard buffer). A low ionic strength buffer was used in these
fluorescence decay measurements ; under these conditions equilibrium
binding studies had shown that all peptide molecules were bound to the
polynucleotide chain. The observed quenching was found to be that expected on the basis of the K_{II} values determined from equilibrium studies
(dissociation of the complexes at high ionic strength allowed us to determine the extent of fluorescence quenching). The lifetimes and relative
contributions of the two components in the fluorescence decay of the
tetrapeptide Lys-Trp-Gly-Lys(OtBu) changed rapidly with temperature
around 10°C, the temperature at which the decay measurements were carried
out. It was therefore difficult to determine whether the observed differences in individual lifetimes and contributions between free and bound
peptide were due to small changes in temperature or whether they reflected real effects of binding. This aspect clearly requires further studies.
However the average lifetime -and therefore the average fluorescence
quantum yield- of the peptide was not affected upon binding to denatured
DNA even though the observed fluorescence quenching was about 85 %. In
the case of native DNA a $\sim$ 10 % increase of the average fluorescence
lifetime was observed upon complex formation (fluorescence quenching was
$\sim$ 40 %). These results mean that the tetrapeptide Lys-Trp-Gly-Lys(OtBu)
behaves quite similarly to Lys-Trp-Lys. The presence of the glycyl residue does not seem to affect the interaction of the tryptophyl residue
with nucleic acid bases. This interaction appears to be mainly determined
by the electrostatic interactions of the α- and ϵ-amino groups of the N-
terminal lysyl residue. The results obtained for the salt dependence of
peptide binding to nucleic acids (see above) showed that these two charges are involved in peptide binding to the polynucleotide backbone.

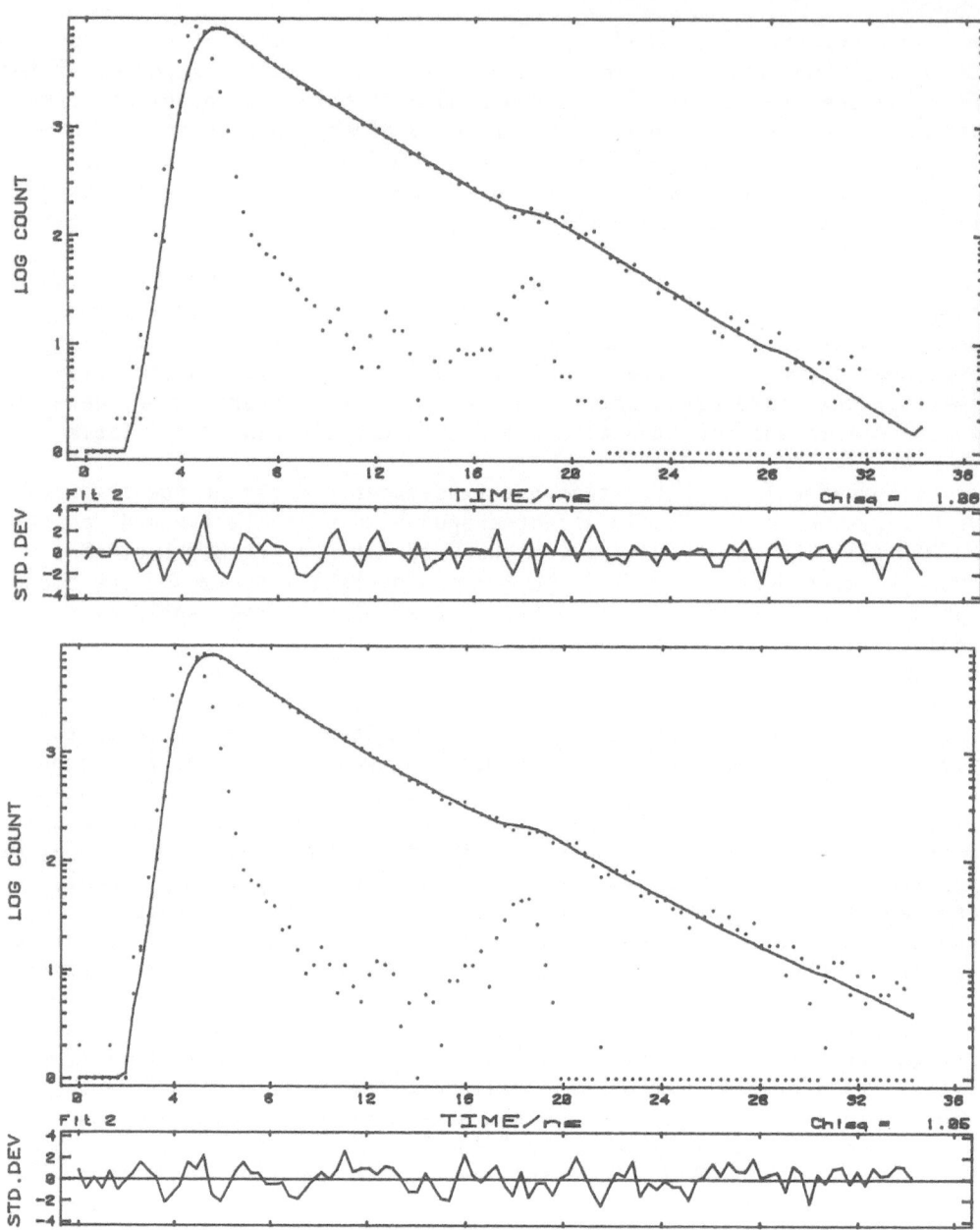

Figure 2 - Fluorescence decay (log I_F vs time in nanoseconds) for Lys-Gly-Trp-Lys(OtBu) in the absence (upper panel) and in the presence (lower panel) of native DNA. The full line represents the calculated curve using a two-exponential decay function $I(t) = \beta_1 \exp -t/\tau_1 + \beta_2 \exp -t/\tau_2$ (see table 3 for the parameters). The flash lamp profile is also given in each panel as well as the standard deviations (below each decay curve).

The results obtained with the tetrapeptide Lys-Gly-Trp-Lys(OtBu) revealed a more complex situation. The fluorescence decay of this pepti- de was the sum of two exponentials but in contrast to the isomeric tetra- peptide discussed above the relative contribution of these two components changed much more slowly with temperature. When complexed with denatured DNA the fluorescence quantum yield of Lys-Gly-Trp-Lys(OtBu) was reduced by ∿ 87 % even though the average fluorescence lifetime was not markedly affected (Table 3). However the relative contribution of the two compo- nents and their lifetimes were modified when compared to the parameters obtained with the free peptide (Table 3). This was also observed with

	τ_1	β_1	τ_2	β_2	$<\tau>$	z	K_I^2/K_I^1	$<K_{II}>$
Free peptide	0.98	0.41	3.05	0.59	2.20	–	–	–
Native DNA	1.57	0.54	3.66	0.46	2.53	–0.15	0.59	0.26
Denatured DNA	1.35	0.58	3.31	0.42	2.17	0.01	0.51	6.65
Poly(U)	1.01	0.74	2.30	0.26	1.35	0.39	0.25	1.65

Table 3 - Fluorescence decay analysis of the tetrapeptide Lys-Gly-Trp- Lys(OtBu) free and bound to different nucleic acids in the standard pH 6 buffer at 10°C. The decay was analysed according to $I(t) = \beta_1 \exp(-t/\tau_1) + \beta_2 \exp -(t/\tau_2)$. The average fluorescence lifetime $<\tau>$ is defined as $<\tau> = (\beta_1\tau_1+\beta_2\tau_2)/(\beta_1+\beta_2)$. The value of z was calculated from equation (12) (or the change in $<\tau>$) and the ratio K_I^1/K_I^2 from equation (13). The average value $<K_{II}>$ was calculated from equations (14) and (15). Compare with the values given in tables 1 and 2 using equation (5) derived from scheme (2) and assuming that complex I has the same fluorescence quantum yield as the free peptide.

native DNA and poly(U) but in addition the average fluorescence lifetime was changed upon binding ; an increase of 15 % was observed with native DNA (fluorescence quenching was ∿ 20 %) ; a decrease of 39 % was obtained with poly(U) (fluorescence quenching was 87 %). In light of these results the previous scheme (equation 2 above) must be modified to account for the possible existence of two conformers in the free peptide (P.1 and P.2). Each conformer can bind electrostatically to the polynucleotide backbone giving two isomeric complexes (I.1 and I.2) in which the aroma- tic ring does not engage any interaction with the nucleic acid bases. Each of these "outside" complexes can then be converted into a stacked complex (II.1 and II.2). The fluorescence lifetimes and the relative po- pulation of the two isomeric "outside" complexes are known from fluores- cence decay measurements as are those of the free peptide. It is not likely that stacked complexes II.1 and II.2 could be interconverted without the involvement of the "outside" complexes.

$$
\begin{array}{ccccc}
\text{P.1 + N} & \underset{}{\overset{K_I^1}{\rightleftharpoons}} & \text{Complex I.1} & \underset{}{\overset{K_{II}^1}{\rightleftharpoons}} & \text{Complex II.1} \\[1em]
\updownarrow & & \updownarrow & & \updownarrow \\[1em]
\text{P.2 + N} & \underset{}{\overset{K_I^2}{\rightleftharpoons}} & \text{Complex I.2} & \underset{}{\overset{K_{II}^2}{\rightleftharpoons}} & \text{Complex II.2}
\end{array}
$$

(4)

In scheme (2) complex II is not emitting fluorescence ; the following relationship (equation 5) was derived assuming that complex I and free peptide have the same fluorescence quantum yield ϕ_F (7)

(5)
$$
\frac{\phi_F}{\phi_F - <\phi>} = \frac{1 + K_{II}}{K_{II}} + \frac{1}{K_I K_{II}} \frac{1}{N_0}
$$

where $<\phi>$ is the average fluorescence quantum yield of the peptide in the presence of nucleic acid at concentration N_0.

If the fluorescence quantum yield of complex I (ϕ_I) is assumed to be different from that of the free peptide (ϕ_F), the following relationship applies

(6)
$$
\frac{\phi_F}{\phi_F - <\phi>} = \frac{1 + K_{II}}{z + K_{II}} + \frac{1}{K_I(K_{II} + z)} \frac{1}{N_0}
$$

where z is the relative change in fluorescence quantum yield of the peptide in complex I as compared to that of the free peptide

$$
(z = \frac{\phi_F - \phi_I}{\phi_F})
$$

When scheme (4) is analysed without making any assumption as to the relative fluorescence quantum yields of complexes I (complexes II are always assumed to be non-fluorescent), an equation analogous to equation (6) can be derived :

(7)
$$
\frac{<\phi_F>}{<\phi_F> - <\phi>} = A + B \frac{1}{N_0}
$$

with

(8)
$$
A = \frac{\left[1 + K_{II}^1 + y(1 + K_{II}^2)\right](1 + ax)}{(1 + ax)\left[1 + K_{II}^1 + y(1 + K_{II}^2)\right] - \alpha(1 + x)(1 + by)}
$$

$$(9) \qquad B = \frac{(1 + x)(1 + ax)}{K_I^1 \left[(1 + ax)\left[1 + K_{II}^1 + y(1 + K_{II}^2)\right] - \alpha(1 + x)(1 + by)\right]}$$

where $\qquad x = \dfrac{[P.2]}{[P.1]}, \qquad y = \dfrac{[I.2]}{[I.1]}, \qquad \alpha = \dfrac{\phi_I^1}{\phi_F^1}, \qquad a = \dfrac{\phi_F^2}{\phi_F^1}, \qquad b = \dfrac{\phi_I^2}{\phi_I^1}$

The values of α, a, b, x, y are determined from the fluorescence decay measurements. $\langle\phi_F\rangle$ is the average fluorescence quantum yield of the free peptide.

$$(10) \qquad \langle\phi_F\rangle = \phi_F^1 \frac{[P.1]}{[P.1] + [P.2]} + \phi_F^2 \frac{[P.2]}{[P.1] + [P.2]} = \phi_F^1 \frac{1 + ax}{1 + x}$$

and $\langle\phi\rangle$ is the overall fluorescence quantum yield of the peptide in the presence of the nucleic acid at concentration N. Quantities represented by [ ] are the concentrations of the indicated species.

The average fluorescence quantum yield for "outside" complexes (I.1 and I.2) is

$$(11) \qquad \langle\phi_I\rangle = \frac{\phi_I^1(1 + by)}{1 + y}$$

The change in fluorescence quantum yield z observed upon formation of complexes I is defined as

$$z = \frac{\langle\phi_F\rangle - \langle\phi_I\rangle}{\langle\phi_F\rangle}$$

and is given by equation (12)

$$(12) \qquad z = 1 - \frac{\alpha(1 + by)(1 + x)}{(1 + ax)(1 + y)}$$

The value of z can also be calculated from the change in average fluorescence lifetimes $\langle\tau\rangle$.

The average value of K_I is defined as

$$\langle K_I\rangle = \frac{1}{N} \frac{[I.1] + [I.2]}{[P.1] + [P.2]}$$

and is given by equation (13)

$$
(13) \qquad \langle K_I \rangle = \frac{K_I^1 + x\, K_I^2}{1 + x} = K_I^1 \frac{1 + y}{1 + x}
$$

The average value of K_{II} is defined as

$$
\langle K_{II} \rangle = \frac{[II.1] + [II.2]}{[I.1] + [I.2]}
$$

and is given by equation (14)

$$
(14) \qquad \langle K_{II} \rangle = \frac{K_{II}^1 + y\, K_{II}^2}{1 + y}
$$

Thus equation (7) can be rewritten as

$$
(15) \qquad \frac{\langle \phi_F \rangle}{\langle \phi_F \rangle - \langle \phi \rangle} = \frac{1 + \langle K_{II} \rangle}{z + \langle K_{II} \rangle} + \frac{1}{\langle K_I \rangle\,(z + \langle K_2 \rangle)}\,\frac{1}{N_0}
$$

which is formally identical to equation (6) with the above definition
for the average values

Plots of

$$
\frac{\langle \phi_F \rangle}{\langle \phi_F \rangle - \langle \phi \rangle}
$$

versus $1/N_0$ should still give a straight line. The ratio of the y-axis
intercept and the slope yields $\langle K_1 \rangle$ ($1 + \langle K_{II} \rangle$) which is the overall
equilibrium constant for the binding of the peptide to the nucleic acid.

The two association constants K_I^1 and K_I^2 can be calculated from
equation (13). The average value $\langle K_{II} \rangle$ can be obtained but not the separated constants K_{II}^1 and K_{II}^2 (unless further assumptions are made). A
complete description of the available results will be published elsewhere.
Table 3 gives the values of z calculated from equation (10) or from the
change in average fluorescence lifetime. The ratio K_I^2/K_I^1 has been calcu- .
lated from equation (13). From the results of table 3, it is apparent
that complex formation involving Lys-Gly-Trp-Lys(OtBu) is accompanied
by a shift of the equilibrium between the two peptide conformers. The
conformer with the longest fluorescence lifetime binds less than the
other one. If complex I is assumed to involve only electrostatic binding
without further interaction of the aromatic ring with nucleic acid bases,
then the above result means that the spatial distribution of positive
charges in conformer 2 is less adapted to the relative position of the

polynucleotide phosphates than it is in conformer 1. It is likely that
the two conformers will have to undergo a conformational change upon
binding to the nucleic acid in order to fit the relative phosphate positions. The polynucleotide backbone is also likely to change slightly
its conformation to allow for the best fit of the charge distributions ;
this is demonstrated by the small changes in circular dichroism which
are induced upon peptide binding even in the absence of any aromatic ring
in the peptide (2). The fluorescence characteristics (lifetimes and quantum yields) of the tryptophyl residue in each conformer are expected to
be sensitive to the position of the aromatic ring with respect to the
amino groups of the peptide. The results obtained with native and denatured DNA (table 3) show that the lifetimes of the two conformers increase
upon formation of complex I. In the complex formed by Lys-Gly-Trp-Lys
(OtBu) with poly(U), conformer 1 has the same fluorescence lifetime as
in the free state whereas conformer 2 has a reduced lifetime. The equilibrium is strongly shifted towards conformer 1 in the outside complex.
The electrostatic association constant is four-fold higher for conformer
1 as compared to conformer 2. Since conformer 1 appears to have a similar conformation (same lifetime) in complex I and in the free state, it
seems to be perfectly adapted to the charge distribution of the polynucleotide phosphates. The reduction in the lifetime of conformer 2 might
have two origins : change in peptide conformation (as described above)
or energy transfer to uracil bases. The second hypothesis seems unlikely
since the critical Förster distances for energy transfer from tryptophan
to nucleic bases are very small (especially in the case of uracil because
the overlap between tryptophan fluorescence and uracil absorption is quite small) (5). Therefore it seems likely that a conformational change in
conformer 2 is responsible for the shortening of its fluorescence lifetime in complex I.

The model described by equation (4) still represents an oversimplification. Even though fluorescence decay measurements reveal the presence of two conformers in the investigated peptides, more than two conformations are expected for the isolated peptides in aqueous solutions.
A shift in equilibrium between these different conformations is likely
to occur upon nucleic acid binding since not all of them will exhibit an
optimum adaptation to the polynucleotide backbone geometry. The polarity
of the peptide and polynucleotide chains should also be taken into account.
A given peptide conformer could bind in the 5' → 3' or 3' → 5' direction
with respect to the polynucleotide backbone. When the peptides contain
two aromatic amino acids as in the pentapeptides mentioned above, several
stacked complexes are expected which have either one or two aromatic
residues stacked with nucleic acid bases.

However the most significant conclusion which can be reached from
the present results (tables 1-3) deals with the formation of the stacked
complex(es) II. The small change of the average fluorescence lifetime
which is observed upon peptide binding to, e.g., denatured DNA (table 3)
is to be compared with the strong reduction of the overall fluorescence
intensity (more than 80 %) ; this points to the very efficient formation
of stacked complexes. Such complexes are not fluorescent and therefore

they are not seen in the fluorescence decay but they play an essential
role in the decrease of the overall fluorescence when the peptide is
bound to the nucleic acid. This holds true even in the few cases when
a decrease in fluorescence lifetime has been detected (see poly(U) in
table 3) : this decrease is far from being sufficient to explain the
overall fluorescence quenching. Stacked complexes are still the major
species in the complexed state.

    Evidence for stacking interactions has also been obtained from NMR
studies and luminescence investigations at low temperature. The proton
resonances of the tryptophyl ring of the investigated peptides are shifted
upfield in the presence of single-stranded poly(A). At 77 K, the luminescence characteristics of the tryptophyl ring are strongly perturbed when
the peptides are bound to poly 5-mercuriuridylic acid. The strong quenching of the tryptophyl fluorescence is accompanied by an enhancement of
the phosphorescence intensity and a drastic shortening (more than three
orders of magnitude) of the phosphorescence lifetime. These changes are
characteristic of the heavy atom effect of the mercury substituent in
the 5th position of the uracil ring and imply a van der Waals contact
of mercury with the aromatic ring of tryptophan (10). Although these
experiments do not unequivocally demonstrate that stacking interactions
are involved, they provide evidence for a close proximity of the polynucleotide bases and the aromatic ring of the peptide.

3. Dynamics of peptide-nucleic acid complexes

    To obtain information on the dynamic state of tryptophan-containing
peptides bound to native or denatured DNA we have used fluorescence anisotropy to analyse the mobility of the aromatic ring in the "outside"
complexes (complexes I in schemes 2 or 4). Stacked complex(es) II do not
emit fluorescence and are not amenable to fluorescence polarization studies. It has already been reported that the fluorescence polarization of
Lys-Trp-Lys is enhanced upon binding to native DNA (11).This holds true
for the tetra and pentapeptides described in the preceding paragraphs.
The anisotropy of the peptide r is related to its rotational correlation
time θ as shown by equation (16)

$$(16) \qquad \frac{1}{r} = \frac{1}{r_0} (1 + \frac{\tau}{\theta})$$

where r_0 is the limit anisotropy reached when the peptide is immobilized,
τ is the fluorescence lifetime of the peptide. The rotational correlation
time in the case of a sphere of volume V depends on viscosity (η) and
temperature (T) according to equation (17)

$$(17) \qquad \theta = \frac{\eta V}{k T}$$

    The calculation of θ can be made from a determination of the fluo-
rescence anisotropy r as a function of the fluorescence lifetime τ. The
latter can be changed by using acrylamide as an external quencher. Acry-

lamide deactivates the excited singletstate of tryptophan through collisions with the excited fluorophore. This leads to a quenching of its
fluorescence and a shortening of its fluorescence lifetime according to
equation (16)

(18) $$\frac{1}{\tau} = \frac{1}{\tau_0} \left(1 + K_{SV}\,[Q]\right)$$

where K_{SV} is the Stern-Volmer constant for quenching by acrylamide at
concentration $[Q]$, K_{SV} is the product of the bimolecular rate constant
k_Q and the fluorescence lifetime in the absence of quencher τ_0.

Fluorescence quenching studies have shown that acrylamide forms a
ground-state complex with tryptophan in addition to the excited-state
interaction. Consequently the fluorescence intensity of the peptide changes with acrylamide concentration $[Q]$ according to the following equation

(19) $$\frac{I_0}{I} = \left(1 + K_{SV}[Q]\right) \exp V[Q]$$

where V is the ground-state association constant. K_{SV} has the same
meaning as in equation (16).

A curve-fitting procedure was used to calculate both K_{SV} and V from
the plot of I_0/I *versus* $[Q]$. From the value of K_{SV}, the fluorescence lifetime τ could then be calculated using equation (18) for each acrylamide
concentration. As shown in table 4 the values of $\overline{K_{SV}}$ at 2°C for the peptide Lys-Gly-Trp-Lys(OtBu) in the absence and in the presence of native
DNA are not markedly different. This result indicates that the tryptophyl
ring is readily accessible in the peptide bound to native DNA. The slight
decrease of K_{SV} can be ascribed to the effect of the DNA double helix on
the solid angle within which the quencher molecule has access to the
excited peptide. A similar result was obtained at 20°C as already reported (9,12).

	$K_{SV}(M^{-1})$	$V\ (M^{-1})$	θ (ps)
KGWK(OtBu)	11	0.5	230
KGWK(OtBu) + native DNA	8.4	0.25	780

Table 4 - Values of K_{SV} (M^{-1}) and V (M^{-1}) obtained at 2°C according to
equation (19) for the quenching by acrylamide of the fluorescence of Lys-
Gly-Trp - Lys(OtBu) and its complex with native DNA. The rotational correlation times θ (in picoseconds) were obtained from a plot of $1/r$ *versus*
τ as shown in figure 3.

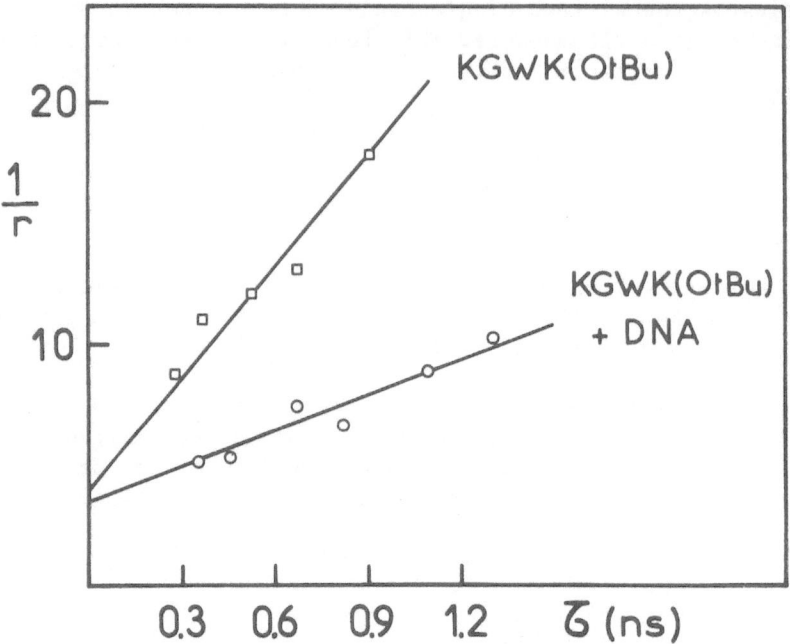

Figure 3 - Plot of 1/r *versus* τ according to equation (16) for the tetrapeptide Lys-Gly-Trp-Lys(OtBu) in the absence (□) or in the presence (O)
of native DNA at 2°C in the standard pH 6 buffer. The fluorescence lifetime τ was changed by adding acrylamide as an external fluorescence quencher (see text).

    A plot of 1/r *versus* τ can then be used to calculate the rotational
correlation time θ according to equation (16). As shown in figure 3 and
table 4 binding of the peptide to DNA is accompanied by an approximately
three-fold increase in the overall rotational correlation time θ. The
values of θ reported in table 4 at 2°C are higher than those previously
reported at 20°C using another method based upon the simultaneous change
in τ and η brought about by addition of glycerol to the solution containing either the free or the bound peptide. The increase in viscosity and
the decrease in temperature can account for the observed differences in
θ values. Preliminary measurements (9) of the decay of the fluorescence
anisotropy have shown that the time dependence of r obeys equation (20)

$$(20) \qquad r(t) \; = \; r_\infty \; + \; (r_0 - r_\infty) \; \exp \; - \frac{t}{\theta}$$

where r_∞ is the anisotropy limit reached at times long as compared to
the fluorescence lifetime. Although r_∞ is small when Lys-Gly-Trp-Lys(Otbu)
is bound to native DNA, its measurable value (0.021) indicates that the
tryptophyl ring is not able to rotate in all directions in space but is

(slightly) restricted to move within a cone whose semi-angle is about
60 degrees (9). This result is expected since the DNA double helix prevents the peptide from free movement in every direction (for the free
peptide $r_\infty = 0$). The θ values calculated from the anisotropy decay are
290 and 750 picoseconds for the free and bound peptides, respectively
(9).

Even though different methods give slightly different values for
rotational correlation times the main conclusion of these dynamic studies
is that the motion of the tryptophyl ring in the peptides bound to native
DNA is only slightly hindered as compared to the free peptide. These results are not compatible with a tryptophyl residue engaged in interactions
with nucleic acid bases in complex(es) I of schemes 2 and 4. In com-
plex(es) II complete fluorescence quenching occurs ; therefore no information on the dynamic state of the tryptophyl residue can be obtained
from fluorescence anisotropy studies. The results presented above deal
only with "outside" complex(es) I ; they are in agreement with schemes
(2) and (4). The rapidity and the amplitude of the motion of the fluorescent tryptophan in complex(es) I excludes that the indole ring is
partially stacked or hydrogen bonded to nucleic acid bases. Only electrostatic interactions of the amino groups with the nucleic acid phosphates
are involved in these complex(es). The modifications of the fluorescence
lifetimes of the two conformers (scheme (4) are therefore likely to be
due to conformational changes induced upon binding to the polynucleotide
backbone.

CONCLUSION

Oligopeptides containing lysyl and tryptophyl residues bind to
polynucleotides and nucleic acids ; two types of interactions (electrostatic and stacking) are involved.

Oligopeptides may adopt different conformations in solution. The
different conformers can bind with different affinities to nucleic acids.
Conformational changes are expected to take place upon binding to allow
the peptide to adapt to the polynucleotide geometry. Also binding to the
nucleic acid is expected to shift the equilibrium between the different
conformers.

Fluorescence steady-state and decay measurements are in agreement
with a binding scheme in which two types of complexes are formed. Both
involve electrostatic interactions ; stacking of the tryptophyl ring
with nucleic acid bases occurs only in the second complex. Fluorescence
anisotropy measurements indicate that the first, purely electrostatic,
complex is characterized by a high mobility of the tryptophyl ring.

Stacking interactions are strongly favored in single-stranded structures (table 2). It is therefore expected that aromatic amino acids play
an important role in the binding of proteins which exhibit a specificity
toward single-stranded nucleic acids (13).

ACKNOWLEDGEMENTS

   This work has been supported in part by the Délégation Générale à la Recherche Scientifique et Technique (81 E 1226). We thank Dr. R. Mayer for a gift of the pentapeptides containing two aromatic residues.

REFERENCES

1. Hélène, C., and Lancelot, G.: 1982, Prog. Biophys. Molec. Biol. 39, pp. 1-68.
2. Hélène, C., and Maurizot, J.C.: 1981, C.R.C. Crit. Rev. Biochem. 10, pp. 213-258.
3. Montenay-Garestier, T., Brochon, J.C., and Hélène, C.: 1981, Intern. J. Quant. Chem. 20, pp. 41-48.
4. Montenay-Garestier, T., and Hélène, C.: 1971, Biochemistry 10, pp. 300-306.
5. Montenay-Garestier, T.: 1975, Photochem. Photobiol. 22, pp. 3-6.
6. Alev-Behmoaras, T., Toulmé, J.J., and Hélène, C.: 1979, Biochimie 61, pp. 957-960.
7. Brun, F., Toulmé, J.J., and Hélène, C.: 1975, Biochemistry 14, pp. 558-563.
8. Record, M.T., Lohman, T.M., and de Haseth, P.: 1976, J. Mol. Biol. 107, pp. 145-158.
9. Montenay-Garestier, T., Fidy, J., Brochon, J.C., and Hélène, C.: 1981, Biochimie 63, pp 937-939.
10. Hélène, C., Toulmé, J.J., and Le Doan, T.: 1979, Nucl. Ac. Res. 7, pp. 1945-1954.
11. Toulmé, J.J., and Hélène, C.: 1977, J. Biol. Chem. 252, pp. 244-249.
12. Behmoaras, T. *et al.*: 1981 in "Intermolecular Forces", B. Pullman (ed.), Reidel, pp. 317-330.
13. Coleman, J.E., and Oakley, J.L.: 1980, C.R.C. Crit. Rev. Biochem. 7, pp. 247-289.

MOLECULAR MECHANISMS FOR THE RECOGNITION OF NUCLEIC ACIDS BY PROTEINS.
PRESENT-DAY AND PREBIOTIC ASPECTS.

Claude HELENE[†§], Gérard LANCELOT[§], Tula BEHMOARAS[†],
Roger MAYER[§], Christian CAZENAVE[†] and Nguyen T. THUONG[§]
†Laboratoire de Biophysique, INSERM U 201, ERA CNRS 951
Muséum National d'Histoire Naturelle, 61, rue Buffon
75005 PARIS, France
and
§Centre de Biophysique Moléculaire, CNRS,
45045 ORLEANS Cedex, France

INTRODUCTION

The binding of proteins to nucleic acids is an essential process in all living systems, allowing cells to build up essential structures (chromatin, ribosomes..) and to regulate the expression of genetic
information. Some of the nucleic acid-protein associates do not involve
specific interactions, i.e., the base sequence of the nucleic acid does
not control the binding process. In many cases the protein is able to
distinguish single-stranded from double-stranded structures. However
there are many nucleic acid-protein complexes which involve very specific
interactions between more or less extended regions of the two molecules.
As in any protein-ligand interactions, there must be a structural complementarity between the binding sites of the two molecules. However this
is not sufficient when a selective recognition of the base or base-pair
sequence is required. The two partners must engage specific interactions
between their functional groups even though the contact points need not
be contiguous in either primary structure.

The study of model systems which has been actively conducted
in our two laboratories during the past ten years has not only provided
information on some molecular mechanisms which control the present-day
recognition of nucleic acids by proteins but it has also revealed some
possibilities for the development of primitive but specific peptide-
polynucleotide associations under prebiotic conditions. These different
aspects will be briefly summarized below ; recent review articles should
be consulted for more in-depth coverage (1,2).

STRUCTURAL COMPLEMENTARITY BETWEEN PROTEINS AND NUCLEIC ACIDS

- Recognition of the DNA double helix

129

C. Hélène (ed.), Structure, Dynamics, Interactions and Evolution of Biological Macromolecules, 129–140.

Several models for the recognition of nucleic acid double helices have been proposed and some of them date back to the early days of the double helix itself. For instance, it was suggested in 1955 that an extended polyarginine chain could wrap around the DNA double helix (3). Then other structural elements of proteins were proposed to be involved in their interactions with nucleic acids. The α-helical N-terminal region of histone 4 was shown to fit quite well within the major groove of DNA (4). A complementarity was demonstrated between a twisted β-sheet structure and the narrow groove of RNA (5) or DNA (6) double helices. More recently, on the basis of the crystal structure of the cro protein from bacteriophage λ a model was built which involves α-helices interacting with the large groove of a right-handed DNA double helix (7). Two such α-helices, one in each subunit of a cro dimer, have a center-to-center distance of 34 Å. Their axes are parallel and inclined at 32° with respect to the line connecting their centers. These two helices can be perfectly accommodated within successive major grooves on one side of the B-DNA double helix (7).

The crystal structure of another DNA-binding protein, the cyclic AMP-binding protein (CAP), involved in the regulation of the transcription of catabolite genes, has led McKay et al. to suggest that α-helices from this protein could fit into the major groove of a left-handed DNA double helix (8). However experiments using circular DNAs containing the CAP binding site have suggested that CAP does not induce a right-to-left transition of the double helix (9). The possibility has been recently suggested that N-terminal regions of α-helices could be utilized by the CAP protein to bind to the major groove of a B-DNA right-handed helix (10).

Recent X-ray crystallographic studies of the operator-binding domain of λ repressor have also emphasized the role that α-helices could play in the recognition of DNA sequences (10). The model proposed on the basis of the crystal structure involves the N-terminal ends of two successive α-helices but none of these α-helices has its axis parallel to the major groove. This is in contrast with the model proposed by Anderson et al. (7) where α-helices are lying along the major groove of the B-DNA double helix. However it should be kept in mind that these models assume that i) the relative positions of the structural elements in both the monomer unit and the dimer inside the protein crystal are identical when the protein dimer binds to its specific site on DNA ; ii) DNA retains the B-conformation in the complex. There is evidence that binding of site-specific proteins to DNA may induce conformational changes in both the protein and the nucleic acid. Crystal structure is not yet available for a protein-nucleic acid complex. The attractive models presented by Anderson et al. (7), McKay and Steitz (8), Pabo and Lewis (10) should be taken as working hypotheses which can be submitted to experimental test. Sequence homologies between DNA-binding domains in different proteins have suggested that a conserved secondary structure (α-helices or α-helix-turn-α-helix) might be the common basis for the recognition of B-DNA base-pair sequences by regulatory proteins (11,12).

In order to test the possible role of α-helices in DNA recognition, a synthetic tetradecapeptide whose sequence is identical to fragment 26-39 of the cro protein has been synthesized in our laboratory by R. Mayer. This peptide contains the sequence which has been postulated by Anderson et al. to fit into the major groove of the DNA double helix (7). When an α-helix is built with this sequence, all hydrophobic residues are found on one side of the helix while hydrophilic residues, including charged amino acids, are located on the other side (Figure 1). In aqueous solution at low ionic strength, the synthetic fragment 26-39 of the cro protein exhibits a disordered structure. Transition towards an α-helical state starts at high ionic strength. Conversion to an α-helix is induced by organic solvents such as hexafluoroisopropanol (HFIP). The transition is cooperative with respect to HFIP concentration and in a mixture HFIP/water 1/4 (v/v) total conversion from a random coil to an α-helix is achieved.

```
      1                              10                                   20
H₂N-Met-Glu-Gln-Arg-Ile-Thr-Leu-Lys-Asp-Tyr-Ala-Met-Arg-Phe-Gly-Gln-Thr-Lys-Thr-Ala-Lys-Asp-Leu
                      ───────────────30──                              40──
      Gly-Val-│Tyr-Gln-Ser-Ala-Ile-Asn-Lys-Ala-Ile-His-Ala-Gly-Arg-Lys│-Ile-Phe-Leu-Thr-Ile-Asn-Ala
      50                              60                                   66
Asp-Gly-Ser-Val-Tyr-Ala-Glu-Glu-Val-Lys-Pro-Phe-Pro-Ser-Asn-Lys-Lys-Thr-Thr-Ala-COOH
```

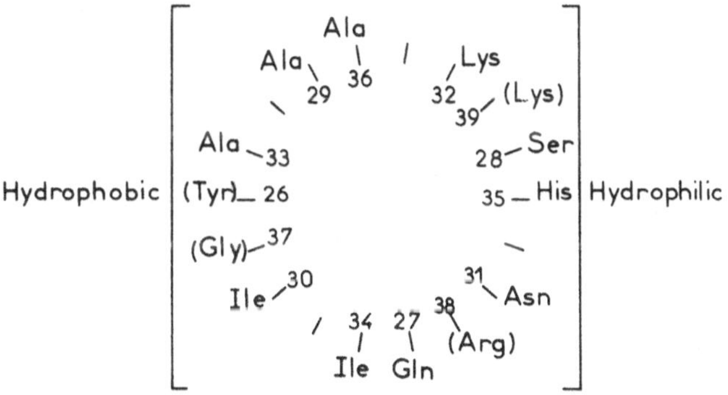

Fig. 1 : *Amino acid sequence of the* cro *protein from bacteriophage λ. The synthetic fragment 26-39 is boxed and the spatial distribution of the side-chains of this fragment in an α-helical conformation is shown on a projection perpendicular to the helix axis. (The amino acids in parentheses are those which are not included in the α-helix in the crystal structure of the* cro *protein (7)).*

In order to investigate the binding of the synthetic 26-39
fragment to DNA double helices short complementary oligodeoxyribonucleo-
tides (10 nucleotides long) have been synthesized including the left
part of the OR3 operator sequence which is selectively recognized by the
cro protein (N.T. Thuong et al., reference 13 and unpublished results).
These oligonucleotides form short double helices in aqueous solutions
at low temperature as revealed by absorption, circular dichroism and
NMR spectroscopy (13,14). Before investigating the interactions between
these short DNA double helices and the cro protein fragment it was
necessary to demonstrate that the double helix could be obtained in the
presence of the α-helix-forming solvent HFIP. In the mixture HFIP/water
1/4 (v/v) the melting temperature of the self-complementary decanucleo-
tide AATTGCAATT is decreased by $\simeq$ 20°C. However it is still possible to
observe the formation of a double helix at low temperature. From the
NMR and CD spectra it can be concluded that a B-like structure is
obtained.

The synthetic 26-39 fragment of the cro protein does interact
with nucleic acids. Preliminary fluorescence, circular dichroism and NMR
data have been obtained for its interaction with short double helices
(G. Lancelot, R. Mayer and C. Hélène, unpublished results). A quenching
of the tyrosine fluorescence is observed upon complex formation. Several
of the proton resonances of both partners are shifted in the complex.
Characterization of the complex structure and of the specificity of the
interaction is presently under way.

- Recognition of single strands

There are many proteins whose role is to bind selectively to
single-stranded nucleic acids (SSB proteins) (15). For example the
gene 32 protein from phage T4 binds to its mRNA (thus regulating its
synthesis at the translational level) and to single-stranded RNA and
DNA with a preference for DNA (whence its role in DNA replication and
repair). Little is known about the structure of SSB proteins and their
mode of binding to single-stranded nucleic acids. The crystal structure
of the gene 5 protein from phage fd has been determined by Mc Pherson
et al. (16). There is no α-helix in this protein ; a long groove
beneath a three-stranded β-sheet structure could be the DNA binding
site. Basic amino acid residues are located in the interior of this
groove while the edges bear several aromatic residues which could be
involved in stacking interactions with the nucleic acid bases.

The specificity of SSB proteins toward single strands is
usually due to a cooperative mode of binding. The most documented case
is the T4 gene 32 protein which binds to isolated double-stranded or
single-stranded sites with nearly the same affinity ; but binding to
single strands only is highly cooperative with a cooperativity parameter
around 10^3 (17). Competition is therefore strongly in favor of single-
stranded structures. Moreover cooperativity amplifies any small difference exhibited by different (short)nucleotide sequences. If the ratio
of the affinities for two isolated sites with different base sequences

is α, the ratio of the binding constants for the two corresponding polynucleotides will be roughly α^n where n is the average cluster size of the cooperatively bound protein. Thus even small differences at the level of oligonucleotides can make a large difference at the level of polymers (18).

The recA protein from *E. coli* plays a very important role in recombination and repair of DNA (19). This protein which binds both to single-stranded and double-stranded nucleic acids exhibits a strong preference for the former above pH 7. Its binding to single strands is highly cooperative (C. Cazenave and C. Hélène, unpublished results). The behavior of the recA protein in this respect looks very similar to that of the gene 32 protein. The strongest affinity in both cases is exhibited by $(dT)_n$ sequences. When *E. coli* DNA replication is blocked, e.g., by thymine starvation, shift to high temperature (in some temperature sensitive mutants) or base damages induced by radiations or chemical substances, the recA protein acquires a protease activity and cleaves the lex A gene product thus derepressing a series of genes (SOS genes) which play an essential role in both DNA repair and mutagenesis (19). Activated recA also cleaves the λ repressor thus leading to prophage induction. The effector required to activate the proteolytic activity of the recA protein is thought to be single-stranded DNA. Due to the cooperativity of recA protein binding to single strands, single-stranded regions of DNA long enough to accommodate a sufficient number of clustered recA protein molecules could act as an effector by displacing recA proteins from their binding sites elsewhere on the DNA. An interesting problem is raised by the interplay of SSB and recA proteins since both bind strongly and cooperatively to single-stranded DNA. Mutations in the ssb gene which codes for the SSB protein lead to defects in SOS induction. This might be due to a role of SSB in the metabolism of the inducing molecules (e.g., wild type SSB could protect single-stranded DNA from degradation by endonucleases) or to its influence on the binding of recA protein to single strands. *In vitro* experiments have shown that SSB competes with recA for binding to limiting amounts of single-stranded DNA but it eliminates the inhibition of recA proteolytic activity when single-stranded DNA is in excess over the stoichiometric recA – ssDNA complex (19). The SSB protein plays also a role in the recombinational activity of recA protein. When SSB is present less recA is required to achieve the same amount of strand exchange between single-stranded DNA and an homologous DNA duplex (36, 37).

SPECIFIC INTERACTIONS INVOLVING AMINO ACID SIDE CHAINS AND NUCLEIC ACID BASES

Interactions between functional groups in protein-nucleic acid associations have been reviewed recently (1). Electrostatic interactions are involved in most cases with lysine, arginine and histidine side chains forming ion pairs with nucleic acid phosphates. The formation of indirect electrostatic bonds mediated by divalent cations such as Zn^{++} has been demonstrated in polypeptide – Zn^{++} – polynucleotide

ternary complexes (20,21). This opens the possibility of bringing
negatively charged regions of proteins close to the DNA bases through
the formation of, e.g., Glu (Asp)... Me^{++}... phosphate bonds. Metal
cations (such as Fe^{2+} which was very abundant under prebiological
conditions, before oxygen evolved) could have played an important role
in "primitive" associations between peptides and polynucleotides.

1) Hydrogen bonding

        The role that hydrogen bonds could play in the recognition
of nucleic acid base sequences by proteins has been emphasized by several authors (22-25). Many amino acid side chains can form two hydrogen
bonds with nucleic acid bases or base pairs. These interactions can
occur in the plane of the base or base pair ("horizontal interaction");
they may also involve two consecutive bases or base pairs in a nucleic
acid ("vertical interactions") (1,26). In the last case, model building
studies have shown that amino acid side chains such as Gln or Asn can
bridge the two chains of a double helix by forming hydrogen bonds with
two bases located on opposite strands and belonging to two consecutive
base pairs (1).

        The recognition code based on hydrogen bonding is degenerate,
i.e., in most cases, a given amino acid side chain may form two hydrogen
bonds with different bases or base pairs. Some selectivity can be
achieved depending on the local structure of the nucleic acid-protein
complex. For example, the side chain of Gln or Asn can form two
hydrogen bonds with a G...C or an A...T base pair ; however this interaction takes place in the minor groove of DNA with a G...C base pair
whereas the major groove is involved for an A...T base pair. Also purines
can be distinguished from pyrimidines in the large groove (A and G)
or the small groove (G) since only purines still possess two hydrogen
bonding groups in the correct position when they are engaged in base
pairing.

        Until now the only specific interaction which has been experi-
mentally demonstrated involves carboxylate anions (the side chain of
Glu or Asp) and guanine (27,28). Among the four bases, only guanine has
two donor groups in the adequate position to form two hydrogen bonds with
a carboxylate anion. This selective interaction has been first demonstrated in an organic solvent (DMSO) (27) and more recently in aqueous
solvents (28). The most remarkable feature of this carboxylate-guanine
pair is its high association constant compared with that of a G...C base
pair. A comparison of these association constants is given in Table 1
for different H$_2$O-DMSO mixtures. Even with a water concentration as high
as 13.7 M the carboxylate-guanine association constant is still about
ten times higher than that of a G...C base pair, even though water is
competing with the two solutes for hydrogen bond formation.

Table 1 : Association constants for the binding of guanine to carboxy-
          late and to cytidine in mixtures of DMSO and H_2O (determined
          from NMR experiments at 295 K, see reference 28).

H_2O concentration	0.6 M	2.7 M	5.7 M	13.7 M
$K_{G...C}$	3.3	2.8	2.3	1.6
$K_{G...Carboxylate}$	110	51	27	18

        The strong association between carboxylate and guanine might
have some relevance not only to protein-nucleic acid associations but
also to prebiotic interactions involved in the establishment of a primitive genetic code.

        2) Stacking interactions

        Evidence for the existence of stacking interactions between
aromatic amino acids and nucleic acid bases has been provided by a study
of peptide binding to nucleic acids (for a review see reference 2).
Stacking is strongly favored in single-stranded nucleic acids and is
likely to be involved in complex formation with single-strand-binding
proteins (SSB).

        A recent study (Colot, Toulmé and Hélène, unpublished results)
has shown that oligopeptides such as Lys-Trp-Lys or Lys-Trp-Gly-LysOtBu
bind to chromatin core particles with nearly the same affinity as to
naked DNA but that the number of binding sites is reduced by $\simeq$ 50 %.
If the binding process is analysed according to the two-step model
previously proposed (equation 1) (29), the values of both K_1 and K_2 are
unchanged in core particles as compared to naked DNA.

$$\text{(1)} \qquad \text{Peptide} + \text{DNA} \overset{K_1}{\rightleftarrows} \text{Complex I} \overset{K_2}{\rightleftarrows} \text{Complex II}$$

        In this model complex II involves both electrostatic interac-
tions (between amino groups of the peptide and phosphate groups of the
nucleic acid) and stacking of the tryptophyl ring with nucleic bases.
In complex I only electrostatic binding is involved. Since K_2 is unchanged in core particles, stacking interactions are as likely to occur in
chromatin as in naked DNA : however half of the binding sites are made
inaccessible due to interactions of the DNA double helix with the histone
core.

        In oligopeptides bound to native DNA no stacking occurs between
tyrosine and nucleic acid bases (30) whereas tryptophan-containing pep-

tides do exhibit such an interaction. Stacking of tryptophan with DNA
bases is expected to give rise to a bending of the double helix ("partial insertion" model). This might be one of the reasons why evolution
has selected proteins containing no tryptophan (histones) to ensure proper compaction of the DNA filament without further distortion of the
double helical structure.

The most favorable situation for stacking interactions occurs
when DNA has been depurinated. Removal of a purine creates a cavity which
has the right size to accommodate the indole ring of tryptophan
(Figure 2). As a matter of fact the K_2 value (equation 1) has been estimated to be around 200 for binding of Lys-Trp-Lys to an apurinic site
as compared to 0.3 for a native site (31). Since electrostatic binding
(K_1) is unchanged the overall association constant of the peptide
$K_1(1 + K_2)$ is increased by more than two orders of magnitude. Selective
recognition of apurinic DNA is therefore achieved by a short peptide as
a result of the insertion of its aromatic ring at the apurinic site.
This interaction could be involved in the binding of specific apurinic
endonucleases whose function involves the selective recognition of
apurinic sites in DNA and the subsequent cleavage of the phosphodiester
backbone which is required as a first step in the repair of apurinic DNA.

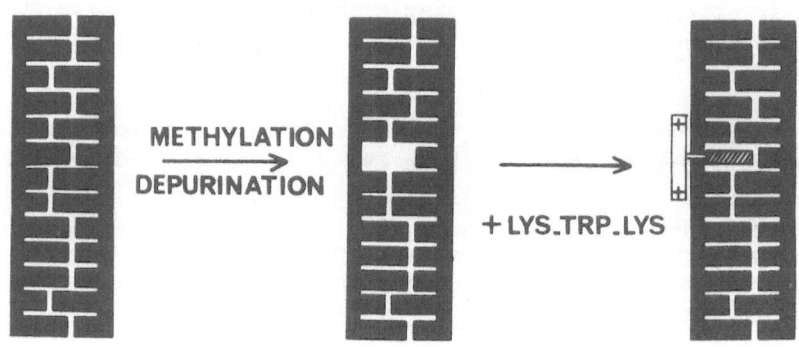

BINDING OF LYS_TRP_LYS TO APURINIC SITES

Figure 2 : *A model for the insertion of the aromatic ring of*
Lys-Trp-Lys in an apurinic site of DNA created by methyla-
tion followed by depurination.

Apurinic sites in DNA are alkali-labile and cleavage of the phosphodiester bond is catalyzed by weak bases. Insertion of the tryptophyl ring of Lys-Trp-Lys into the apurinic cavity brings two lysyl side chains and the α-amino group of the peptide close to the DNA backbone. Therefore it could be expected that these amines might catalyze the cleavage of the phosphodiester bond if they had the proper orientation with respect to the DNA backbone. As a matter of fact when the peptide-apurinic DNA complex is incubated at 37°C such a cleavage is observed (32-34).

Removal of a purine from DNA creates an aldehydic function on the sugar moiety. If the aldehyde is reduced by sodium borohydride no cleavage is observed. The α or ε amino groups of the peptide could form a Schiff base with this aldehyde. Using sodium cyanoborohydride to reduce and trap the Schiff base (cyanoborohydride does not reduce aldehydes) it was shown that less than 10 % of the cleaved sites had resulted from Schiff base formation (T. Behmoaras, unpublished results). It is therefore likely that the cleavage of the phosphodiester bond results from a β-elimination reaction following abstraction of a proton in the 2'-position of the sugar by one of the amino groups of the peptide.

PREBIOTIC ASPECTS OF PEPTIDE-NUCLEIC ACID COMPLEXES

We have shown above that a simple tripeptide such as Lys-Trp-Lys is endowed with two remarkable properties : selective recognition of apurinic sites ; cleavage of the phosphodiester bond at these apurinic sites. It behaves as a mini-enzymatic system with a recognition site (Trp) and a catalytic site (Lys). Another remarkable property of this tripeptide should also be recalled : when its complex with UV-irradiated DNA is submitted to UV radiations a photosensitized splitting of pyrimidine dimers is observed (35). Thus Lys-Trp-Lys mimics both apurinic endonucleases and photoreactivating enzymes. The mechanisms of the two activities exhibited by this peptide may be different from those involved in the present-day enzymatic processes. β-elimination does not seem to be the major route utilized by most apurinic endonucleases ; the electron transfer reaction which appears responsible for the sensitized splitting of pyrimidine dimers involves another chromophore than tryptophan in photoreactivating enzymes. Nevertheless it is an attractive hypothesis that enzymes such as apurinic endonuclease or photoreactivating enzymes could have evolved from simple peptides containing lysyl and tryptophyl residues.

It must also be kept in mind that stacking interactions of aromatic amino acids and nucleic acid bases are strongly favored in single-stranded nucleic acids (1,2). Most-if not all-proteins which exhibit a specificity toward a particular nucleic acid structure or a DNA base sequence also bind in a non-specific way to any nucleic acid. When oligopeptides containing basic and aromatic residues are bound to nucleic acids, basic residues form electrostatic bonds with nucleic

acid phosphates and are responsible for non-specific binding ;
aromatic residues are involved in stacking interactions which ensure
specific binding to single strands. Therefore simple peptides containing basic and aromatic residues might have been the precursors of
single-strand binding proteins during the course of evolution.

ACKNOWLEDGMENTS :

        We wish to thank Drs T. Montenay-Garestier, F. Toulmé and
J.J. Toulmé for stimulating discussions and communication of unpublished
results. This work was supported by CNRS, INSERM (ATP 77 79 109 and
CRL 794 033 2 and 812 032), DGRST (81 E 1226), Ligue Nationale Française contre le Cancer and Fondation pour la Recherche Médicale.

REFERENCES :

1. HELENE C. and LANCELOT G. (1982) Prog. Biophys. Molec. Biol. 39,
   pp. 1-68.

2. HELENE C. and MAURIZOT J.C. (1981) CRC Crit. Rev. Biochem. 10,
   pp. 213-258.

3. FEUGHELMAN M. et al., (1955), Nature 175, 834-838.

4. SUNG M.T. and DIXON G.H. (1970) Proc. Natl. Acad. Sci. USA, 67
   pp. 1616-1623.

5. CARTER C.W. and KRAUT J. (1974), Proc. Natl. Acad. Sci. USA, 71
   pp. 283-287.

6. CHURCH G.M., SUSSMAN J.L. and KIM S.H. (1977), Proc. Natl. Acad. Sci.
   USA, 74, pp. 1458-1462.

7. ANDERSON W.F., OHLENDORF D.H., TAKEDA Y. and MATTHEWS B.W. (1981)
   Nature, 290, pp. 754-758.

8. McKAY D.B. and STEITZ T.A. (1981), Nature 290, pp. 744-749.

9. KOLB A. and BUC H. (1982), Nucl. Ac. Res. 10, pp. 473-485.

10. PABO C.O. and LEWIS M. (1982) Nature 298, pp. 443-447.

11. STEITZ T.A., OHLENDORF D.H., McKAY D.B., ANDERSON W.F. and
    MATTHEWS B.W. (1982), Proc. Natl. Sci. USA 79, pp. 3097-3100.

12. SAUER R.T., YOCUM R.R., DOOLITTLE R.F., LEWIS M. and PABO C.O.
    (1982), Nature 298, pp. 447-451.

13. THUONG N.T., CHASSIGNOL M., LANCELOT G., MAYER R., HARTMANN B., LENG M. and HELENE C. (1981), Biochimie 63 pp. 775-784.

14. LANCELOT G., MAYER R., THUONG N.T., CHASSIGNOL M. and HELENE C. (1981),Biochimie 63 pp. 785-790.

15. HELENE C., TOULME J.J. and MONTENAY-GARESTIER T. (1982) in "Topics in Nucleic Acid Structure", S. Neidle Ed. Part 2, Mc MILLAN Publishers Ltd.

16. Mc PHERSON A., JURNAK F.A., WANG A.H.J., MOLINEUX I. and RICH A. (1979),J. Mol. Biol. 134, pp. 379-400.

17. KOWALCZYKOWSKI S.C., LONBERG N., NEWPORT J.W. and VON HIPPEL P.H. (1981),J. Mol. Biol. 145, pp. 75-104.

18. NEWPORT J.W., LONBERG N., KOWALCZYKOWSKI S.C. and VON HIPPEL P.H. (1981),J. Mol. Biol. 145, pp. 105-121.

19. LITTLE J.W. and MOUNT D.W. (1982),Cell. 29, pp. 11-22.

20. HELENE C. (1975), Nucl. Ac. Res. 2, pp. 961-969.

21. BERE A. and HELENE C. (1979), Biopolymers 18, pp. 2659-2672.

22. BRUSKOV V.I. (1975), Molek. Biol. (USSR) 9, pp. 304-309.

23. SEEMAN N.C., ROSENBERG J.M. and RICH A. (1976), Proc. Natl. Acad. Sci. USA 73, pp. 804-808.

24. HELENE C. (1976), Studia Biophysica 57, pp. 211-222.

25. HELENE C. (1977), FEBS Letters 74, pp. 10-13.

26. WOODBURY C.P., HAGENBUCHLE O. and VON HIPPEL P.H. (1980); J. Biol. Chem. 255, pp. 11534-11546.

27. LANCELOT G. and HELENE C. (1977), Proc. Natl. Acad. Sci. USA, 74, pp. 4872-4875.

28. LANCELOT G. and MAYER R. (1981),FEBS Letters 130, pp. 7-11.

29. BRUN F., TOULME J.J. and HELENE C. (1975), Biochemistry 14, pp. 558-563.

30. MAYER R., TOULME F., MONTENAY-GARESTIER T. and HELENE C. (1978), J. Biol. Chem. 254, pp. 75-82.

31. BEHMOARAS T., TOULME J.J. and HELENE C. (1981), Proc. Natl. Acad. Sci. USA 78, pp. 926-930.

32. BEHMOARAS T., TOULME J.J. and HELENE C. (1981), Nature 292, pp. 858-859.

33. PIERRE J. and LAVAL J. (1981), J. Biol. Chem, pp. 10217-10220.

34. BEHMOARAS T., TOULME J.J. and HELENE C. (1981) C.R. Acad. Sci. Paris 293, pp. 5-8.

35. HELENE C. and CHARLIER M. (1977), Photochem. Photobiol. 25, pp. 429-434.

36. McENTEE K., WEINSTOCK G.M. and LEHMAN I.R. (1980), Proc. Natl. Acad. Sci. USA 77, pp. 857-861.

37. SHIBATA P., DASGUPTA C., CUNNINGHAM R.P. and RADDING C.M. (1980), Proc. Natl. Acad. Sci. USA 77, pp. 2606-2610.

THE *LAC* REPRESSOR. STRUCTURE AND INTERACTIONS

Bernard Barbier, Michel Charlier, Françoise Culard, Maurice
Durand, Jean-Claude Maurizot and Manfred Schnarr
Centre de Biophysique Moléculaire, C.N.R.S.,
1A, avenue de la Recherche Scientifique,
45045 Orléans Cedex, France.

ABSTRACT

Recent results from our laboratory, on the structure and the interactions of *lac* repressor with the DNA are reviewed. Informations on the
shape and the spatial organization of the *lac* repressor, its tryptic core
and the headpieces were obtained by small angles neutron scattering experiments. The conformation and stability of the headpiece against thermal denaturation and tryptic hydrolysis as a function of salt concentration were also studied. Interaction between non operator DNA and the
repressor or the headpieces were investigated using the photochemical
crosslinking method, circular dichroism, fluorescence quenching, thermal
denaturation of the nucleic acid. Similitudes and differences between
the headpiece and the repressor behaviours are analysed. The specific
interaction between the repressor and DNA fragments containing the operator sequence were studied using circular dichroism. The interactions
between the headpiece and the *lac* operator were also studied by fluorescence and circular dichroism. All these investigations lead to a "low
resolution" model of the *lac* repressor and its complexes with nucleic
acids.

FOREWORD

Since the formulation of the concept of operon structure, expression
and regulation by Jacob and Monod in 1961 (1), the operator-repressor
system of the *lac* operon has been extensively studied by numerous methods.
At the beginning, these studies were essentially carried out using bacterial genetics. The isolation of the *lac* repressor by Gilbert and Müller-
Hill in 1966 (2) and the ensuing building of overproducing strains have
opened the way to biochemical studies. Towards the middle of the seven-
ties, the biochemical knowledge of the system was large enough to develop a biophysical approach. This paper is dedicated to Professor Charles
Sadron pioneer of molecular biophysics, for his eightieth birthday. We
present here some recent results we obtained on the *lac* system using
biophysical methods.

C. Hélène (ed.), Structure, Dynamics, Interactions and Evolution of Biological Macromolecules, 141–153.

INTRODUCTION

The *lac* repressor from *E.coli* is a protein, which binds very tightly a precise part of the DNA, the operator, and can block the expression of the structural genes of the lactose operon. The inducer, *in vivo* the allolactose, binds the repressor and decreases its affinity constant for the operator by about three orders of magnitude. This decrease is large enough to lead, *in vivo*, to repressor release, and thus allows the transcription of the genes z, y and a. The repressor can also bind non operator DNAs, a process which is commonly called the "non specific binding". The affinity constant corresponding to this interaction is smaller by several orders of magnitude than that for the specific operator-repressor binding. Nevertheless in the genome of *E.coli*, the non operator DNA considerably exceeds the only operator sequence, so that the biological role of the non specific binding becomes essential (3-5).

It is now well admitted that the repressor is a protein organized in several domains. The genetic analysis of numerous mutants by the groups of B. Müller-Hill (3) and J. Miller (6) showed unambiguously that the N-terminal part of the protein is involved in the binding process with the nucleic acids. K. Weber and coworkers (7) have shown that the domains can be obtained using proteolytic cleavage (trypsin, chymotrypsin, clostripain) in particular conditions of ionic strength.

In these conditions, one obtains a tetrameric core containing four times the amino acids 60 to 360, and four N-terminal peptides, the headpieces, containing amino acids 1-51, 1-56 or 1-59, depending on the enzyme and the conditions used.

The inducers bind the core with the same affinity as the native repressor, and the headpieces interact with nucleic acids. Several studies have shown that the conformation of the isolated core is the same as it is in the native repressor. For instance, we have found that the induced circular dichroism of effectors bearing an orthonitrophenyl group is the same when they are bound on the core and on the native repressor (8). The environment of the binding site of the inducer is thus similar in the core and in the repressor. Using the hydrogen-tritium exchange kinetic method, we reached alike the conclusion that the conformation of the core is the same when it is isolated and when it is in the repressor (9).

In this paper, we wish to present some of our recent results dealing with the structural organization of the domains, and with the interactions between the repressor or its isolated domains and the nucleic acids.

STRUCTURAL STUDIES

Small angle neutron scattering (ILL Grenoble, Collaboration with G. Zaccaï)

A study has been carried out by this method on the native repressor

and its domains. It provides immediate informations about the shape and
the structural organization of the molecule. Figure 1 shows the Guinier
plots obtained for the repressor, the core and the headpiece. From the
radius of gyration R_G of these three particles, we can deduce several
informations :

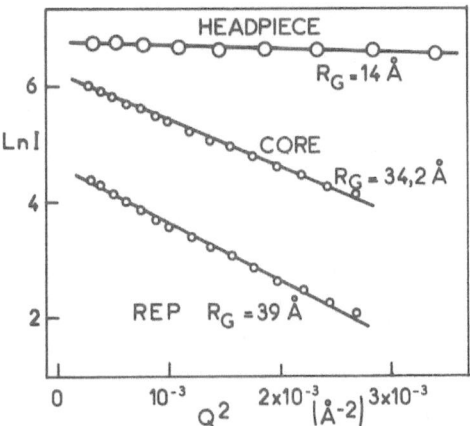

Figure 1 - Guinier plots of the neutron scattering data for solutions
of repressor, core and headpiece in deuterated buffer (> 98 % D_2O).

  - The radius of gyration of the repressor (39 Å in D_2O buffer and
42.5 Å in H_2O buffer is particularly high for a protein of a molecular
weight of 158000. This value shows that scattering matter is located far
from the center of mass of the molecule, and consequently, that the molecule exhibits an elongated shape. The analysis of scattering data at
larger angles confirms this result (10).

  - The radius of gyration of the core (34.5 Å) leads to the same
conclusions.

  - The radius of gyration of the headpiece is 14 Å. Only a very elongated shape for this peptide can explain this value. One can calculate
that the height of an equivalent cylinder (same volume and same radius
of gyration) would be 45 Å. For an equivalent probate ellipsoid, the
great axis would be 59 Å in length.

  Using the measured radii of gyration of the repressor, of the core
and of the headpiece, and applying the parallel axes theorem, we calculated the distance d between the center of mass of the repressor and the
center of mass of each headpiece :

$$M R_G^2 = M' R_G'^2 + 4 \, m r_G^2 + 4 \, m d^2$$

M, M' and m are the scattering masses of the repressor, the core

and the headpiece respectively, and R_G, $R_G^!$ and r_G the corresponding
radii of gyration. We found d = 61 ± 10 Å. This value implies an outward
localization of the headpiece in the *lac* repressor.

Several other arguments (some of which will be discussed below)
indicate that the headpieces must be arranged in pairs. This leads us to
propose for the *lac* repressor the model drawn in the figure 2.

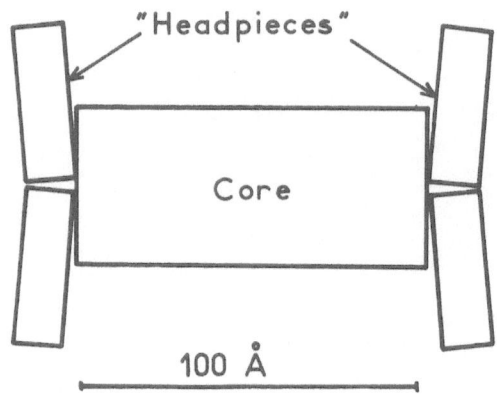

Figure 2 - Low resolution model for the *lac* repressor as deduced from
the scattering measurements.

Structural stability of the headpiece

Intact headpieces are obtained by proteolysis of the repressor at
high ionic strength (1 M Tris). When the hydrolysis is performed at low
ionic strength, the N-terminal part is cleaved in several fragments. This
led us to analyse the conformational stability of the headpiece as a
function of the ionic strength and of the temperature (12).

The CD signal of the headpiece decreases strongly when increasing
the temperature. This is due to the complete vanishing of the helicoidal
structure of the headpiece. This phenomenon can be analysed as a two
state process, with a very low cooperativity. One of the most striking
aspects of this phenomenon is its strong dependence on the ionic strength.
The temperature of the midpoint of the denaturation increases from 37°C
in absence of salt to 68°C in 1 M NaCl. Such a variation is markedly
larger than that commonly observed for proteins (13).

We have also studied the kinetic of trypsinolysis of the headpiece
at various ionic strength. This can be followed by CD since the hydrolysis leads to a complete vanishing of the secondary structure. Figure
3 shows the kinetic curves at various ionic strengths. They indicate pseudo first order kinetics, which strongly depend on the salt concentration.

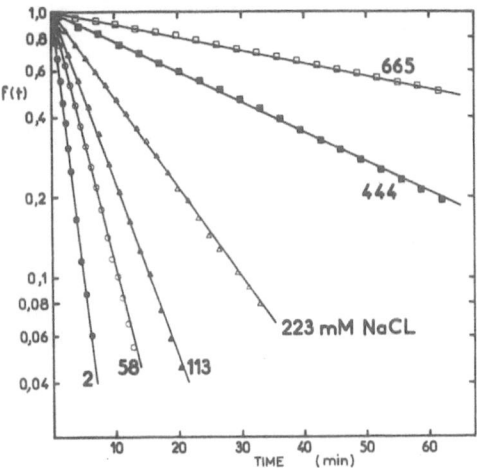

<u>Figure 3</u> - Kinetic curves of the tryptic digestion of the headpiece at
various ionic strengths. f(t) is the relative variation of the CD signal
at 220 nm. (From reference 12, with kind permission of Elsevier Biomedi-
cal Press).

This result is in agreement with the hypothesis that only the denatured
form of the headpiece is a substrat for trypsin.

    We put forward the idea that this outstanding dependence of the
headpiece stability with the ionic strength could be related to the fact
that it has to bind the DNA. It would correspond to a particular geometrical arrangement of the positive charges of the headpiece, which might
be complementary of the set of the phosphate negative charges of the DNA
molecule.

    It has been shown by Schlotman and Beyreuther (14) that for a great
number of i⁻d mutants (mutants with altered operator-binding activity)
there was a very fast endogenous degradation of the N-terminal part of
the *lac* repressor by intracellular proteases. Our results can explain
this observation. One may think that the "missense" mutations in this
N-terminal part of the protein just act by lowering the stability of the
headpiece, thus allowing a faster proteolytic hydrolysis than for the
wild-type repressor.

    It must be pointed out that our results were obtained on the isola-
ted headpiece. When it is included in the *lac* repressor supplementary
stability may arise from the interaction with the core. We have shown
that in the *lac* repressor the headpieces stabilize the core against urea
denaturation (27) and it is probable that the stabilization is mutual.
However, the fact that the proteolysis of the *lac* repressor leaves the
headpiece uncleaved when performed at high ionic strength indicates that

the stabilization of the headpiece by neutral salt also occurs when it is
part of the native repressor.

INTERACTION WITH THE NON-OPERATOR DNA. NON SPECIFIC BINDING

Photochemical studies

It is possible to covalently cross-link repressor to 5-BrUracil substituted DNA by near UV irradiation (λ > 295 nm) of the complex. Using
proteolytic cleavage after crosslinking, we have demonstrated that the
repressor is linked to DNA through the headpieces. Moreover, we were
able to recrosslink a repressor previously crosslinked to DNA, after
a DNase digestion of the excess DNA. This fact shows that the repressor
has two separate binding sites for the non-operator DNA (Figure 4), and
is in full agreement with the model of the protein shown on the figure 2.

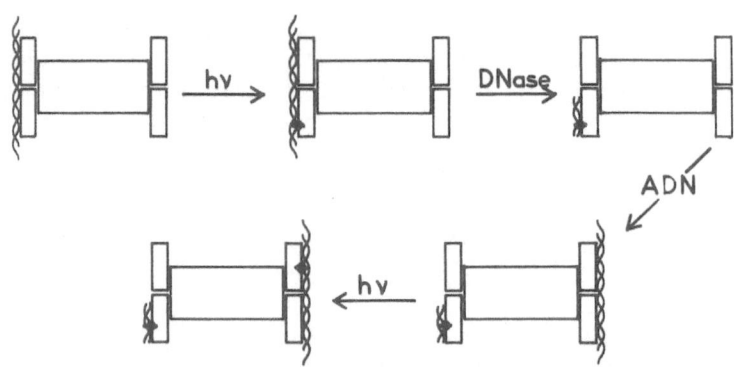

Figure 4 - Synopsis of the experiment of double photochemical crosslink
of repressor and 5-BrUracil substituted DNA.

Interactions between the headpiece and non-operator DNA

The headpieces interact with DNAs, leading to a conformational
change of the nucleic acid. The CD changes observed for binding on
E.coli DNA, poly d(AT), or poly d(GC) are qualitatively the same than
those observed for the binding of *lac* repressor on the same nucleic acids
(15-17).

In the presence of DNA, the headpiece fluorescence due to the
tyrosyl residues is quenched by about 60 %. This quenching can be used
to quantitatively analyse the binding of the headpiece on DNAs. This
analysis is made according to the theory developed by McGhee and von
Hippel (18), which takes into account the fact that the protein covers
several base pairs on the DNA and that each base pair is the beginning
of a possible site for the protein. Figure 5 shows the experimental

binding isotherms for the headpiece 1-51 and *E.coli* DNA, at various
ionic strengths. On the same figure are plotted the theoretical curves
calculated according to McGhee and Von Hippel. The best fit is obtained
for a site extent of three base pairs.

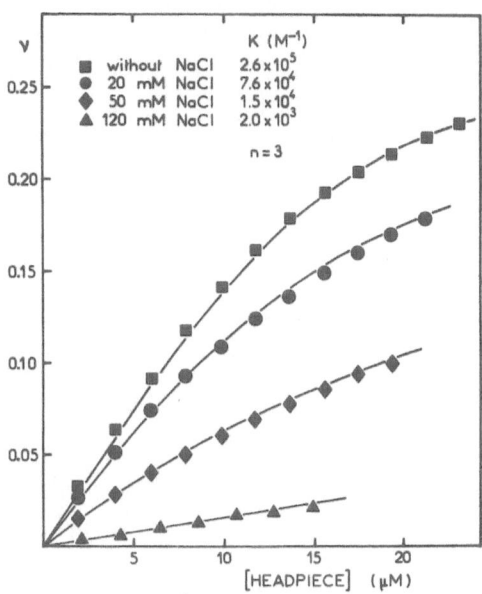

Figure 5 - Isotherms of binding for the complex non-operator DNA-
headpiece. ν is the number of bound headpiece per base pairs. Continuous
lines are calculated according to McGhee and Von Hippel (18) using a
site extent of 3 base pairs and the indicated values of the constants.

        This value is far more shorter than that we have determined for the
repressor (15 base pairs). The values we found for the association constants are also far more lower than those observed for repressor at identical ionic strength (16,17).

        The variation of the association constant with the ionic strength
has been analysed to determine the number of ion - pairs formed on the
binding process (19). A maximum number of three ion-pairs is found. This
value is considerably lower than the 12 ± 2 ion - pairs observed for the
non specific binding of the repressor (20).

        The site extent and the number of ion - pairs formed might led to
think that the repressor interact with the DNA through four headpieces.
However, the headpieces location determined by neutron scattering makes
this hypothesis irrelevant. Moreover, in this case, the association constant for the repressor should obey to the following relationship :

$$K_{REP} = (K_{Headpiece})^4$$

and thus should be equal to 10^{12} M^{-1} instead of the observed 10^7 M^{-1} at
0.1 M salt concentration. All these observations more likely reflect the
fact that the geometrical arrangement of the headpieces on the DNA are
not similar when they interact as isolated headpiece or as headpieces
in native repressor, in spite of the similitude of the conformational
changes of DNA in both cases. In good agreement with this assumption is
our observation that isolated headpieces cannot be photocrosslinked to
5-Br-Uracil substituted DNA, although they are complexed.

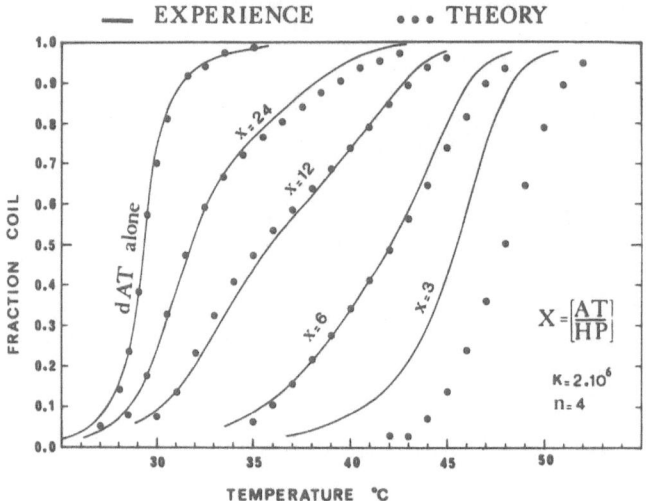

Figure 6 - Denaturation curves of poly d(AT) in the presence of headpieces. The theoretical points are calculated according to McGhee (21)
using the indicated parameters.

The interaction between the headpieces and DNAs can also be studied
using thermal denaturation experiments. The figure 6 shows that the
headpiece binding induces a thermal stabilization of the poly d(AT). The
greater the headpiece amount, the more important this stabilization effect. It is possible, using the theory of McGhee (21), to calculate the
theoretical profiles of the denaturation curves, assuming that the denatured DNA does not bind the headpieces. The DNA is treated in the infinite homogeneous Ising model approximation, and all calculations are
done by Lifson's method of sequence generating functions. An acceptable
fit is obtained only if one considers that four base pairs of poly d(AT)
are protected against thermal denaturation by the headpiece. This result
is in good agreement with the fluorscence results. We get the same agreement for the association constant.

REPRESSOR—OPERATOR INTERACTION. SPECIFIC BINDING

This study has been carried out using two different restriction fragments of DNA bearing the operator sequence. The first one, a short fragment of 2x29 bases contains in its mid part 21 base pairs of the operator
sequence. The second one is a long fragment of 203 base pairs and the
operator sequence represents only about 10 % of the total DNA. Moreover,
this later fragment carries the O3 pseudo operator sequence, and the CAP
and RNA polymerase binding sites (figure 7).

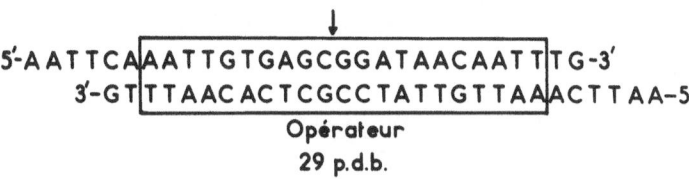

Opérateur
29 p.d.b.

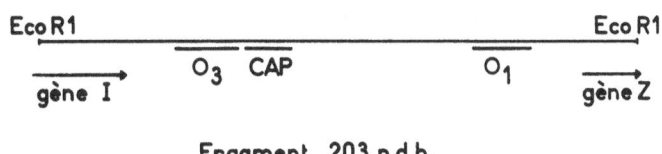

Fragment 203 p.d.b.

Figure 7 - The two DNA fragments bearing the operator sequence used in
the specific binding experiments.

Circular dichroism study

The CD spectrum of the short DNA operator fragment is typical of a
classical B form (22). When repressor is added to the operator fragment
there is an increase of the CD signal in the wavelength range of 260-300
nm. The observed spectral change could, in principle, be due either to a
change in the conformation of the protein or of the DNA fragment. However
as it has been several times discussed (15-17,22,23), one can exclude
that an eventual conformational change of the protein makes a significant contribution to the CD change in this wavelength range. As shown by
the difference CD spectrum (figure 8), this change is different from that
observed for the non specific binding (15-17,23). The final conformation
of the DNA fragment does not correspond to any classical form of the DNA
yet described (A,C,Z,Ψ).

This result has been confirmed with the 203 b.p. fragment. In this
later case, one can distinguish two steps for the repressor interaction.
The first one is the specific binding on the operator sequence, exhibiting the spectral change described above. The second one is the non specific binding on the other DNA regions of the fragment and the spectral
change is then the same as observed for a non operator DNA-repressor

complex (Culard and Maurizot, in press).

With the short fragment, another important information is obtained from the CD experiment. We evidenced a complex of one repressor:two operators, by analysing the titration curves at different wavelengths. The existence of this complex RO$_2$ has been confirmed by polyacrylamide gel electrophoresis, small angle X rays scattering, and gel chromatography (22). This means that the repressor has at least two binding sites for the operator. The model presented in figure 2 fully meets this requirement.

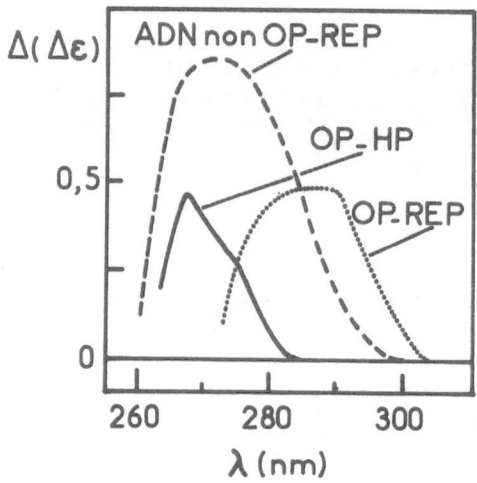

Figure 8 - Difference CD spectra between the complexes and the DNA alone.

Small angle X rays scattering studies

These experiments have been done in collaboration with A. Tardieu, at LURE, using the synchrotron radiations. We measured the radius of gyration of the complex RO$_2$, and found 51 Å. Assuming classical values for the electronic densities of the DNA and the protein, and using the parallel axis theorem (see above), we determined the distance between the two operator fragments bound to the repressor. This value is 130 ± 20 Å, in good agreement with the proposed model of repressor (figure 2) if we consider that the binding occurs through two headpieces located at the extremities of the repressor molecule.

Does the core bind the operator ?

When the core is added to the operator, we do not observe any CD change of the DNA signal. There are two possible explanations for this fact :
i) The core binds the operator, but does not modify the DNA conformation. Such a phenomenon is highly unlikely, and would be, at our know-

ledge, an unique exemple of DNA-protein interaction without any conformational change of the DNA.

ii) The core affinity for the operator is too small to give a complex in our experimental conditions. This would imply that the affinity
constant is smaller than 10^5 M^{-1}. This value is also the upper limit determined by Fried and Crothers (24) by polyacrylamide gel electrophoresis.
But it disagrees with the value of 10^7 M^{-1} proposed by O'Gorman and co-
workers (25). Nevertheless, it seems us that the more likely proposition
is the second one. Moreover, one must point out that the absence of
interaction between the operator and the isolated core does not imply
the absence of contact points between the operator and the core *in situ*
in the native repressor.

Operator - headpieces interactions

The headpiece binding modifies the CD spectrum of the operator, indicating a conformational change of the DNA (figure 8). The spectral change
depends on the amount of added headpiece. Until a ratio of two headpieces
per operator is reached, a first type of change is observed, different
from that observed for the headpiece-non operator DNA binding. It is also
different from that induced by the repressor-operator binding. In a second
step, for a ratio headpiece to operator greater than 2, the observed conformational change is the same as for the non specific DNA-headpiece
interaction. The interaction can also be studied by fluorescence spectroscopy. The quenching of the fluorescence of the headpiece tyrosines is the
same in the binding to operator and to non operator DNA. The saturation
curve shows that until a ratio of two headpieces per operator, all the
headpieces added to the *lac* operator fragment are bound. For more headpiece added a curvature is seen and the binding curve reaches a plateau
corresponding to four headpieces bound on the short operator fragments.
In our experimental conditions the linearity of the first part of the
curve implies that the binding constant is larger than 5×10^7 M^{-1}. This
is at least two orders of magnitude larger than the value found for the
binding to non operator DNA.

These results, in good agreement with those obtained by Ogata and
Gilbert (26) about the methylation patterns, show that two headpieces can
specifically bind the operator sequence. But they show that the headpieces
are not identical to the repressor in changing the operator DNA conformation. This later point does not obviously appear in the Ogata and Gilbert
results.

CONCLUSION

From all our results, a model for the repressor and its interactions
with the DNA emerges.

The repressor is a long shaped molecule, with the headpieces located
by pairs at the extremities of the molecule. These headpieces are long
shaped, their α structure content is high (33) and their stability is

strongly ionic strength dependent. In the native protein, the core is
stabilized by the headpieces interactions, under the dependence of the
inducer binding (27). The interaction of repressor with operator or non
operator DNA implies directly a pair of headpieces (figure 9).

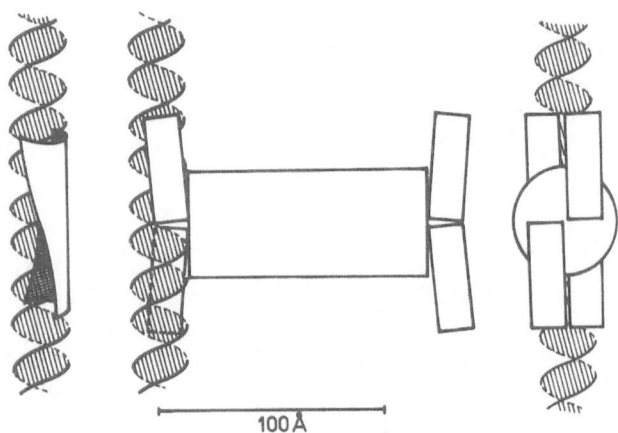

100 Å

Figure 9 - Schematic drawing of the repressor and of its interaction with
the operator. In the left part is shown the DNA B helix with the surface
of contact with the repressor (from references 26,30-32). Midpart : side
view of the complex. Right part : front view of the complex. Figure reprinted from reference 11 (with permission of Journal of Molecular Biology, Copyright Academic Press, Inc., London).

     The free headpieces interact with the DNAs in modifying their confor-
mation. Nevertheless, the conformational changes are different from those
induced by the repressor, for the non specific as well as for the specific interaction. In spite of the mobility of the headpieces in the repressor (28,29), the core imposes geometrical constraints in their relative
position. These constraints do not allow them the same contacts with the
DNA when they are isolated or *in situ* in the repressor. However, the isolated headpieces keep their ability to specifically recognize the operator
sequence.

     If the proposed model can be taken as a progress in the knowledge of
the operator-repressor functioning, it is still a "low resolution" approach
of the system. Our main goal is now to improve the resolution.

ACKNOWLEDGEMENTS

     We thank Alain Gervais for preparing the repressor used in these
studies, and Professor Claude Hélène who initiated this work. Part of this
work was supported by the DGRST (contract n° 81E1213), the Fondation pour
la Recherche Médicale Française, and the Comité du Loiret de la Ligue
Nationale Française contre le Cancer.

REFERENCES

1. Jacob, F. and Monod, J.:1961, J. Mol. Biol. 3, 318.
2. Gilbert, W. and Müller-Hill, B.:1966, Proc. Natl. Acad. Sci. USA 56, 1891.
3. Müller-Hill, B.:1975, Prog. Mol. Biol. 30, 227.
4. Bourgeois, S. and Pfahl, M.:1976, Adv. Protein. Chem. 30, 1.
5. Kao-Huang, Y., Revzin, A., Butler, A., O'Connor, P., Noble, D. and Von Hippel, P.H.:1977, Proc. Nat. Acad. Sci. USA 74, 4228.
6. Miller, J.:1979, J. Mol. Biol. 131, 249.
7. Weber, K. and Geissler, N.:1978, dans "The Operon", J. Miller and W. Reznikoff eds, Cold Spring Harbor, p. 155.
8. Maurizot, J.C. and Charlier, M.:1977, Eur. J. Biochem. 79, 395.
9. Ramstein, J., Charlier, M., Maurizot, J.C., Szabo, A.G. and Hélène, C.:1979, Biochem. Biophys. Res. Commun. 88, 124.
10. Charlier, M., Maurizot, J.C. and Zaccaï, G.:1980, Nature 286, 423.
11. Charlier, M., Maurizot, J.C. and Zaccaï, G.:1982, J. Mol. Biol. 153, 177.
12. Schnarr, M. and Maurizot, J.C.:1982, Biochim. Biophys. Acta 702, 155.
13. Von Hippel, P.H. and Wong, K.Y.:1965, J. Biol. Chem. 240, 3909.
14. Schlotmann, M. and Beyreuther, K.:1979, Eur. J. Biochem. 95, 39.
15. Maurizot, J.C., Charlier, M. and Hélène, C.:1974, Biochem. Biophys. Res. Commun. 60, 951.
16. Butler, A.P., Revzin, A. and Von Hippel, P.H.:1977, Biochemistry 16, 4769.
17. Durand, M. and Maurizot, J.C.:1980, Biochimie 62, 503.
18. McGhee, J.D. and Von Hippel, P.H.:1974, J. Mol. Biol. 86, 469.
19. Record, M.T., Anderson, C.F. and Lohman, T.M.:1978, Quart. Rev. Biophys. 11, 103.
20. De Haseth, P., Lohman, T.M. and Record, M.T.:1977, Biochemistry 17, 4783.
21. McGhee, J.D.:1976, Biopolymers 15, 1345.
22. Culard, F. and Maurizot, J.C.:1981, Nucleic Acids Res. 9, 5175.
23. Spodheim-Maurizot, M. and Maurizot, J.C.:1980, Cancer Letters, 11, 21.
24. Fried, M. and Crothers, D.M.:1981, Nucleic Acids Res. 9, 6505.
25. O'Gorman, R.B., Dunaway, M. and Matthews, K.S.:1980, J. Biol. Chem. 255, 10100.
26. Ogata, R. and Gilbert, W.:1979, J. Mol. Biol. 132, 709.
27. Schnarr, M. and Maurizot, J.C.:1981, Biochemistry 20, 6164.
28. Buck, F., Rüterjans, H. and Beyreuther, K.:1978, FEBS Letters 96, 335.
29. Wade-Jardetzky, N., Bray, R.P., Conover, W.W., Jardetsky, O., Geisler, N. and Weber, K.:1979, J. Mol. Biol. 128, 259.
30. Jobe, A., Sadler, J.R. and Bourgeois, S.:1974, J. Mol. Biol. 85, 231.
31. Goeddel, D.V., Yansura, D.G. and Caruthers, M.H.:1978, Proc. Natl. Acad. Sci. USA 71, 172.
32. Siebenlist, U., Simpson, R.B. and Gilbert, W.:1980, Cell, 20, 269.
33. Schnarr, M., and Maurizot, J.C.:1982, Eur. J. Biochem., in the press.

ROLE OF MESSENGER RNA SPECIFIC SECONDARY STRUCTURE IN THE CONTROL OF
GENE EXPRESSION.

Sylvain BLANQUET, Guy FAYAT, Jean-François MAYAUX, Mathias
SPRINGER, Marianne GRUNBERG-MANAGO.
Laboratoire de Biochimie, Ecole Polytechnique, 91128 Palaiseau
Cedex, France.
Institut de Biologie Physico-Chimique, 13, rue Pierre et Marie
Curie, 75005 Paris, France.

ABSTRACT : Prokaryotic messenger RNA can form specific secondary structures involved in transcription termination. These structures can be located within as well as at the end of an operon. In the case of several
aminoacid biosynthetic operons, termination sites are found between the
promoter and the structural part of the operon. Regulation of operon
expression can be achieved by controlling transcription termination
versus read-through at these termination sites. The control involves
the coupling of RNA translation to the transcription process. The present
review summarizes the evidence in favour of the same kind of control
mechanism in the expression of *E.coli* phenylalanyl-tRNA synthetase
operon.

Nearly twenty-three years ago, it was proposed that each ribonucleic
acid chain consisted of a number of imperfect hairpin-like helices similar in character to those of deoxyribonucleic acid, linked by flexible,
unstructured regions (1). Such secondary structures in transfer RNA and
ribosomal RNA molecules are now well proved (2,3). Stems and loops are
thought to account for the highly specific functional properties of
these RNA species and also to explain their stability regarding ribonuclease action.

In contrast, the turnover rate of messenger RNA is very high (4,5).
For this reason, in the early history of studies on gene expression,
messenger RNA, at least the prokaryotic one, was believed to have no
secondary structure, thus control signals of gene expression were
searched at the DNA level.

Then, several studies on the expression of the single-stranded RNA
genomes from bacteriophages R17, f2 and MS2 revealed that the secondary
(or tertiary) structure of the phage RNA was essential for the regulation of initiation of protein synthesis *in vitro* (6-8). There is recent
evidence that translation of prokaryotic messenger RNA is controlled by
the position of the initiator codon in the secondary structure (9). In
addition the accessibility of the Shine and Dalgarno's sequence to the

C. Hélène (ed.), Structure, Dynamics, Interactions and Evolution of Biological Macromolecules, 155–165.
Copyright © 1983 by D. Reidel Publishing Company.

3'-end of 16S rRNA is also an important feature for ribosome attachment
to initiation sites (10). On the other hand, secondary structure of
mRNA appears to be involved in nuclease resistance and processing (11-
12).

Secondary (or tertiary) structure of mRNA was also shown to play
a role in transcription termination (13). In general, the transcription
termination sites consist of a G/C-rich inverted repeat followed of a
stretch of U residues. This inverted repeat gives the mRNA the potential
ability to form a stem and loop type structure. After this particular
RNA structure has been synthesized, RNA polymerase stops transcription.

During the past 10 years, it has become evident that transcription
termination sites can be located within as well as at the end of an
operon. Such termination sites have been shown to be involved in the
control of gene expression (14-16). Regulation can be achieved by controlling the levels of transcriptional read-through (i.e. antitermina-
tion) at these termination sites. In particular, transcription termination versus read-through can be controlled by the coupling of RNA translation to the transcription process. This mode of control has been recently shown to occur in certain of the aminoacid biosynthetic operons
(17-18). In the well-documented case of the tryptophan (trp) operon,
control is known to occur, in part, through transcription termination
at a site defined as the attenuator. This site is located between the
trp promoter and the structural part of the operon (15 and references
therein).

According to Yanofsky's model, elongation of the trp operon mRNA
beyond the attenuator is controlled by the secondary structure of the
transcript corresponding to the region between the promoter and the attenuator (leader RNA). This leader RNA was shown to contain an open
reading frame of 14 codons corresponding to a hypothetical 14-residue
peptide (leader peptide) containing two adjacent tryptophan residues.
The secondary structure of the leader RNA depends on ribosome movements
on that open reading frame. Ribosome position on the leader RNA is regulated by the level of aminoacylated $tRNA^{Trp}$ in the cell. Low Trp-
$tRNA^{Trp}$ levels (i.e. tryptophan starvation) will cause ribosome to stall
at the Trp codons of the leader peptide reading frame. The stalling will
in turn create a leader RNA secondary structure causing transcription
read-through at the attenuator and thus derepression of the operon.
High $Trp-tRNA^{Trp}$ levels will affect translation of the leader peptide
and thus the structure of leader RNA as to increase termination at the
attenuator.

The 3'-end of the trp leader RNA can form two alternative and exclusive stem and loop structures. The stem and loop structure nearest
to the 3'-end of leader RNA displays all the features of a transcription
termination site. The other alternative stem and loop structure (anti-
terminator) of the leader mRNA precludes formation of the termination
structure. Under tryptophan-limiting conditions where ribosomes are
stalled at the tandem of UGG codons, the anti-terminator stem and loop

structure can form at the expense of the terminator structure and transcriptional read-through increases. In the presence of tryptophan, i.e.
of charged tRNA^{Trp}, ribosomes move through the leader peptide-coding
RNA sequence to the UGA codon. This movement hinders formation of the
anti-terminator stem and loop structure. The terminator structure then
forms and transcriptional read-through decreases.

The basic features of this mechanism of regulation, called attenuation,are strongly supported by the recent studies of Yanofsky and
others (19,20).

Attenuation control of operon expression is a simple economical
mechanism involving only common cell components and general features of
the processes of transcription and translation. It seems to operate in
the studied cases of a number of aminoacid biosynthetic operons in prokaryotes such as his, pheA, leu, thr and ilv (21-27).

We will discuss now recent work in our groups suggesting the same
kind of mechanism for the regulation of $E.coli$ phenylalanyl-tRNA synthetase genetic expression.

Aminoacyl-tRNA synthetases.

Aminoacyl-tRNA synthetases are essential enzymes for protein synthesis. These enzymes catalyse the esterification of each aminoacid to
its cognate tRNA(s) to form aminoacyl-tRNAs. They are thus involved in
the previously discussed regulation of aminoacid biosynthetic pathways
as well as in several other cellular processes (for reviews see 28,29).

During the last decade, the regulation of the expression of bacterial aminoacyl-tRNA synthetases themselves has been extensively
studied (reviewed in 30,29). For most studied aminoacyl-tRNA synthetases, transient derepression of synthesis is observed under conditions
of starvation of the cognate aminoacid. In the case of three $E.coli$
aminoacyl-tRNA synthetases among which phenylalanyl-tRNA synthetase (31),
long-term derepression is obtained. On the other hand, the $E.coli$ aminoacyl-tRNA synthetases have been shown to be subject to metabolic regulation, i.e. their cellular concentration increases in parallel with growth
rate of the bacteria (32).

Only recently, studies have been initiated in order to explain the
features of aminoacyl-tRNA synthetase regulation in precise molecular
terms (33,34).

Phenylalanyl-tRNA synthetase and the pheS-pheT operon.

Phenylalanyl-tRNA synthetase (PheRS) is a tetrameric enzyme of the
$\alpha_2\beta_2$ type (35). The α and β subunits have molecular ratios of 94 K and
37 K, respectively. It has been shown that the two genes, pheS and pheT,
encoding PheRS subunits are clustered on the $E.coli$ chromosome with two
other genes involved in protein translational machinery (36-38). These

two other genes, thrS and infC, code for threonyl-tRNA synthetase and
initiation factor IF3, respectively (Figure 1).

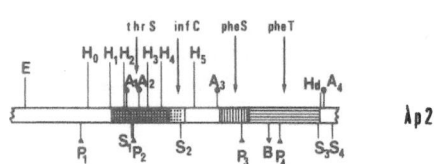

Figure 1. Physical map of the *E.coli*
DNA insert of λp2. The positions of
the four genes thrS, infC, pheS and
pheT are taken from data in (39,49,
40). The position of the restriction
sites are indicated as E = EcoRl,
H = HpaI, A = AvaI, P = PstI, S =
SacII, B = BamHl, Hd = HindIII. Sites are numbered from left to right.
The EcoRl to HindIII fragment of λp2
was cloned in pBR322 giving pBl (39).
Additional *E.coli* DNA is present on
either side of this fragment.

    The pheS, pheT, thrS and infC genes were originally cloned on a
single λ transducing phage, λp2 (36-38) and recloned in plasmid DNA
pBR322 to give the recombined plasmid pBl (39). The genes were physically localised on pBl DNA and it has been shown that all were transcribed
in the same direction, from thrS to pheT. The four genes are transcribed
as three separate transcription units : the first unit corresponds to
thrS, the second to infC and the third to both pheS and pheT (40,41).
The two genes encoding PheRS subunits thus form a multicistronic operon,
pheS-pheT.

Studies on the regulation of the pheS-pheT operon.

    The regulatory region of the pheS-pheT operon is located on the
left-hand side of the AvaI3 restriction site (A3 on Figures 1 and 2),
immediately preceding pheS on the endonuclease restriction site map of
λp2 or of plasmid pBl. That the structural part of the operon is located
on the right side of this AvaI3 site is evidenced by substituting DNA
on the left of AvaI3 by foreign DNA containing a foreign promoter such
as the arabinose promoter or that of thrS (42,43). In each case a PheRS
with the same molecular ratio as wild type PheRS was synthesized. On
the other hand, a deletion located on the left-hand side of AvaI3 eliminates expression of both the products of pheS and pheT (44). In addition,
PheRS expression from a mutated pBl DNA plasmid carrying a 1.3 kilo-base
insertion on the left-hand side of AvaI3 is strongly decreased (45).

    All these observations support the conclusion that the DNA region
upstream to the AvaI3 restriction site is necessary to *in vivo* expression of PheRS.

    In order to study the effect of phenylalanine deprivation on PheRS
expression, we have fused the promoter region of pheS-pheT operon, i.e.
DNA on the left hand side of AvaI3, to lac structural genes (46,42).

A plasmid (pMF3) was constructed by inserting the AvaI_2-AvaI_3 fragment from pB1 into the unique XmaI site of pMC306 (47). This XmaI site immediately precedes lacZ, the structural gene of β-galactosidase (Figure 2).

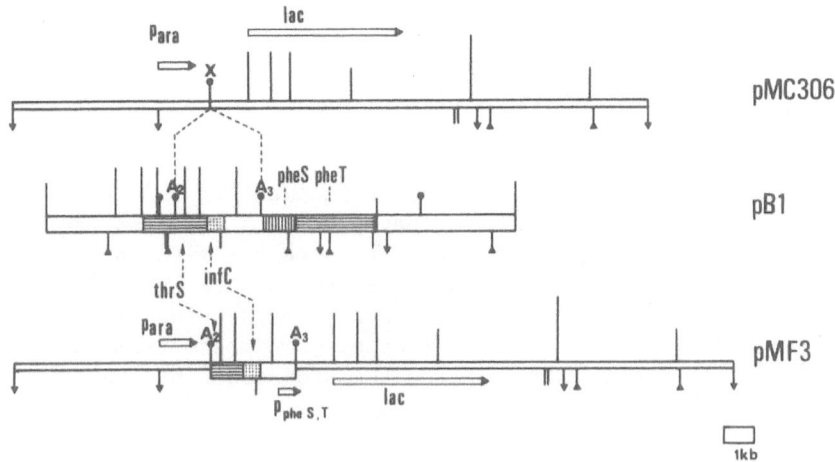

Figure 2. Physical map of pMC306 and pMF3 (taken from (42)). The structure of pMC306 is taken from (47). Genes and restriction sites of pB1 are indicated as in λp2 (Figure 1). X is the unique XmaI site of pMC306. P_{ara} indicates the arabinose BAD promoter. lac indicates β-galactosidase structural gene. $\text{P}_{pheS,T}$ indicates the promoter for the pheS- pheT operon.

It was firstly shown that the plasmid pMF3 synthesizes a constitutive level of β-galactosidase which reflects the normal functioning of the pheS-pheT promoter. In order to starve the cell for phenylalanine, two phenylalanine requiring *E.coli* strains were used to host plasmid pMF3. One (a leaky phenylalanine auxotroph) is mutated in the gene pheA coding for the principal enzyme of the phenylalanine biosynthetic pathway. The other (also a phenylalanine leaky auxotroph) is modified in pheS, the gene coding for the small subunit of PheRS. When pMF3 was hosted in any one of these two strains, starvation for phenylalanine was shown to cause a 7-10 fold increase of β-galactosidase activity. The response of the pheS-pheT promoter to phenylalanine starvation is specific since histidine starvation (induced by 3-amino-1,2-4-triazole addition to growth medium (48)) has no effect on β-galactosidase synthesis. On the other hand, phenylalanine starvation has no effect on β-galactosidase expression in the same pheA strain containing pMC306, the parental plasmid of pMF3, with lac under control of the arabinose promoter.

This experiment shows that the DNA region on the left hand side of Aval₃ carrying the pheS-pheT promoter region, positively responds to phenylalanine starvation by transcriptional derepression. This observation is in agreement with the earlier study of Nass and Neidhardt (31), who showed phenylalanine starvation to cause derepression of phenyl-alanyl-tRNA synthetase synthesis in a bradytroph *E.coli* strain.

The effect of phenylalanine deprivation on the pheS-pheT promoter in either a pheA or a pheS *E.coli* strain can be explained by the deprivation affecting directly or indirectly the cellular ratio of charged to uncharged tRNA^{Phe}.

Nucleotide sequence analysis of pheS-pheT.

In order to gain more information on the expression of pheS-pheT operon at the molecular level, a DNA fragment of 3.3 Kb of plasmid pBl carrying infC, pheS and the beginning of pheT has been sequenced (49, 42, Fayat et al. in preparation). In agreement with the genetic studies (40,41), the structural part of pheS is found to start 87 nucleotides downstream from the center of the Aval₃ restriction site (Figure 3). The structural part of pheT is located downstream from and adjacent to pheS, with 14 nucleotides between the stop codon of pheS and the start codon of pheT.

```
TCGTTTCAACGCCATCAAAACATTGACTTTTTATCGCCGTAGCCTTT

        p2          p2'              MetAsnAlaAlaIlePheArgPhePhePheTyrPhe
TCAATAAAGGTCTTTTGAAGAGTAACCAAAAGGTAACGCAAGCAATGAATGCTGCTATTTTCCGCTTCTTTTTTTACTTT

SerThr  Hin fI                                                _____ t2 _____
AGCACCTGAATCCAGGAGGCTAGCGCGTGAGAAGAGAAACGGAAAACAGCGCCTGAAAGCCTCCCAGTGGAGGCTTTTTT

     MetArgValLeuLysAsnLysLysProValSerSerGlyValIleValAlaAlaArgLeuAspThrAlaSerAla
TGTATGCGCGTTTTGAAAAATAAAAAGCCTGTCTCATCAGGTGTAATTGTCGCAGCCAGACTGGACACGGCCAGCGCG

GluLysProGluArgIle                          Ava I(3)
GAGAAACCGGAGCGTATTTAAGTACGTGAGAATTTCGAGCACAGCCCGGGACCAAAATGGCAAGTAAAATAGCCTGATGG

Hpa II
     Dde I                                            MetSerHisLeuAlaGlu
ATAGGCTCTAAGTCCAACGAACCAGTGTCACCACTGACACAATGAGGAAAACCATGTCACATCTCGCAGAA
```

Figure 3. Nucleotide sequence of the regulatory region of the pheS-pheT operon (taken from (42)). The positions of the restriction sites HinfI, HpaII, Aval₃, DdeI are shown. The dyad symmetry corresponding to the transcription terminator t2 is indicated by lines above the sequence, p2 and p2' corresponding to the mapped start points of transcription are shown by horizontal arrows, the length of which indicates the uncertainty in mapping. The pribnow box-like sequences TTCAATA and TCTTTTC and their corresponding -35 regions, TTGACT and TTATCG, respectively, are shown by boxes around the sequences. The aminoacid composition of the 14 residue phenylalanine-rich peptide is shown above the DNA sequence, as is the aminoacid composition of the 31 residue peptide devoid of phenylalanine located after the proposed attenuator DNA sequence, t2. The N-terminal sequence of the pheS structural gene is shown at the 3' end of the DNA sequence.

Upstream to the AvaI₃ restriction site, where the regulatory region for PheRS expression is located, the DNA sequence indicates a rho-independent transcription termination site (13) (t₂ on Figure 3). Examination of the DNA sequence shows that the stem and loop structure corresponding to the terminator is in competition with another stem and loop structure (Figure 4). Upstream to these alternative and exclusive RNA structures is a short open reading frame coding for a 14-residues-peptide containing 5 phenylalanines, 3 of which are adjacent. The coding sequence corresponding to the putative phenylalanine-rich leader peptide is preceded by several possible Shine and Dalgarno-type sequences capable of pairing with 16 S rRNA (10).

This organization of a leader peptide followed by alternative RNA structures, one of which is a transcription terminator, strongly recalls the basic features of the mechanism of attenuation in aminoacid biosynthetic operons (15).

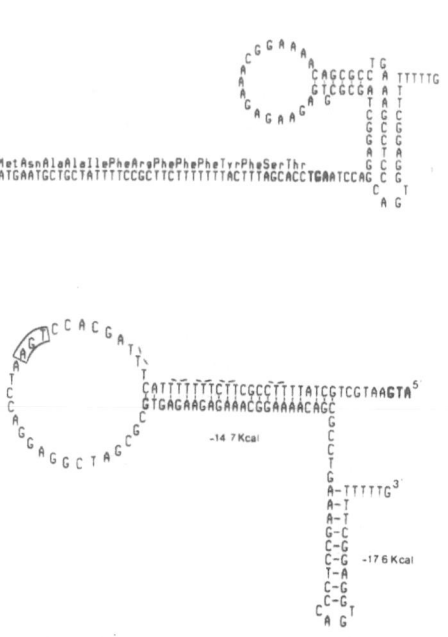

Figure 4. Possible alternate secondary structures of the RNA in the region of the transcription terminator, t2. In the top scheme, the DNA sequence comprising the leader peptide coding sequence and the two alternative base paired stems of the attenuator-like arrangement are indicated. The bottom scheme shows another secondary structure of mRNA in this region involving possible base pairs between mRNA of the leader peptide coding sequence and the large loop of the upper structure (see text for details). The free energies associated with this structure are given. The phenylalanine codons of the leader peptide are indicated by parentheses and the termination codon by a box (taken from (42)).

In vitro transcription of the pheS-pheT regulatory region.

The regulatory region of the pheS-pheT operon is the source *in vitro* of several transcripts of variable lengths and intensities (42, Fayat et al. in preparation). The origins and lengths of the transcripts were found using DNA restriction fragments of variable lengths derived from different parts of the pheS-pheT regulatory region. The main conclusion is that the DNA region encoding the putative "leader peptide" is covered by several transcripts of 144 ± 5 nucleotides originating from several adjacent start points (p_2 and p'_2) located upstream to the leader peptide initiation codon (Figure 3). These start points are immediately preceded by possible Pribnow box and −35 regions (50,13) shown on Figure 3. This location of the start points of the transcripts covering the DNA region of the leader peptide is confirmed by recent S1-nuclease mapping experiments of the 5'ends of RNA's synthesized *in vitro* from pB1. Each of the four adjacent T's in the start region p_2 appear to be the major start points of transcription.

Transcription starting from these origins can either terminate at the terminator (t_2) located after the leader peptide-coding RNA sequence and in front of the AvaI$_3$ site or proceed through this transcription terminator into the structural genes of the operon. That the transcripts of 144 ± 5 nucleotides are capable of extending into the pheS structural region was shown by monitoring the *in vitro* transcription experiments using ITP instead of GTP. This has been shown to promote read-through of other attenuators due to the weak base pairing between I and C in the synthesized RNA transcript (51).

CONCLUSION.

The genetic and structural studies reviewed here strongly indicate that expression of the pheS-pheT operon involves an attenuation mechanism mediated by the level of aminoacylation of tRNA^{Phe}.

Under normal growth conditions, ribosomes translate the leader RNA encoding the leader peptide and stop at the UGA stop codon. This position of the ribosome prevents formation of the first stem and loop structure, i.e. the antiterminator, in the mRNA. The second structure corresponding to the terminator forms and premature termination of transcription occurs (Figure 5). Transcription termination would be equally obtained if the ribosome dissociates from mRNA at the level of the stop codon of the leader peptide reading frame. In this case, another stem and loop structure involving the region of mRNA encoding the leader peptide and the region downstream can form before that RNA polymerase has synthesised the RNA sequence of the antiterminator. When formed, this stem and loop structure (shown on Figure 4) precludes the formation of the antiterminator.

In the case of cells starved of phenylalanine, a ribosome translating the leader peptide would stall at one of 5 phenylalanine codons

(Figure 5). RNA polymerase continues to elongate mRNA transcription beyond this region and the antiterminator structure can form before that synthesis of the terminator RNA sequence has been completed.

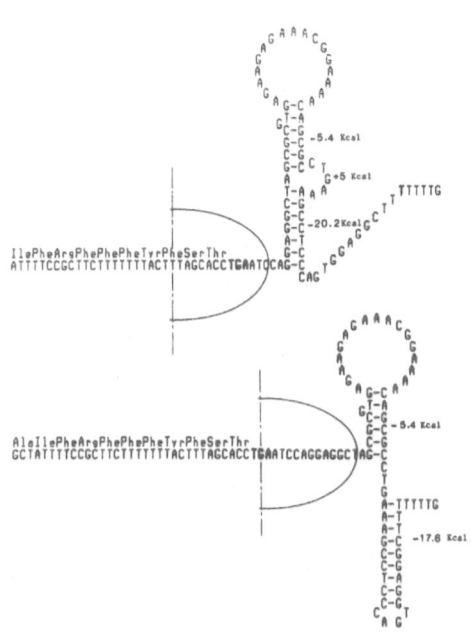

Figure 5. Possible alternate RNA secondary structures thought to regulate transcription termination at the attenuator-like structure of the pheS-pheT operon. The top scheme represents the situation where cells are starved of phenylalanine. The bottom scheme shows the case of non-starvation. According to Yanofsky (15), it is proposed that a ribosome stalled over one of the codons of the transcript covers about 10 more nucleotides on the downstream region of the transcript, indicated in the figure by the half oval structure (taken from (42)).

A difference between the structure of the leader region of pheS-pheT operon and that of aminoacid biosynthetic operons is the indication by the DNA sequence of an open reading frame, downstream from the terminator, with an initiation triplet preceded by a GGAGG Shine and Dalgarno's sequence making part of the terminator stem. This open reading frame indicates a 31-aminoacid peptide characterized by the absence of phenylalanine residues (Figure 3). One possible role for the translation of such a coding sequence could be that coupled with transcription it insures maximum RNA elongation in this region reducing premature transcription termination due to polarity. Another interesting possibility is that the hypothetical peptide is involved in controlling PheRS expression or phenylalanine metabolism.

Another important feature is that phenylalanyl-tRNA synthetase activity is only stimulated two-and-half-fold in a pheA bradytroph strain submitted to phenylalanine deprivation (31). In contrast, a 10-fold derepression of β-galactosidase activity was measured in pheA or pheS cells carrying plasmid DNA pMF3. One possible explanation for the relatively small derepression of PheRS compared to that of β-galactosidase when the two pheS-pheT and lac operons are under control of the same regulatory DNA region, could be the occurence of an abnormally high level of phenylalanine codons in the structural part of pheS (42). The corresponding phenylalanine residues are mostly clustered in the C-terminal half of the polypeptide. Thus extreme phenylalanine deprivation although stimulating transcription at the level of attenuator structure might induce ribosome stalling within the pheS structural gene and hence transcription termination due to polarity. In this case premature termination of transcription within pheS would counter-react with the positive effect of phenylalanine starvation exerted on mRNA read-through at the level of the attenuator. A 10-fold derepression of β-galactosidase activity when lac is under control of pheS-pheT promoter region might reflect the normal tuning of the attenuator in the absence of polarity effects during transcription of the structural part of the operon.

ACKNOWLEDGEMENTS : This work was supported by the following grants : Centre National de la Recherche Scientifique (L.A. n° 240, G.R. n° 18), Délégation Générale à la Recherche Scientifique et Technique (Conventions 80.E.0872 and 81.E.1207).

REFERENCES.

1 - Doty,P., Boedtker,H., Fresco,J.R., Haselkorn,R. and Litt,M.: 1959, Proc.Natl.Acad.Sci.USA, 45, pp.482-486.
2 - Rich,A. and Rajbhandary,U.L.: 1976, Ann.Rev.Biochem.45, pp.805-860.
3 - Stiegler,P., Carbon,P., Ebel,J.P. and Ehresmann,C.: 1981, Eur.J.Biochem.120, pp. 487-495.
   - Branlant,C., Knol,A., Machatt,M.A., Pouyet,J., Ebel,J.P., Edwards, K. and Kössel,H.: 1981, Nucleic Acids Res.9, pp.4303-4324.
4 - Volkin,E. and Astrachan,L.: 1956, Virology 2, pp.149-161.
5 - Gros,F., Gilbert,W., Hiatt,H.H., Kurland,C.G., Risebrough,R.W. and Watson,J.D.: 1961, Nature 190, pp.581-585.
6 - Lodish,H.F.: 1970, J.Mol.Biol.50, pp.689-702.
7 - Steitz,J.A.: 1969, Nature 224, pp.957-964.
8 - Fiers,W., Contreras,R., Duerunck,S., Haegeman,G., Iserentant,D., Mereegaert,J., Min-Jou,W., Molemans,S., Raeymaekers,A., Van den Berghe, A., Volckaert,G., Ysebaert,M.: 1976, Nature 260, pp.500-507.
9 - Hall,M.N., Gabay,J., Debarbouillé,M. and Schwartz,M.: 1982, Nature 295, pp.616-618.
10 - Shine,J. and Dalgarno,L.:1974, Proc.Natl.Acad.Sci.USA 71, pp.1342-1346.
11 - Saito,H. and Richardson,C.C.: 1981, Cell 27, pp.533-542.
12 - Gegenheimer,P. and Apirion,D.: 1981, Microbiol.Rev.45, pp.502-541.
13 - Rosenberg,M. and Court,D.: 1979, Ann.Rev.Genet.13, pp.319-353.
14 - Adhya,S. and Gottesman,M.: 1978, Ann.Rev.Biochem.47, pp.967-996.
15 - Yanofsky,C.: 1981, Nature 289, pp.751-758.
16 - Ward,D.F. and Gottesman,M.E.: 1982, Science 216, pp.946-951.
17 - Kasai,T.: 1974, Nature 249, pp.523-527.
18 - Bertrand,K., Korn,L.J., Lee,F., Yanofsky,C.: 1977, J.Mol.Biol.117, pp.227-247.
19 - Stroynowsky,I. and Yanofsky,C.: 1982, Nature 298, pp.34-38.
20 - Stroynowsky,I., Van Cleemput,M. and Yanofsky,C.: 1982, Nature 298, pp.38-41.

21 - Johnston,H.M., Barnes,W.M., Chumley,F.G., Bossi,L. and Roth,J.: 1980, Proc.Natl.Acad.Sci.USA 77, pp.508-512.

22 - Gardner,J.F.: 1979, Proc.Natl.Acad.Sci.USA 76, pp.1706-1710.

23 - Lawther,R.P. and Hatfield,G.W.: 1980, Proc.Natl.Acad.Sci.USA 77, pp.1862-1866.

24 - Nargang,F.E., Subrahmanyam,C.S. and Umbarger,H.E.: 1980, Proc.Natl. Acad.Sci.USA 77, pp.1823-1827.

25 - Keller,E.B. and Calvo,J.M.: 1979, Proc.Natl.Acad.Sci.USA 76, pp.6186-6190.

26 - Zurawski,G., Gunsalus,R.P., Brown,K.D. and Yanofsky,C.: 1981, J.Mol.Biol.145, pp.47-73.

27 - Gowrishankar,J. and Pittard,J.: 1982, J.Bacteriol.150, pp.1130-1137.

28 - Schimmel,P.R. and Soll,D.: 1979, Annu.Rev.Biochem.48, pp.601-648.

29 - Neidhardt,F.C., Parker,J. and Mc Keever,W.: 1975, Ann.Rev.Microbiol.29, pp.215-250.

30 - Morgan,S.D. and Soll,D.:1978, Progr.Nucl.Acid.Res. and Molec.Biol. 21, pp.181-207.

31 - Nass,G. and Neidhardt,F.C.:1967,Biochim.Biophys.Acta 134,pp.347-359.

32 - Neidhardt,F.C., Bloch,P.L., Pedersen,S. and Reeh,S.: 1977, J.Bacteriol.129, pp.378-387.

33 - Putney,S.D. and Schimmel,P.: 1981, Nature 291, pp.632-635.

34 - Hall,C.V. and Yanofsky,C.: 1981, J.Bacteriol.148, pp.941-949.

35 - Fayat,G., Blanquet,S., Dessen,P., Batelier,G. and Waller,J.P.: 1974, Biochimie 56, pp.35-41.

36 - Springer,M.,Graffe,M. and Hennecke,H.: 1977, Proc.Natl.Acad.Sci. USA 74, pp.3970-3974.

37 - Hennecke,H., Springer,M. and Böck,A.: 1977, Molec.Gen.Genet.152, pp.205-210.

38 - Hennecke,H., Böck,A., Thomale,J. and Nass,G.: 1977, J.Bacteriol. 131, pp.943-950.

39 - Plumbridge,J.A., Springer,M., Graffe,M., Goursot,R. and Grunberg-Manago,M.: 1980, Gene 11, pp.33-42.

40 - Springer,M., Plumbridge,J.A., Trudel,M., Graffe,M. and Grunberg-Manago,M.: 1982, Molec.Gen.Genet.186, pp.247-252.

41 - Plumbridge,J.A. and Springer,M.: 1980, J.Mol.Biol.144, pp.595-600.

42 - Springer,M., Plumbridge,J.A., Trudel,M., Grunberg-Manago,M., Fayat, G., Mayaux,J.F., Sacerdot,C., Dessen,P., Fromant,M. and Blanquet,S.: 1982, in Translational/Transcriptional Regulation of Gene Expression (Grunberg-Manago,M. and Safer,B., eds) Elsevier, in press.

43 - Plumbridge,J.A. and Springer,M.: 1982, J.Bacteriol., in press.

44 - Springer,M., Graffe,M. and Grunberg-Manago,M.: 1979, Molec.Gen. Genet.169, pp.337-343.

45 - Plumbridge,J.A. and Springer,M.: 1982, J.Bacteriol., in press.

46 - Trudel,M.: 1982, Thèse de Doctorat de l'Université Paris VI.

47 - Casabadan,M.J. and Cohen,S.N.: 1980, J.Mol.Biol.138, pp.179-207.

48 - Hilton,J.L., Kearney,P.C. and Ames,B.N.: 1965, Arch.Biochem.Biophys. 112, pp.544-547.

49 - Sacerdot,C., Fayat,G., Dessen,P., Springer,M., Plumbridge,J.A., Grunberg-Manago,M. and Blanquet,S.: 1982, EMBO J.1, pp.311-315.

50 - Pribnow,D.: 1975, Proc.Natl.Acad.Sci.USA 72, pp.784-789.

51 - Lee,F. and Yanofsky,C.: 1977, Proc.Natl.Acad.Sci.USA 74, pp.4365-4369.

THE ORGANIZATION OF PLANT VIRAL GENOMES

H. Guilley, R.K. Dudley, L. Hirth, G. Jonard, G. Lebeurier,
J. Menissier and K.E. Richards
Institut de Biologie Moléculaire et Cellulaire, Laboratoire de
Virologie, 15, rue Descartes, 67084 Strasbourg cedex, France

INTRODUCTION

Most plant viruses are simple in structure being composed uniquely
of a protein envelope and a nucleic acid. Our laboratory has been drawn
to the study of these viruses because they can be considered as simple
model systems for the comprehension of a number of fundamental problems
of molecular biology such as, for example, the nature of protein-nucleic
acid interactions, the organization of the viral genome and the expression of this genome.

Plant viruses may be divided into two major groups : those possessing an RNA genome and those possessing a DNA genome with the RNA viruses being more common. In general, the RNA genome of plant viruses is of
positive polarity, that is, it possesses a message that can be directly
translated into protein. In most cases the genetic information is contained either in a single RNA molecule which is polycistronic (monopar-
tite viruses) or on several RNA molecules which may be either mono- or
polycistronic (multipartite viruses). Tobacco mosaic virus (TMV) is the
best studied member of the former group (for a review, see reference 1).
DNA -containing plant viruses are less numerous and their study has only
been undertaken fairly recently. There are single-stranded DNA containing
viruses (for example the geminiviruses) and those containing double-
stranded DNA (e.g. the caulimoviruses). The best studied representative
of the caulimoviruses is cauliflower mosaic virus (CaMV).

In the last few years, there has been heightened interest in this
virus because it is possible to envisage the construction, using all or
part of this DNA, of a vector for gene transfer in higher plants (for a
review, see reference 2).

C. Hélène (ed.), Structure, Dynamics, Interactions and Evolution of Biological Macromolecules, 167–175.
Copyright © 1983 by D. Reidel Publishing Company.

TOBACCO MOSAIC VIRUS (TMV)

Genetic organization of TMV RNA

    The genetic information of TMV is contained on an RNA molecule of
about 6400 nucleotides (Figure 1). The viral RNA contains three cistrons
which can encode for proteins of 165000 daltons (110000 daltons), 30000
daltons and 17500 daltons (3,4). The last protein corresponds to the
viral coat protein. Other strategic regions have also been localized on
the RNA molecule, including the initiation site for viral morphogenesis
(5,6) situated about 1000 nucleotides from the 3'OH extremity, which is
interesting in itself as it can be specifically aminoacylated with histidine by histidyl-tRNA synthetase (7).

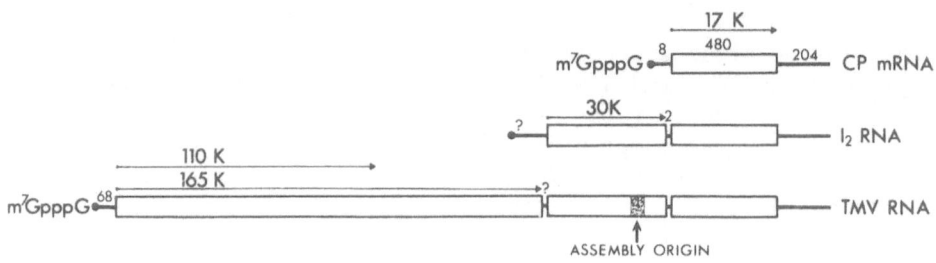

Figure 1 - Genetic organization of TMV vulgare and its subgenomic mRNAs.
Cistrons are represented by open rectangles. Arrows indicate cistrons
functional in protein synthesis and the molecular weights of the proteins
synthesized. Length of nontranslated regions are given in nucleotides if
known. Drawing is not to scale.

The *in vitro* translation of the viral RNA

    When TMV RNA is introduced into an *in vitro* proteosynthetic system
only the first cistron is expressed : one observes synthesis of proteins
of 165000 daltons and 110000 daltons (the 165000 d protein is an extension
of the 110000 d protein by suppression of a termination codon (4). The
two other proteins are not expressed from the genomic RNA. Thus, the genomic RNA, which is polycistronic, behaves like a monocistronic messenger.
The proteins encoded by the silent interior cistrons are obtained from
subgenomic RNAs, produced by an unknown mechanism from the genomic RNA
and which are found in the infected plant cell but which are not necessarily encapsidated (Figure 1) (3,8,9). It thus appears to be a fairly
general rule for RNA plant viruses, that only the first cistron of the
viral RNA is expressed. We thus became interested in the sequence at the
5' extremity of different cistrons in order to understand why certain of
these regions are more or less well recognized by ribosomes while others
are not. Analysis of these sequences has led us to the following conclusions (10) :
    1) The 5' noncoding region of many plant viral RNAs is characterized

by the presence of a "cap" m/Gppp and by a deficiency of guanosine residues.
2) the 5' noncoding region of subgenomic RNAs is shorter than the corresponding part of the genomic RNA.
3) the 5' noncoding region of subgenomic RNAs contains a triplet which probably represents the termination codon of the preceding cistron on the genome. It is thus probably that, in a polycistronic RNA, the 3' coding portion of cistron n-1 overlaps the 5' noncoding region of the subgenomic RNA corresponding to cistron n (11).

The 5' terminal part of the internal cistrons must be masked in some way on the polycistronic RNA so that these cistrons are only expressed from the subgenomic RNAs. On the other hand, the 5' terminal regions of the genomic and subgenomic RNAs which can fix a ribosome during translation initiation have few elements in common other than a capped 5' extremity and an AUG initiation codon. These observations, along with results obtained by other workers, have permitted Kosak to propose her scanning model for translation initiation of eucaryotic mRNAs (12).

The mechanism of TMV morphogenesis

Early experiments allowed us to show that specific regions of TMV RNA can interact specifically with viral coat protein, in particular with an aggregate of 34 coat protein molecules (the double disk) (13,14). These regions have been sequenced and one of them, called P1, situated about 1000 nucleotides from the 3'OH extremity, corresponds to the viral morphogenesis initiation site (6). These results, along with microscopic observations of reassembling TMV particles, permitted us (15) to propose a mechanism for TMV morphogenesis (Figure 2). The RNA hairpin corresponding to the P1 sequence is inserted between the two layers of the double disk provoking the planar disk structure to dislocate and to form a two-turn helix. Elongation then proceeds in the 3' to 5' sense along the RNA by the stacking and dislocation of disks on the initiation complex. During this process, the 5' terminal part of the RNA is pulled through the central channel of the assembling particle. The addition of protein to the 3' terminal portion of the RNA is probably the last step in the assembly process.

Aminoacylation of TMV RNA

A large number of plant viral RNAs are capable of fixing an amino acid at their 3'OH extremity in the presence of the appropriate aminoacyl-tRNA synthetase. Thus TMV RNA can be aminoacylated with histidine (7, and R. Giege, personal communication). The structure of the 3' termini of these RNAs is not, however, identical to that of the corresponding tRNA (11) but certain domains of sequence, found in the homologous tRNA and thought to be important for synthetase recognition, exist in the plant viral RNA sequences. The ability to be aminoacylated may be linked with the control of the viral RNA replication process but no experimental evidence for this hypothesis has yet been brought forth.

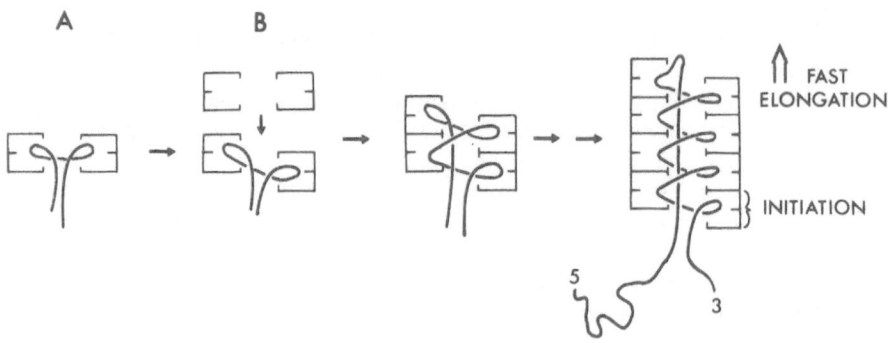

Figure 2 - First steps in TMV self-assembly. (A) Insertion of the assembly origin sequence between layers of a disk from inside. (B) Transformation to helix and addition of a second disk.

CAULIFLOWER MOSAIC VIRUS (CaMV)

Description of the virus

     CaMV is a virus with icosahedral symmetry. The virion is composed of
a single type of subunit of about 42000 d (16) and a molecule of double
stranded DNA. Neutron scattering studies of the virion have shown that
the DNA is bound to the inner surface of the capsid shell so that the
center of the particle is empty (17). No histones are associated with
the DNA thus distinguishing CaMV from the papovaviruses.

Structure of the genome

     The DNA of CaMV has a molecular weight of 5.10^6 daltons and contains
about 8000 base pairs. Two types of DNA can be extracted from virions :
circular molecules (80 %) and linear molecules (20 %). We have established the complete sequence of two strains of viral DNA : the Cabb-S
strain and D/H, with 8024 and 8016 base pairs, respectively (18,19).

     - The sequence discontinuities
     Volovitch and collaborators (20) have shown that the encapsidated
DNA of CaMV is characterized by the presence of three single-strand sequence interruptions situated at well-defined sites on the DNA of almost
all strains of CaMV. The existence of these interruptions can be demonstrated by gel electrophoresis of the viral DNA after denaturation by heat or
alkali. One of these interruptions called $\Delta 1$, is in the α-strand, the
strand which is transcribed *in vivo* ; the other two interruptions $\Delta 2$ and
$\Delta 3$ are in the complementary strand. Contrary to expectation, sequence analysis has shown that the discontinuities are not nicks or gaps in one
strand of the DNA but instead correspond to three-stranded structures,
where the 5' and 3' ends of the interrupted strand have short redundant
sequences (21). The sequence at the 5' end of each discontinuity is homo-

genous but the 3' ends are heterogenous in length. For the moment, the role of the interruptions is unknown.

   - Genetic organization of CaMV
   Analysis of the distribution of termination codons in each strand of the DNA has led to the following conclusions :

       1) The α-strand, that which contains Δ1, has no significant coding capacity
       2) The complementary strand, that which contains two interruptions, is almost entirely coding except for a region of 1000 nucleotides around Δ1.

   These conclusions are in agreement with the results obtained by us and by others which show that only the α-strand is transcribed *in vivo*.

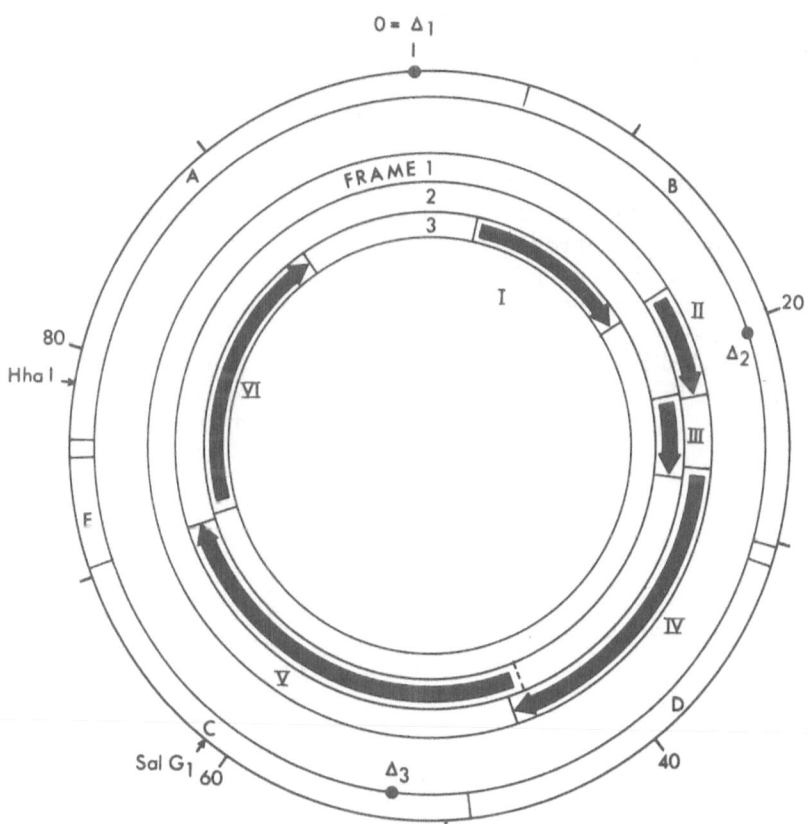

Figure 3 - Coding regions on CaMV DNA. Inner circles give the position on the DNA α-strand of the six long open reading frames described in the text. The outer circle gives the position of EcoRI fragments A-E and the three discontinuities (Δ) in virion DNA.

Six coding regions have been localized (Figure 3) with lengths of 1017, 501, 408, 1503, 2082 and 1626 nucleotides corresponding to polypeptides of 38 kd, 18 kd, 15 kd, 57 kd, 79 kd and 61 kd, respectively. The six regions overlap one another only slightly if at all. Preliminary eviden-ce suggests that the viral genes are not mosaic in nature (spliced), at least for the coding regions.

- Identification of genes on the viral DNA
    If we assimilate the six aforesaid coding regions with potential genes, we can attempt to correlate them with the two proteins of viral origin which have been identified so far. These two proteins are, on one hand, the coat protein (42000 d) and, on the other hand, the viral inclu-sion body protein (61000 d) (22), this latter protein being isolated from the inclusion bodies which accumulate in virus-infected cells. The sequence analysis reveals that only gene IV can code for a protein for which the central part has an amino acid composition comparable to that of coat protein. It is hence, probable that the 57000 d polypeptide, expressed by gene IV, represents a precursor of the coat protein. The viroplasm protein, on the hand, has been shown conclusively to derive from gene VI (23,24). The products encoded by the other genes, if they exist, have not yet been identified.

In vivo and *in vitro* transcription of CaMV DNA

The next important question is how the various viral polypeptides are expressed from the six potential genes. The DNA could be transcribed either as a long polycistronic messenger or as several monocistronic species, each with its own promotor. To distinguish between these possi-bilities polyadenylated RNA was prepared from infected leaves, fractio-nated on an agarose gel after denaturation, transferred to DBM paper, and, finally, hybridized with radioactive viral DNA probe. Four products of transcription were thus identified and their extremities have been mapped precisely on the genome by S1 nuclease mapping (Figure 4) (25).

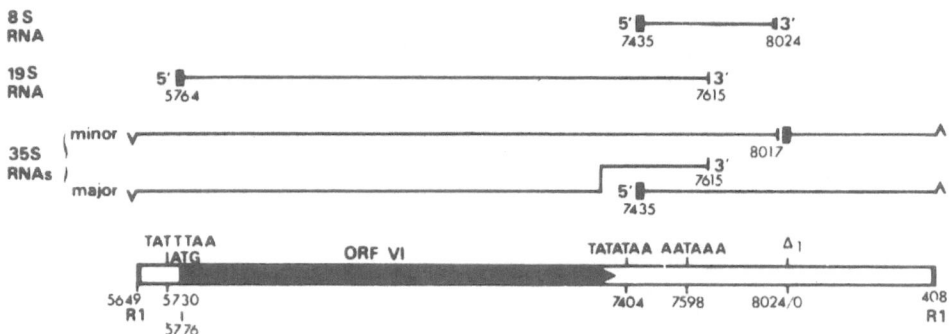

Figure 4 - Coordinates of the four viral transcripts described in the text.

The transcripts are :

1) A major 35S RNA whose 5' extremity is located at nucleotide (nt) 7435 and its 3' extremity at nt 7615. This RNA corresponds to the total transcription of the genome with an overlap of 180 nucleotides at the two extremities.
2) A minor 35S RNA which starts and ends at Δ1.
3) A 19S RNA whose 5' extremity is at nt 5764 and whose 3' extremity is at nt 7615.
4) An 8S RNA with 5' extremity at nt 7435 and 3' extremity at Δ1.

Upstream of the 5' end of the 19S and major 35S RNAs there are sequences ressembling a "TATA box", which is thought to form part of the eucaryotic transcription promotor signal. We have tested these sequences for promotor activity by transcription *in vitro*, using a Hela cell extract (26). The results show that both sequences are capable of specifically initiating transcription *in vitro* at the same point as for the authentic *in vivo* RNAs indicating that both species are primary transcripts. These observations also suggest that the signals governing transcription of animal DNA and that of plants are probably very similar. The various viral transcripts have been introduced into an *in vitro* translation system but only the 19S RNA has messenger activity. It encodes a protein of 61 kd which has been shown to correspond to the viral inclusion body structure protein referred to above (27). On the other hand, efforts to obtain translation of the 35S RNAs have failed. In summary, we have shown that CaMVDNA contains at least two promotors. Other signals, such as transcription termination and polyadenylation, have also been detected. At this stage of our work, however, a point remained unclear ; in fact, transcription of RNA from DNA containing a discontinuity is unlikely as it is difficult to imagine how the RNA polymerase II molecule could copy over Δ1. For this reason, we have sought and found within the cytoplasm of infected plant a supercoil viral DNA species (containing no interruptions) which undoubtedly, serves as substrate for transcription. The supercoiled DNA is not encapsidated (28,29).

Infectivity of native and cloned CaMV DNA. Phenomena of ligation and recombination

The DNA extracted from CaMV particles is infectious, whether it is circular or linear. By way of contrast, this same DNA cloned into pBR322 (in any insertion site) is non infectious. The infectivity is restored if the DNA is excised from the plasmid by the same restriction enzyme used for insertion. CaMV DNA possesses two BamHI restriction sites. The two BamHI fragments, inoculated separately, are not infectious. But, if the mixture of restriction fragments is inoculated, weak infectivity is restored. This result indicates that restriction fragments may be ligated together in the infected plants cytoplasm (30).

As already mentioned, clones of CaMV DNA inserted in different sites of plasmid pBR322 are not infectious. If, however, these clones are inoculated in mixtures of two, systemic symptoms appear and viral DNA (but not plasmid DNA) can be isolated. In this case, it is evident that recombination events have occurred in the plant cell, representing the first example of such phenomena for the higher plants (31).

REFERENCES

1. Hirth, L., and Richards, K.E.: 1981, Adv. Virus Res. 26, pp 145-199.
2. Hohn, T., Richards, K.E., and Lebeurier, G.: 1982 in "Current topics in microbiology and immunology" Editors : Hofschneider, P.H. and Goebel, W., Springer Verlag 96, pp 193-236.
3. Hunter, T.R., Hunt, T., Knowland, J., and Zimmern, D.: 1976, Nature (London) 260, pp 759-764.
4. Pelham, H.R.B.: 1978, Nature (London) 272, pp 469-471.
5. Zimmern, D.: 1977, Cell 11, pp 463-482.
6. Jonard, G., Richards, K.E., Guilley, H., and Hirth, L.: 1977, Cell 11, pp 483-493.
7. Oberg, B., and Philipson, L.:1972, Biochem. Biophys. Res. Commun. 48, pp 927-932.
8. Siegel, A., Montgomery, V.H.I., and Kolacz, K.: 1976, Virology 73, pp 363-378.
9. Bruening, G., Beachy, R.N., Scalla, R., and Zaitlin, M.: 1976, Virology 71, pp 498-517.
10. Briand, J.P., Keith, G., and Guilley, H.:1978, Proc. Natl. Acad. Sci. USA 75, pp 3168-3172.
11. Guilley, H., Jonard, G., Kukla, B., and Richards, K.E.:1979, Nucleic Acids Res. 6, pp 1287-1308.
12. Kozak, M.: 1978, Cell 15, pp 1109-1123.
13. Butler, P.J.G., and Klug, A.: 1971, Nature New Biology 229, pp 47-50.
14. Guilley, H., Jonard, G., Richards, K.E., and Hirth, L.: 1975, Eur. J. Biochem. 54, pp 135-144.
15. Lebeurier, G., Nicolaeiff, A., and Richards, K.E.: 1977, Proc. Natl. Acad. Sci. USA, 74, pp 149-153.
16. Alani, R., Pfeiffer, P., and Lebeurier, G.: 1979, Virology 93, pp 188-197.
17. Chauvin, C., Jacrot, B., and Lebeurier, G., and Hirth, L.: 1979, Virology 96, pp 640-641.
18. Franck, A., Guilley, H., Jonard, G., Richards, K.E., and Hirth, L.: 1980, Cell 21, pp 285-294.
19. Balazs, E., Guilley, H., Jonard, G., and Richards, K.E.: 1982, Gene, in press.
20. Volovitch, M., Drugeon, G., and Yot, P.:1978, Nucl. Acids. Res. 5, pp 2913-2925.
21. Richards, K.E., Guilley, H., and Jonard, G.: 1981, FEBS Letters 134, pp 67-70.
22. Alani, R., Pfeiffer, P., Whitechurch, O., Lesot, A., Lebeurier, G., and Hirth, L.: 1980, Ann. Virol. (Inst. Pasteur) 131E, pp 33-53.
23. Odell, J.T., Dudley, R.K., and Howell, S.H.: 1981, Virology 111, pp 377-385.
24. Covey, S.N., and Hull, R.: 1981, Virology 111, pp 463-474.
25. Guilley, H., Dudley, R.K., Jonard, G., Balazs, E., and Richards, K.E.:

   Cell, in press.
26. Manley, J.L., Fire, A., Cano, A., Sharp, P.A., and Gefler, M.L.:
    1980, Proc. Natl. Acad. Sci. USA 77, pp 3855-3859.
27. Odell, J.T., and Howell, S.H.: 1980, Virology 102, pp 349-359.
28. Menissier, J., Lebeurier, G., and Hirth, L.: 1982, Virology 117, pp
    322-328.
29. Olszewski, N., Hagen, G., and Guilfoyle, T.J.: 1982, Cell 29, pp
    395-402.
30. Lebeurier, G., Hirth, L., Hohn, T., and Hohn, B.: 1980, Gene 12, pp
    139-146.
31. Lebeurier, G., Hirth, L., Hohn, B., and Hohn, T.: 1982, Proc. Natl.
    Acad. Sci. USA 79, pp 2932-2936.

STRUCTURE, EVOLUTION AND INTERACTIONS OF RIBOSOMAL RNAs

J.P. Ebel, C. Branlant, P. Carbon, B. Ehresmann, C. Ehresmann,
A. Krol and P. Stiegler.
Laboratoire de Biochimie, Institut de Biologie Moléculaire et
Cellulaire, 15, rue René Descartes 67000 Strasbourg (FRANCE).

The ribosomes are multimolecular particles where protein synthesis takes place. This complex process is achieved through a coordinated interaction between the ribosomes and the macromolecules involved : mRNAs, tRNAs and translational factors. Understanding of the functional role of the ribosome requires detailed knowledge of its structure and of the spatial relationship between its constituent molecules. Despite the complexity of the its structure, the ribosome is the subcellular particle where the largest amount of information is available, as the primary structure of the 3 RNAs and of the the 52 proteins of E.coli ribosome is completely determined.

The ribosomal RNAs play a fundamental role in ribosome structure and function. For a long time, only the structural function was considered : they represent the backbone around which the ribosomal proteins are assembled in the course of ribosome assembly. But more recently it appeared that they may also play a functional role, as several regions within the RNAs have been found to interact directly with the macromolecules involved in protein synthesis.

In our laboratory, we have concentrated our efforts on two major problems : i) the determination of the primary and secondary structure of the two large ribosomal RNAs, 16S and 23S RNAs ; ii) the study of RNA-protein interactions or neighbourhoods in the ribosomal subunits. This paper describes our recent results in these fields.

A. STRUCTURAL ORGANIZATION OF THE LARGE RIBOSOMAL RNAs

Whereas the nucleotide sequence of the small E.coli 5S RNA has been established in 1968 in Sanger's group (1), the primary structures of the two large E.coli 16S and 23S RNAs were only determined ten years later between 1978 and 1980 on the corresponding genes in Noller's group (2,3) and directly on the RNAs in our own laboratory (4,5,6).

Completion of the nucleotide sequences of these RNAs has allowed investigation of the secondary structure of the nucleotide chains. It is obvious that the the structural and functional role of the ribosomal RNAs can only be understood if the folding of these RNAs within the ribosome is known.

C. Hélène (ed.), Structure, Dynamics, Interactions and Evolution of Biological Macromolecules, 177–193.

I. CRITERIA USED TO BUILD SECONDARY STRUCTURE MODELS FOR THE E.COLI RIBOSOMAL RNAs

The total number of the theoretical base pairings and their topological combinations are so immense in the case of large RNAs like 16S or 23S ribosomal RNAs that restriction must be introduced. This was achieved by integrating in computer programs all available experimental data on the topography of these RNAs within the ribosomal subunits. The criteria used to build the secondary structure models of E.Coli 16S and 23S RNAs we proposed (7,8,9) were the following :

1. Accessibility to ribonucleases

- of the free 16S and 23S RNAs in solution ;
- of these RNAs within reconstituted complexes between the RNA and one or several ribosomal proteins ;
- of the RNAs within either the intact 30S or 50S subunits or partially unfolded subunits or within 70S ribosomes.
Two types of ribonucleases were used :
- single strand specific ribonucleases like T1, pancreatic or S1 ribonucleases
- a double strand specific ribonuclease extracted from **Naja oxiana** cobra venom.
These ribonucleases were very useful as secondary structure probes for an accurate mapping of residues located in either single stranded or helical regions. In addition they provided information on the most exposed RNA regions, thus giving rise to a precise picture of the RNA surface topography.
The accessibility data used for the building of our secondary structure models are referenced in (8) and (9).

2. Accessibility to chemical reagents

To complete these enzymatic data, we also used results from Noller's group on the reactivity towards kethoxal modification which modifies guanylic residues in single stranded regions (10, 11). These accessibility studies were performed in either active or inactive subunits or in 70S ribosomes.

3. Experimental evidence for long range RNA-RNA interactions

Specific long range RNA-RNA interactions could be detected by mild ribonuclease hydrolysis of ribosomal RNA-protein complexes. This was achieved by isolation of the digested complexes under non-denaturing conditions preserving the RNA-RNA interactions, followed by a deproteinization of the complex and by a selective dissociation of the interacting RNA fragments (12). The analysis of these fragments showed the presence of complementary sequences widely separated in the primary structure, which were good candidates for RNA-RNA long distance interactions.

4. Comparative sequence analysis

Several complete small and large ribosomal subunit RNA sequences of prokaryotic, eukaryotic or organellar ribosomes are now available (see next paragraph), as well as numerous partial sequences (listed in (8), (9)). They were used for comparative sequence analysis.

Examination of these sequences showed that a large number of secondary structure elements proposed in the secondary structure models for E.Coli 16S and 23S RNAs are conserved in the corresponding RNAs of the other species : the replacement of one base in one strand of a helical region is compensated by the presence of a complementary base in the opposite strand, the secondary structure being thus maintained. The finding of such compensatory base changes in the structure of other ribosomal RNAs can be considered as a strong evidence for the reality of many helical structures in the E.coli model.

5. Computer assistance

Computer analysis was used to determine the base paired regions, taking into account the classical thermodynamic data. The total number of potential base pairings was restricted by integration of the accessibility data into the computer program. In addition, computer assistance was used to characterize in the various ribosomal RNAs sequence homologies and secondary structure conservation through compensatory base changes. The programs used are described in (8) and (9).

II. THE 16S RIBOSOMAL RNA MODEL

Figure 1 shows the characteristics of the model we propose for the 1642 nucleotides long E.coli 16S RNA. The figure indicates all the accessibility data to the various single stranded and double-stranded specific ribonucleases and to kethoxal. The helices which are supported by phylogenetic evidence are indicated by bars. It must be noted that some other helices are preserved by strict nucleotide sequence conservation. The model is characterized by the following features :

- Almost 49 per cent of the nucleotides are base paired, thus forming 57 double helical regions. 44 of these helices are conserved in other rRNA molecules through coordinate base changes.

- The folding of the 16S RNA chain is mainly directed by 5 long-range interactions, e.g. base pairing between sequences widely separated in the primary structure : i) 27-37/546-555 ; ii) 563-569/879-885 ; iii) 925-932/1383-1390 ; (IV) 945-954/1224-1234, and (V) 983-989/1214-1220.

These base pairings have a structural key-role since they delineate 4 distinct structural domains, which have also been defined by topographical studies (Figure 1).

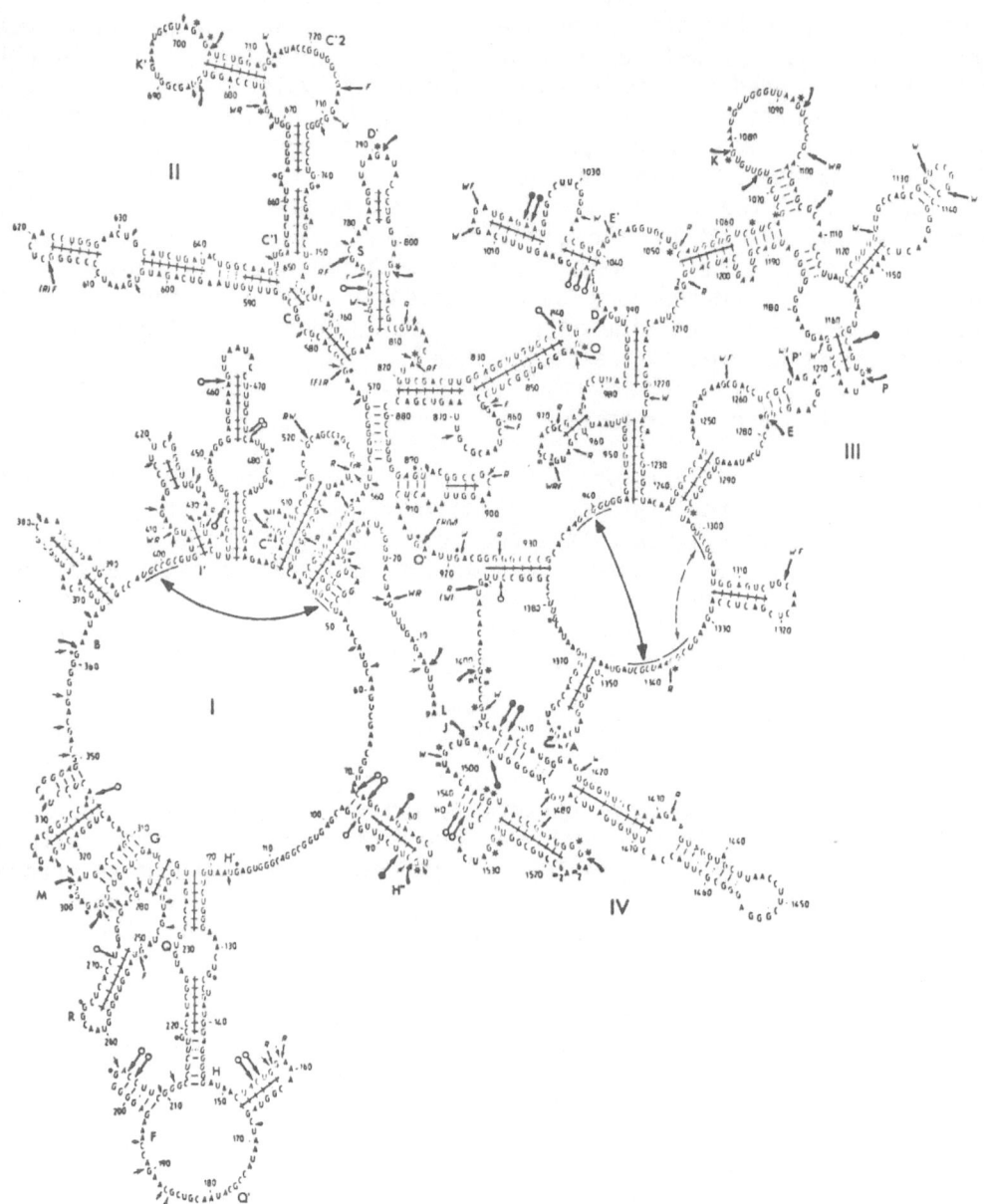

Figure 1 – SECONDARY STRUCTURE MODEL OF THE E.coli 16S RNA

Points of nuclease attack are indicated by arrows : (➤) RNase A ; (→) RNase T₁ in
isolated protein RNA-complexes ; (➤) common major RNase T₁ cuts
      (➡) major, (→) medium, (–) minor T₁ RNase cuts   (•➤)
primary, (○➤) secondary, (○➤) tertiary cobra venom RNase cuts in 30S subunits.
Kethoxal-modified G residues are indicated by (*) in active subunits   and (•)
in inactive 30S subunits .   Helices supported by phylogenetic evidence are indicated
by heavy lines. Letters denote the different sections of the 16S rRNA primary structure.

In this model of the 16S RNA folding a few number of secondary structure motifs are merely tentative (8). Actually insufficiency of experimental evidence or even contradictory sets of data do not allow any definite conclusion. Some other base pairing possibilities may also exist between RNA regions that are still single stranded in our model.

The model we propose (7, 8) is close to that proposed by Woese et al. (13). It also shares large similarities with another model proposed by Brimacombe (13). However several differences essentially based on accessibility data and phylogenetic data can be noted between our model and the two ones models (8). In some cases they may reflect possible conformational changes, as already suggested by Herr et al. (10) and Brimacombe (13).

III. THE 23S RIBOSOMAL RNA MODEL

Figure 2 shows the characteristics of the secondary structure model (in a schematic representation) we propose for the 2902 nucleotides long E.coli 23S RNA (9). This model has been built using the same criteria as for 16S RNA. As for 16S RNA the various single- and double-stranded regions of 23S RNA are supported by accessibility data to various ribonucleases and to kethoxal. For a large number of helices phylogenetic evidence is brought. The folding of 23S RNA is, as with 16S RNA, directed by several long-range RNA-RNA interactions, which delineate 7 distinct structural domains. Interestingly the 5'- and 3'-extremities (1-8/2984-2902) are base-paired.

Here too, a limited number of secondary structure motifs are merely tentative. This is particularly true for domain IV in the central region of 23S RNA, where several alternative base-pairings are possible.

Recently two other secondary structure models have been proposed (15, 16). There is a large agreement between these two models and ours. However Noller's model (16) displays some differences in domaine IV and in particular an additional long distance interaction between a sequence of domain IV and one of domain II. Brimacombe's model (15) has taken into account information brought by ultraviolet irradiation induced intramolecular RNA-RNA crosslinks. This model differs from ours by small differences in the base pairing in all the domains.

IV. SEQUENCE AND SECONDARY STRUCTURE HOMOLOGIES BETWEEN E.COLI 16S AND 23S RNAS AND THE RIBOSOMAL RNAS FROM DIFFERENT SOURCES : EVIDENCE FOR COMMON SECONDARY STRUCTURE MODELS

We investigated the extent of both nucleotide sequence and secondary structure conservation which might exist between the E.coli 16S and 23S RNAs and the following complete nucleotide sequences of small and large ribosomal subunit RNAs, covering diverse types of prokaryotes, eukaryotes and organelles :

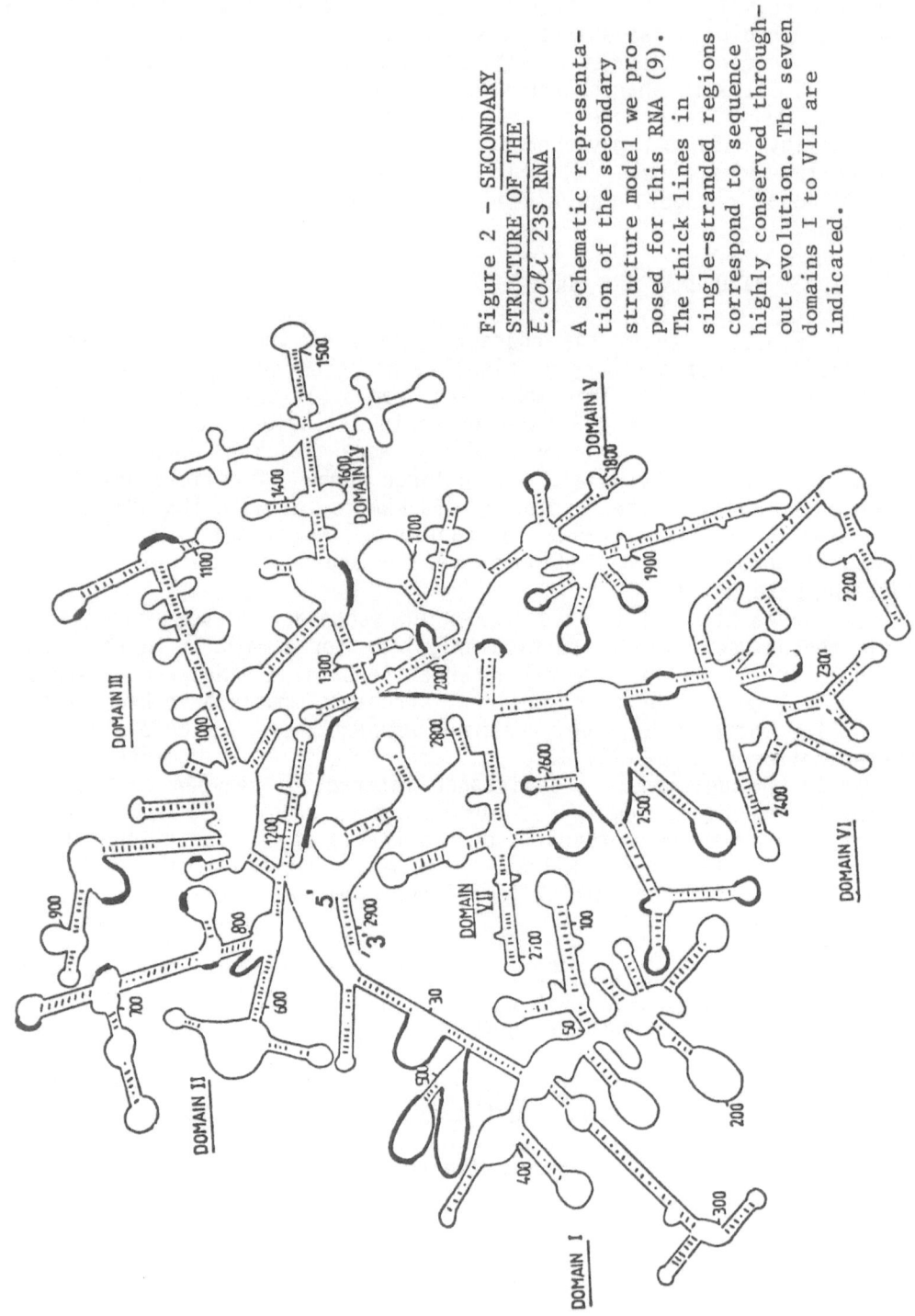

Figure 2 - SECONDARY STRUCTURE OF THE *E.coli* 23S RNA

A schematic representation of the secondary structure model we proposed for this RNA (9). The thick lines in single-stranded regions correspond to sequence highly conserved throughout evolution. The seven domains I to VII are indicated.

        - in the case of small subunit ribosomal RNAs :
* prokaryotic RNAs :
     Proteus vulgaris 16S RNA (17)
* eukaryotic cytoplasmic RNAs :
     Yeast 18S RNA (18)
     Xaenopus laevis 18S RNA (19)
* organelle's RNAs :
     Zea mays chloroplastic 16S RNA (20)
     Yeast mitochondrial 15S RNA (21)
     Mouse mitochondrial 12S RNA (22)
     Human placenta mitochondrial 12S RNA (23)
        - in the case of large subunit ribosomal RNAs :
* eukaryotic cytoplasmic RNAs :
     Yeast 26S RNA (24)
* organelle's RNAs :
     Zea mays chloroplastic 23S RNA (25)
     Mouse mitochondrial 16S RNA (22)
     Human placenta mitochondrial 16S RNA (23)

        All the RNA molecules examined could be folded into secondary
structure schemes that brought to light a remarkable preservation of many
structural motifs as well as strong sequence conservations compared to the
E.coli molecule (26, 9).
        The homology is particularly striking in the case of the
prokaryotic **Proteus vulgaris** 16S RNA, where 93 per cent of sequence
homology is found compared to E.**coli** 16S RNA. Another example of
remarkable homology is found between the two chloroplastic **Zea Mays** 16S
and 23S RNAs and the prokaryotic E.**coli** 16S and 23S RNAs. In this case,
the fact that nearly superimposable secondary structures can be deduced
provides an even more convincing proof for the prokaryotic nature of
chloroplastic ribosomes than their 70-75 per cent sequence homology. In
the case of the eukaryotic yeast and **Xaenopus laevis** RNAs, despite their
larger size and an overall lower sequence homology, similar secondary
structure schemes can be drawn, where many motifs are conserved. The
mitochondrial ribosomal RNAs were of particular interest since great
variability is observed in both their size and base composition : the
mitochondrial 12S and 16S RNAs from mouse (22) and human placenta (23)
have significant shorter chain length than their bacterial counterpart,
and a rather low G + C content. However, here too, remarkable sequence
homology and secondary structure conservation were observed. (For a
detailed study, see (26) and (9).
        Another interesting observation is that in all the RNAs studied
the homologous regions are interrupted by domains varying in both length
and nucleotide sequence. Furthermore the various folding schemes proposed
highlight a striking constancy in size and secondary structure of several
domains. This is illustrated in both the small and large subunit ribosomal
RNAs.

1. Sequence and secondary structure homologies in small subunit ribosomal RNAs

Figure 3 presents the base paired elements which appear to be common among all the small subunit RNA molecules studied (26). It can be seen that the principal long distance RNA-RNA interactions which delineate, in E.coli 16S RNA, the four well defined domains are conserved, suggesting a common basic organization. But there are also variable domains (A-G) where extensive divergence both in base composition and in chain length is observed. These variable domains are all located in finite areas in each structural domain.

The RNA sequences analyzed here not only exhibit a common basic structural organization but also display single nucleotide residues or sequences which are strictly conserved at equivalent positions. These residues are good candidates for beeing invariant nucleotides. Most of them (85%) appear not to be base-paired. They are distributed in each structural domain but are particularly clustered in the single stranded sequences flanking motif 36 near the 3' extremity in domain IV. They are also concentrated in the 3'-part of domain II, in the center of the molecule. Invariant nucleotides may be involved in tertiary structure interactions as they are in tRNA molecules. Another possible role could be their active participation in ribosome funtion. Indeed, there is strong evidence that both the central and 3'-terminal domains contain sequences that are exposed on the surface of the subunit and located on the interface of the two subunits in the E.coli ribosome (9, 27, 28), some of which making contact with the large subunit (10) (Fig.5). Furthermore one of these highly conserved regions in the 3'-domain of the E.Coli RNA appears to be in contact with the anticodon of the tRNA when it is bound at the ribosomal P site (29).

In conclusion, extensive phylogenetic variability appears to be permitted but seems to be restricted to distinct parts of the molecule. The intrinsic molecular pecularities of each species might be expressed in these variable domains. In this way, the chloroplastic RNA appears to be more closely related to its bacterial counterpart than the mitochondrial RNAs. The homology between E.coli 16S RNA and yeast cytoplasmic 18S RNA is even greater than that of E.coli with the various mitochondrial RNAs. Phylogenetic variability must also be related with changes in the mechanism of protein synthesis. For instance, as the mechanism of initiation is concerned, both the cytoplasmic and mitochondiral small subunit RNAs lack the mRNA binding Shine and Dalgarno sequence (30) which is found near the 3'-extremity of both bacterial and chloroplastic RNAs.

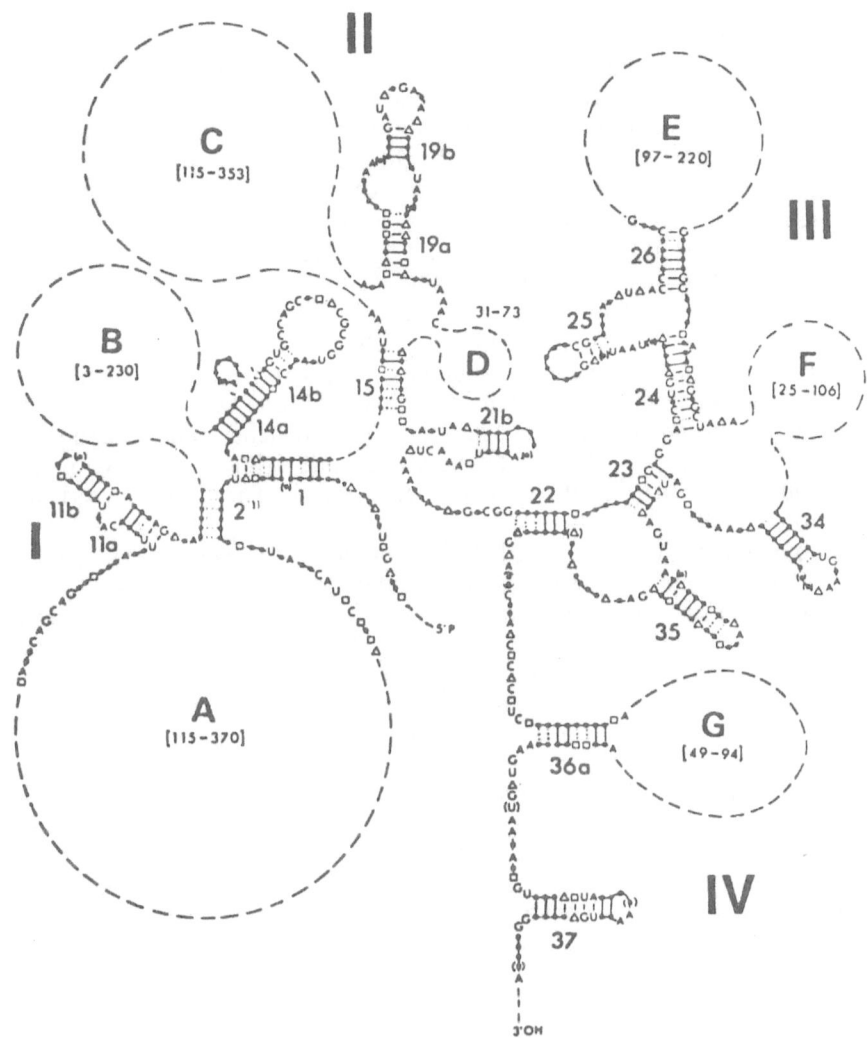

Figure 3 - COMMON BASIC STRUCTURAL ORGANIZATION OF THE SMALL
RIBOSOMAL SUBUNITS RNAs. Only secondary structure motifs which are
common to all the RNAs studied are shown. The helices are numbered
according to (12). Nucleotides are symbolized by filled circles and
base pairs by bars. Dashed bars indicate that base-pairing between
the relevant nucleotides is not possible in every case. The four
structural domains (I-IV) are shown. Variable domains (A-G) are
symbolized by a dashed line, their size in nucleotide residues
varying between the two extreme values indicated in parentheses.
Invariant nucleotides, found in homologous positions in all RNA
sequences, are indicated. Semi-invariant residues are symbolized
(□) for purines and (△) for pyrimidines. In some cases, a single
residue was inserted or deleted and these are indicated in parentheses.

2. Sequence and secondary structure homologies in large subunit ribosomal RNAs

The models we built for a prokaryotic, a cytoplasmic, a chloroplastic and two mitochondrial large RNAs all display the same basic organization with remarkable sequence and secondary structure homologies (9, 24). As mentioned before the degree of homology between prokaryotic and chloroplastic RNAs is particulary high.

It should be noticed that in the chloroplast ribosomes from higher plants, the counterpart of the 3'-terminal region of bacterial 23S RNA is a small additional RNA species : the 4.5S rRNA (31). In the same way, the cytoplasmic 5.8S rRNA is the counterpart of the 5'-terminal region of bacterial 23S RNA [Fig. 4] (24). The genes for 5.8S and 26S rRNAs are separated by a stretch of DNA which is transcribed but eliminated during the maturation process, leading to two separated RNA species 5.8S and 26S rRNAs. Such an observation suggests that in the course of evolution an insertion has occurred in the gene coding for the large rRNA. The inserted DNA is transcribed but eliminated by processing enzymes. The same explanation may be given for the existence of chloroplastic 4.5S RNA.

Occurrence of additional DNA sequences in the genes coding for the large rRNAs is not restricted to these two cases. Indeed, intron sequences are observed in genes for some mitochondrial, chloroplastic and cytoplasmic large rRNAs. Interestingly, these intron sequences are all located in the DNA region coding for domains V and VI (9).

Domain V and VI are precisely the domains which are the most conserved throughout evolution, both in primary and secondary structure. As is the case for 16S RNA, the degree of conservation of the large rRNA is not constant all along the molecule. Some of the domains : III and more particulary V and VI are more conserved. We observed that the degree of conservation is in direct relation with the degree of accessibility within the subunit : in other words, the RNA regions which are accessible at the surface of the subunit are more evolutionarily conserved. On the other hand we previously found that the domains I, II, IV and VII, where we observed a great variability in size and primary structure, constitute a compact area within the subunit (32).

Finally domains V and VI probably are part of the functional site of the 50S subunit : i) these two domains contain most of the posttranscriptionally modified nucleotides of **E.coli** 23S RNA (9) ; ii) results of kethoxal modification suggest that domain VI is at the subunit interface (33) ; iii) two ponctual mutations in domain VI lead to chloramphenicol resistance (34) ; iv) a covalent linkage obtained between 23S RNA and a puromycin derivative is probably located in domain VI (35).

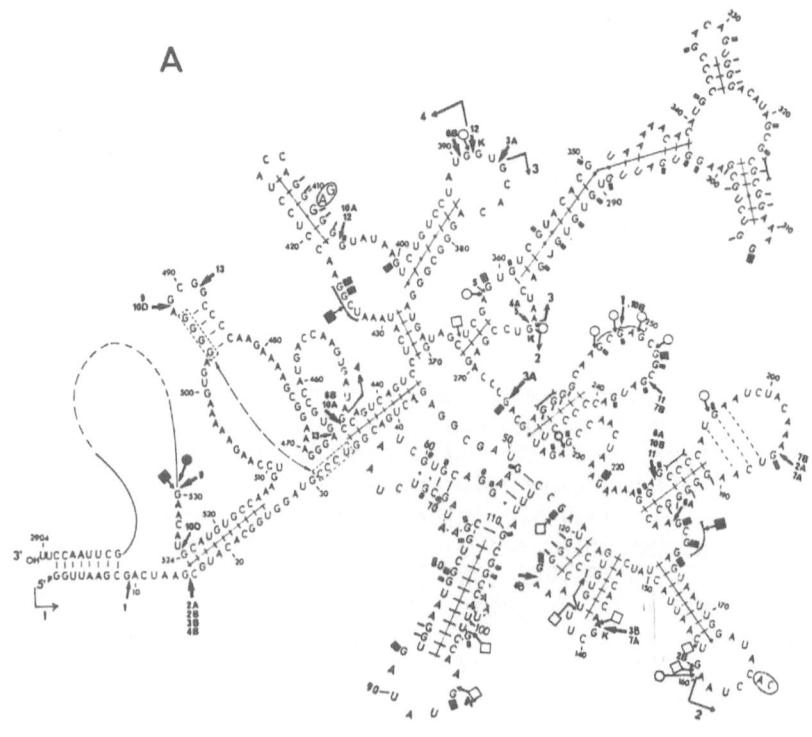

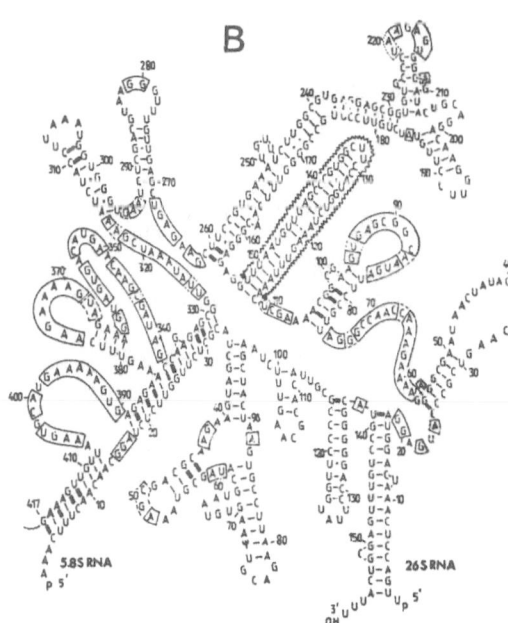

Figure 4. – <u>DOMAIN I OF *E.coli*</u> <u>23S RNA (A) AND ITS COUNTERPART</u> <u>IN YEAST CYTOPLASMIC RIBOSOMES (B).</u>

In yeast cytoplasmic ribosomes domain I is constituted by 5.8S RNA and the 417 nucleotides at the 5'-end of 26S RNA. Nucleotides in single-stranded regions which are conserved as compared to *E.coli* 23S RNA are boxed with full lines. Conserved base pairs are indicated by thick bars, and additional sequences in yeast RNA which are absent in *E.coli* RNA are denoted by ⌇⌇⌇ .

B. RNA-PROTEIN INTERACTIONS OR NEIGHBOURHOODS IN THE RIBOSOMAL SUBUNITS

Ribosomal RNAs represent the backbone around which ribosomal proteins are attached during ribosome assembly. This fundamental role can only the explained by specific interactions between both ribosomal RNAs and proteins. The determination of the contact regions reflecting either direct interaction or close proximity of the ribosomal RNA and protein is essential for understanding the ribosome architecture.

The approaches we have used in our laboratory to characterize the contact areas between ribosomal RNAs and proteins are the followings.

1. Protection experiments

The first techniques we used consisted to form a synthetic complex between an isolated ribosomal RNA and a specific ribosomal protein which interacts directly with the RNA, to digest this complex by a ribonuclease and to characterize the RNA fragments which are protected by the protein against the ribonuclease digestion.

This method gave valuable information about the positioning of the proteins along the RNA molecules. With several proteins like S8 and S15 in their complexes with 16S RNA (36) (see Fig.5) or L1, L20 and L23 in the complex with 23S RNA (47), fragments of small size were characterized. But with other proteins, like S4 in its interaction with 16S RNA (36) and L24 in its interaction with 23S RNA (46), very large areas of RNA were found in the remaining complex after ribonuclease hydrolysis. This was due to RNA-RNA interactions. Only a few fragments directly interact with the protein, the other ones being present in the complex because of a compact secondary or tertiary structure due to short or even long range RNA-RNA interactions. This approach was actually very useful for the detection of these long distance interactions (12). Therefore it turned out that the protection approach did not allow in several cases an accurate information about the precise interaction site between the RNA and the protein.

2. Crosslinking experiments

We therefore developed techniques which involve the formation of covalent crosslinks between RNA and proteins, using either direct ultraviolet irradiation or chemical bifunctional reagents. After ribonuclease digestion, the oligonucleotides covalently bound to the ribosomal proteins were characterized.

These techniques were first applied to reconstituted RNA-protein complexes (38, 39). But as the location of the ribosomal protein along the RNA could undergo some changes in the course of the ribosome assembly, they were also applied to the intact ribosomal subunits (40, 41).

Ultraviolet irradiation allows crosslinking of RNA and protein, when the two partners are in close contact. The oligonucleotides and peptides crosslinked in these conditions are good candidates to be interaction regions. This technique has given excellent results with RNA-protein reconstituded complexes. For instance protein S4 was found to be mainly crosslinked to fragments all located in domain I : fragment 428-505 and to a lesser extent to regions 301-323 and 530-556. In the case of protein S20, only one RNA fragment (213-296) was found to be crosslinked, also within domain I (38, 39).

In the case of ribosomal subunits, direct ultraviolet irradiation led, with doses which do not lead to unfolding of the subunit, to the crosslinking of only a limited number of proteins. In the case of the 30S subunit, only protein S7 was linked to 16S RNA to an extent allowing the characterization of the crosslinked oligonucleotide. In our hands (40), it was found to be oligonucleotides 1261-1266, whereas BRIMACCMBE et al. (41), using the same technology characterized oligonucleotide 1233-1240, close to the preceeding one, in domain III.

These results show that direct ultraviolet irradiation of the ribosomal subunits is not a convenient technique for a general mapping of the proteins along the RNA molecules within the ribosomal subunits.

For this reason, we were obliged to develop crosslinking methods using chemical bifunctional reagents. The use of these reagents was much more successful, but it must be emphasized that, due to the length of the reagent, this crosslinking approach does not give the precise position nor the nature of the interacting residues. Nevertheless it gives valuable information about RNA-protein neighbourhood and therefore about the topographical arrangement between these components in the ribosome.

Three reagents have first been used :

$$N_3\text{—}\langle O \rangle\text{—CONH–CH}_2\cdot \overset{+}{C}=NH_2\cdot Cl$$
$$\overset{|}{OC_2H_5}$$

4-azidobenzoylamino acétimidate
hydrochloride (42)

4-azido-3,5 dichloro
-2,6 difluoropyridine
(43)

4 azido-2,3,5,6
tetrafluoropyridine
(43)

(1)   (2)   (3)

These reagents all possess one function able to substitute protein amino-groups. The greater inertness of nucleic acids can be overcome by the presence, as the second function, of an azido-group, which can be photoactivated yieding an unstable nitrene thus allowing reaction with a vicinal nucleic acid. Furthermore, since the reaction of nitrene is very unspecific, this enhances chance to react with the RNA molecule.

Recently another reagent has been used (9) :

$$(4) \qquad CH_3\text{-}CH_2\text{-}N=C=N\text{-}CH_2\text{-}CH_2\text{-}CH_2\text{-}\overset{H}{\underset{CH_3}{N^+}}\text{-}CH_3$$

1-ethyl-3-dimethylaminopropyl carbodiimide (44).

This reagent can be used to activate side chain carboxyl groups in ribosomal proteins and provoke formation of amide bonds between the activated glutamyl- and aspartyl- residues of these proteins and amino groups in ribosomal RNA (44).

All these reagents have been used to induce crosslinks within the 30S ribosomal subunit. With all reagents, proteins S4, S7 and S9 have been crosslinked to 16S RNA. With reagents 1 and 4, proteins S5, S12 and S18 were also crosslinked. All these proteins must be in close proximity to 16S RNA within the 30S subunit. Ribonuclease hydrolysis of the subunit allows the isolation of the proteins with the oligonucleotide covalently attached and the characterization of the latter. Using reagent 4 a complex could be isolated containing protein S12, which is not a primary binding protein, and T1 ribonuclease oligonucleotide which could be analyzed and located in the 16S RNA structure in position 1316-1322 in domain III (44).

Another approach using chemical crosslinking with reagents 1 to 3 within the ribosomal subunit was worked out in our laboratory (45). A ribosomal protein is isolated, modified with the reagent and reintroduced within the ribosomal subunit. The azido-function of the reagent is then photoactivated to induce its reaction with the RNA to form a covalent link between this RNA and this unique protein. This technique has been applied to protein S1 which is not a primary binding protein. The crosslinked oligonucleotide was in position 861-889 in domain II (45).

Location of the ribosomal proteins along the ribosomal RNAs

First it must be emphasized that the results given by these various techniques are in good agreement. In no case, a protein which has been located in one domain of secondary structure by one technique has been found in another domain by a second technique.

The location in the secondary structure model of 16S RNA of the 30S ribosomal proteins which we have studied is shown in Fig.5.

The location in the secondary structure model of E.Coli 23S RNA of some ribosomal proteins from the 50S subunit is, also known : domain I for protein L24 (46), domain III for protein L20, domain IV for protein L23 and domain VI for protein L1 (47).

The location of these proteins in the secondary structure models of the 16S and 23S RNAs gives information which might be useful for the characterization of the folding of these RNAs in the tertiary structure and for functional implications. As an example protein S12 has been shown to be located in the immediate neighbourhood of protein S4 by chemical crosslinking (48) and neutron scattering (49). It is also known that streptomycin dependence caused by mutation of S12 can be reversed by a mutation in S4 (50). Since S4 has been shown to associate with the 5'-terminal region of 16S RNA (domain I) and since the fragment crosslinked to protein S12 is found near the 3'-end of 16S RNA in domain III, these results suggest that these domains approach each other closely within the 30S subunit. Topological conclusions can be drawn from the crosslinking data between ribosomal RNAs and the other proteins presented here and from the available data concerning the localization of these ribosomal proteins within the ribosomal subunits. All there features may thus be of fundamental importance for the understanding of the tertiary conformation of the ribosomal RNAs within the ribosome.

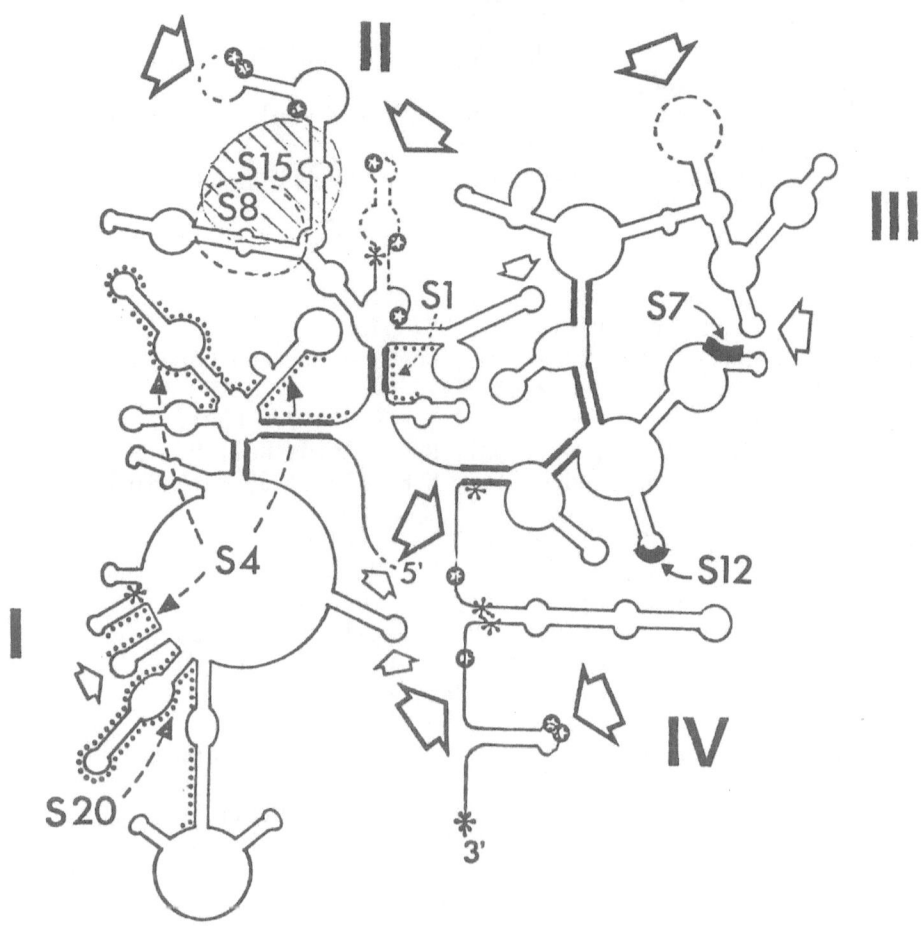

Figure 5 – ACCESSIBLE REGIONS AND PROTEIN CONTACT AREAS IN
16S RNA WITHIN THE 30S SUBUNIT.

⬦ Regions highly accessible to ribonucleases and to kethoxal
at the surface of the 30S subunit.

✳ Sites accessible to cobra venom ribonuclease in the 30S
subunit, but protected in the presence of the 50S subunit.

⊕ Sites accessible to kethoxal in the 30S subunit, but
protected in the presence of the 50S subunit.

REFERENCES

1. Brownlee, G.G., Sanger, F., and Barrel, B.G. (1968) J. Mol. Biol. 34, 379-412.
2. Brosius, J., Palmer, M.L., Kennedy, P.J., and Noller, H.F. (1978) Proc. Natl. Acad. Sci USA 75, 4801-4805.
3. Brosius, J., Dull, T.J., and Noller, H.F. (1980) Proc. Natl. Acad. Sci. USA 77, 201-204.
4. Carbon, P., Ehresmann, C., Ehresmann, B., and Ebel, J.P. (1978) FEBS Lett. 94, 152-156.
5. Carbon, P., Ehresmann, C., Ehresmann, B., and Ebel, J.P. (1979) Eur. J. Biochem. 100, 399-410.
6. Branlant, C., Krol, A., Machatt, M.A., and Ebel, J.P. (1979) FEBS Lett. 107, 177-181.
7. Stiegler, P., Carbon, P., Zuker, M., Ebel, J.P., and Ehresmann, C. (1980) C.R. Acad. Sci. Paris 291, 937-940.
8. Stiegler, P., Carbon, P., Zuker, M., Ebel, J.P., and Ehresmann, C. (1981) Nucl. Acids Res. 9, 2153-2172.
9. Branlant, C., Krol, A., Machatt, M.A., Pouyet, J., Ebel, J.P., Edwards, K., and Kossel, H. (1981) Nucl. Acids Res. 9, 4303-4324.
10. Herr, W., Chapman, N.M., and Noller, H.F. (1979) J. Mol. Biol. 130, 433-449.
11. Herr, W., and Noller, H. (1979) Cell 18, 55-60.
12. Ehresmann, C., Stiegler, P., Carbon, P., Ungewickell, E., and Garrett, R.A. (1980) Eur. J. Biochem. 103, 439-446.
13. Woese, C.R., Magrum, L.J., Gupta, R., Siegel, R.B., Stahl, D.A., Kop, J., Crawford, N., Brosius, J., Guttel, R., Hogan, J.J., and Noller, H.F. (1980) Nucl. Acids Res. 8, 2275-2293.
14. Brimacombe, R. (1980) Biochem. Int. 1, 162-171.
15. Glotz, C., Zwieb, C. Brimacombe, R., Edwards, K., and Kossel, H. (1981) Nucl. Acids Res. 9, 3287-3306.
16. Noller, H., Kop, J., Wheaton, V., Brosius, J., Gutell, R., Kopylov, A., Dohme, F., Herr, W., Stahl, D., Gupta, R., and Woese, C. (1981) Nucl. Acids. Res. 9, 6167-6189.
17. Carbon, P., Ebel, J.P., and Ehresmann, C. (1981) Nucl. Acids Res. 9, 2325-2333.
18. Rubtsov, P.M., Musakhanov, M.M., Zakharyev, V.M., Krayev, A.S., Skryabin, K.G., and Bayev, A.A. (1980) Nucl. Acids Res. 8, 5779-5794.
19. Salim, M., and Maden, E.H. (1981) Nature 291, 205-208.
20. Schwarz, Zs., and Kossel, H. (1980) Nature 283, 739-742.
21. Sor, F., and Fukuhara, H. (1980) C.R. Acad. Sci. Paris, Ser. D. 291, 933-936.
22. Van Etten, R.A., Walberg, M.W., and Clayton, D.A. (1980) Cell 22, 157-170.
23. Eperon, I.L., Anderson, S., and Nierlich, D.P. (1980) Nature 286, 460-467.
24. Veldman, G.M., Klootwijk, J., de Regt, C.H.F., Planta, R.J., Branlant, C., Krol, A., and Ebel, J.P. (1981) Nucl. Acids Res., 6935-6952.

25. Edwards, K., and Kossel, H., (1981) Nucl. Acids Res. 9, 2853-2869.
26. Stiegler, P., Carbon, P., Ebel, J.P., and Ehresmann, C. (1981) Eur. J. Biochem. 120, 487-495.
27. Chapman, N.M., and Noller, H.F. (1977), J. Mol. Biol. 109, 131-149.
28. Vassilenko, S.K., Carbon, P., Ebel, J.P., and Ehresmann, C. (1981) J. Mol. Biol. 152, 699-721.
29. Ofengand, J., Liou, R., Kohut, J., Schwartz, R., and Zimmermann, R.A. (1979) Biochemistry 18, 4322-4332.
30. Shine, J., and Dalgarno, L. (1974) Proc. Natl. Acad. Sci. USA 71, 1342- 1346.
31. Machatt, M., Ebel, J.P., and Branlant, C. (1981) Nucl. Acids Res. 9, 1533-1549.
32. Branlant, C., Krol, A., Sriwidada, J., and Ebel, J.P. (1977) J. Mol. Biol. 116, 443-467.
33. Herr, W., and Noller, H. (1979) Cell 18, 55-60.
34. Dujon, B. (1980) Cell 20, 185-197.
35. Greenwell, P., Harris, R. and Symons, R. (1974) Eur. J. Biochem. 49, 539-554.
36. Ungewickel, E., Garrett, R.A., Ehresmann, C., Stiegler, P., and Fellner, P. (1975) Eur. J. Biochem. 51, 165-180.
37. Ehresmann, B., Millon, R., Backendorf, C., Golinska, B., Olomucki, M., Ehresmann, C., and Ebel, J.P. (1980) in Biological implications of protein-nucleic acid interactions, Adam Mickiewicz University Press, Poznan, 63-75.
38. Ehresmann, B., Backendorf, C., Ehresmann, C., and Ebel, J.P. (1977) FEBS Letters 78, 261-266.
39. Ebel, J.P., Ehresmann, B., Backendorf, C., Reinbolt, J., Tritsch, D., Ehresmann, C., and Branlant, C. (1978) FEBS Proc. Meet. 43, 109-119.
40. Ehresmann, B., Backendorf, C., Ehresmann, C., Millon, R., and Ebel, J.P. (1980) Eur. J. Biochem. 104, 255-262.
41. Zwieb, C., and Brimacombe, R. (1979) Nucleic Acids Res. 6, 1175-1190.
42. Millon, R., Olomucki, M., Le Gall, J.Y., Golinska, B., Ebel, J.P., and Ehresmann, B. (1980) Eur. J. Biochem. 110, 485-492.
43. Millon, R., Ebel, J.P., Le Goffic, F., and Ehresmann, B. (1981) Bioch. Biophys. Res. Comm. 101, 784-791.
44. Chiaruttini, C., Expert-Bezançon, A ., Hayes, D., Thurlow, D.L. and Ehresmann, B. (1982) Nucl. Acids Res., in the press.
45. Golinska, B., Millon, R., Backendorf, C., Olomucki, M., Ebel, J.P., and Ehresmann, B. (1981) Eur. J. Biochem. 115, 479-484.
46. Branlant, C., Sriwidada, J., Krol, A., and Ebel, J.P. (1977) Eur. J. Biochem. 74, 155-170.
47. Branlant, C., Sriwidada, J., Krol, A., Ebel, J.P., Sloof, P., and Garrett, R. (1975) FEBS Letters 52, 195-201.
48. Traut, R.R., Lambert, J.M., Boileau, G., and Kenny, J.W. (1980) in Ribosomes, Chambliss, G., Graven, G.R., Davies, J., Davis, K., Kahan, L., and Nomura, M., Eds. pp. 89-110.
49. Ramakrishnan, V.R., Yabuki, S., Sillers, I.Y., Schindler, D.M., Engelman, D.M., and Moore, P.B. (1981) J. Mol. Biol. 153, 739-760.
50. Dabbs, E.R., and Wittmann, H.G. (1976) Mol. Gen. Genet. 149, 303-309.

LIPOSOME-MEDIATED GENE TRANSFER *IN VIVO*. UPTAKE AND EXPRESSION OF THE PREPROINSULIN I GENE BY RATS AND MICE.

Claude Nicolau and Philippe Soriano
Centre de Biophysique Moléculaire, C.N.R.S.
45045 Orléans Cedex, France

The use of liposomes for gene transfer into eucaryotic cells *in vitro* indicated that these carriers are quite appropriate for the introduction of DNA molecules into cells (1-4). Interesting insights have been obtained by studying the different pathways of liposome uptake by the cells (5) ; the efficiency of nuclear uptake of the exogeneous DNA was correlated with the mechanism of entry of the liposomes and the cell cycle stage of the target cells (6,7). A number of other questions arise from these studies, e.g. concerning the pathway of intracellular migration of DNA to the cell nucleus, interactions of the DNA-loaded liposomes with lysosomes, the possible interaction of those liposomes with mitochondria, etc...

The efficiency of the gene transfer -and expression- does not, however, exceed that of other more conventional methods (2,4,5). It appears therefore that the great potentialities of the liposome-mediated gene transfer technology lie in its *in vivo* applications, since none of the other gene transfer methods, e.g. coprecipitation with Ca^{2+} salts, fusion using polyethylene glycol or microinjection into nuclei can be used *in vivo*, in live animals.

Recent studies have shown that, when injected intravenously, liposomes are taken up essentially in the liver and the spleen, by the macrophages of the reticulo-endothelial system (8-11). Preparation of liposomes with long latencies of the entraped molecules, when injected intravenously (12), allows the transport of substantial amounts of such molecules to the cellular site of liposome uptake, i.e. splenic macrophages and liver Kupffer cells. These observations suggest that DNA-loaded liposomes, when injected intravenously, would be phagocyted by the same cells. The availability of a recombinant plasmid encoding rat preproinsulin I, p(gRl 9.4) (13), gave us the opportunity to try to transport and to express this gene in recipient animals (14).

C. Hélène (ed.), Structure, Dynamics, Interactions and Evolution of Biological Macromolecules, 195–206.
Copyright © 1983 by D. Reidel Publishing Company.

I - LIPOSOME DISTRIBUTION IN THE RAT FOLLOWING INTRAVENOUS INJECTION

In order to precisely locate the natural target cells of the i.v. injected liposomes, we prepared liposomes loaded with ^{111}In-bleomycin. The time course and the distribution of these "γ-emitting liposomes" in the animal organism can be visualized by external γ-camera imaging.

The images obtained show, between injection time and the 20th minute after injection, a very rapid and intensive uptake of the ^{111}In-bleomycin-liposomes by the liver. A weak urinary elimination is observed as well (14). The radioactivity in the bladder is due probably to the destruction of liposomes in the liver and the release of the label. Similar observations were made with ^{99}Tc-labeled liposomes (15). 6 hours after the inoculation, the rats were killed and the different organs were individually visualized with the γ-camera.

In order to check whether the liposomes were entirely responsible for this organ distribution of bleomycin, the same assay was repeated with free ^{111}In-bleomycin and with [111]In Cl$_3$. The latter was performed in order to check the pattern of distribution of the free ions, which could have resulted for the potential dissociation of ^{111}In-bleomycin.

With these controls the distribution pattern differs strongly from that observed with the liposome-encapsulated label. For ^{111}In-bleomycin, 10 s after inoculation the image shows a high heart radioactivity corresponding to the phase of tracer distribution. 2 minutes after injection the bleomycin is already found in liver and in urine. The urinary excretion is particularly strong 5 minutes after injection and 20 minutes after injection practically the entire radioactivity is found in the urinary bladder. The absence of kidney radioactivity corresponds to the asymptotic phase of the plasmatic clearance. The [111]In Cl$_3$ distribution differs in so far from that of ^{111}In-bleomycin that the urinary excretion is faster and even more significant. Not even a transient liver uptake is observed (16).

The organ activities were measured individually 20 minutes after injection ; 42 % and 15 % of the radioactivity were found in the liver and bladder, respectively, following liposome administration. When ^{111}In bleomycin was injected alone, these figures were 10 % and 82 %. In the case of ^{111}In injection, no radioactivity could be detected in the liver and 95 % were found in the bladder. Moreover, the kinetics of organ uptake and of urinary excretion of the radioactivity are completely different with the liposome-encapsulated ^{111}In-bleomycin, ^{111}In-bleomycin and [111]In Cl$_3$ (Fig. 1) (14,16). The data presented indicate that intravenously injected, large liposomes are taken up nearly essentially by the liver and the spleen of the injected animals. When these liposomes contain high amounts of cholesterol, e.g. molar ratios of phospholipid-cholesterol $\sim$ 1, they are very stable in the blood for at least 18 hours (12). The kinetics of the liposome uptake by the liver is fast ($t_{1/2} \sim 2$ min) and once in the liver, the radioactivity transported by liposomes decreases very slowly over periods of time (Fig. 1a).

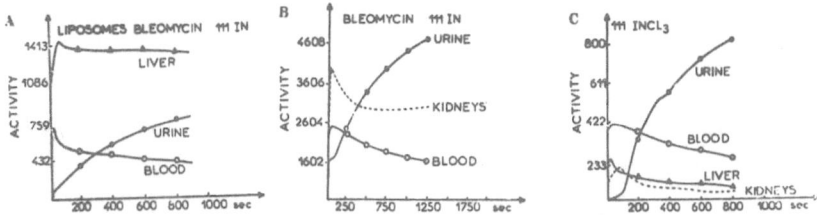

Figure 1 - Kinetics of the organ uptake of ^{111}In-bleomycin following i.v. injection to rats.   a) Liposome encapsulated ^{111}In-bleomycin ; b) Free ^{111}In-bleomycin ; c) ^{111}In Cl$_3$. Experimental methods as in reference 14.

II - *IN VIVO* EXPRESSION OF THE LIPOSOME-ENTRAPED RAT INSULIN GENE

The DNA segment encoding rat preproinsulin I which we have used in this study contains, in addition to the insulin coding and intervening sequences, several kilobases of flanking sequences on each side of the gene (13). This suggests that regulatory signals are also included within the insulin plasmid and that the gene might be functional when transferred *in vivo* after its entrapment in liposomes.

The results concerning the distribution of liposome-encapsulated ^{111}In-bleomycin indicated that, should the liposome-encapsulated insulin gene penetrate any cells, the target cells would be essentially in the liver and the spleen. Knowing now that the percentage of the total lipo-some-encapsulated radioactivity injected was decreasing very little in the target organs (liver and spleen) during the first 6 hours, animals were inoculated with the liposome-encapsulated insulin gene, liposome-encapsulated pBR 322 plasmid, "empty" liposomes and free p(gRl 9.4) plas-mid and were assayed 6, 10, 12 and 18 hours after inoculation for the expression of the gene (i.e. insulin level and blood glycemia).

The treated animals (5 in each of 2 independent experiments) and all the controls (2x2 for each type) had their blood glucose and blood, splenic and hepatic insulin assayed. Blood specimen were collected from the tail vein 6, 10, 12 and 18 hours after injection, using heparinized microtubes. The glycemia was measured immediately by the glucose-oxydase Microtest (Mercktest-Glucose, Merck, Darmstadt, FRG). Immunoreactive blood insulin was assayed by the method of Okajima and Ui on 0.1 ml plas-ma samples (17). Rat insulin (Novo Industries, Denmark) was used as a standard.

Table 1 indicates that injection of the liposome encapsulated insu-lin gene has a significant effect on the glycemia and insulin level in the animals having received it. Glycemia in the "treated animals" shows a significant decrease 6 hours after injection which corroborates quite well with the increase in the insulin blood level in these animals. At the same time, 6 hours after injection, the insulin found in the liver

Table 1 - Glycemia blood, liver and spleen insulin in control animals and in rats injected i.v. with the liposome-encapsulated insulin gene*.

Animals	Glycemia, mg glucose/100 ml blood ± SD					Blood insulin μU/ml ± SD					Liver[1] and spleen[2] insulin μU/g tissue ± SD				
	0 hr	6 hrs	10 hrs	12 hrs	18 hrs	0 hr	6 hrs	10 hrs	12 hrs	18 hrs	0 hr	6 hrs	10 hrs	12 hrs	18 hrs
Controls															
- "Empty" liposomes (4)	110±2	105±2	118±2	112±2	110±2	38±6	40±5	32±8	42±7	43±8	85±13[1], 118±15[2]	87±15[1], 112±20[2]	n.d.[**]	n.d.	84±12[1], 127±20[2]
- Free insulin gene (4)	106±2	109±2	n.d.	104±2	n.d.	41±5	45±5	n.d.	51±7	n.d.	n.d.[**], n.d.[**]	n.d.[**], n.d.[**]	n.d.[**], n.d.[**]	n.d.[**], n.d.[**]	n.d.[**], n.d.[**]
- Liposome-encapsulated pBR322 plasmid (4)	115±3	113±3	n.d.	107±7	n.d.	45±5	36±5	n.d.	46±7	n.d.	89±12[1], 125±20[2]	82±13[1], 121±18[2]	n.d.	n.d.	82±14[1], 123±20[2]
- Non injected (4)	103±2	107±2	116±2	110±2	103±2	40±6	43±5	45±6	50±5	45±7	n.d.[**], n.d.[**]	n.d.[**], n.d.[**]	n.d.[**], n.d.[**]	n.d.[**], n.d.[**]	n.d.[**], n.d.[**]
Liposome-encapsulated insulin-gene (9)	108±3	72±6	86±5	129±6	138±9		61±8	42±6	24±6	30±5	82±10[1], 115±18[2]	205±22[1], 242±20[2]	n.d.	n.d.	105±15[1], 143±20[2]

a) For each animal 2 determinations were performed and the SD was calculated for the average value per animal.

**) Since little variations of the blood glycemia and blood insulin level were observed during 18 hours with these controls, the assays of liver and spleen insulin were not performed.

and spleen is twice the amount in the controls. This insulin, found in the liver and spleen of control animals is the blood pancreatic insulin which is bound on the insulin receptors present on these cells (18-19). An interesting feature in Table 1 is the relatively short time of increased insulin synthesis observed with the "treated" animals either because of the ejection of the gene by the host cells or "quenching" of this gene. The glycemia starts increasing 10 hours after inoculation and 12 hours after injection it is higher than the control values, whereas the blood insulin decreases below the control values. The liver and spleen insulin have reached almost their normal values again, 18 hours after injection.

Another feature which emerges from Table 1 is the relative stability of all these values in the controls. Whatever systems were injected, i.e. free insulin-gene, empty liposomes or liposome-entraped pBR 322 gene, the glycemia and insulin values varied little over 18 hours. The results presented raise some questions :

1) How is it possible that non-specialized cells (apparently splenic macrophages and liver Kupffer cells, which take up the major fraction of i.v. injected liposomes (8-11)) express and secrete insulin in the blood ?

2) Would it be possible to send the gene to other cells as well - eventually perhaps target it to the Langerhans islets ?

3) Could the exogeneous genes be detectable in the spleen and liver cell by the Southern blotting techniques ?

We tried to provide answers, at least partly, to these questions :

1) In pancreatic beta cells insulin is first synthesized as an inactive precursor which is later activated intracellularly by the converting enzymes. Our natural target cells are non insulinogenic and thus probably do not possess the specialized converting enzymes present in the pancreatic beta cells (20). They are, however, quite rich in lysosomal proteases. Should the insulin gene be transcribed and translated in the splenic macrophages and in the Kupffer cells it seems quite conceivable that the signal peptide is removed by a peptidase present in these cells and that random processing occurs in the Golgi apparatus. Even if the proteases present in the pancreas beta cells, which specifically process insulin, are absent from splenic macrophages and Kupffer cells the processing of part of the proinsulin might occur.

It is quite difficult to explain though, how the secretion by the macrophages of the newly synthesized insulin takes place -macrophages have secreting properties- lymphokine secretion is well known- but this does not mean that a new protein, synthesized transiently by the cells must be secreted. We are investigating the possibility of insulin secretion by genetically "transformed" macrophages using monoclonal anti rat insulin antibodies and assaying this expression -the increased blood insulin levels- with such antibodies in other animal species.

In order to answer this intriguing question a suggestion has been recently made (M.Poznansky, personal communication, July 1982). According to this suggestion, the splenic and hepatic insulin should bind on the macrophage (and Kupffer cell) insulin receptors, so that the insulin requirements for these receptors are met. More pancreatic insulin would then be available in the blood and this could explain the physiological effects observed and answer the question of secretion. Again, the radio-immunoassay of the blood insulin with monoclonal anti rat insulin antibodies in other animal species having received i.v. the p(gR1 9.4) gene might help to elucidate these aspects.

2) Despite the insulin expression in the spleen and hepatic cells we were well aware that neither macrophages nor Kupffer cells would qualify as target cells for gene expression *in vivo*, should this expression be a lasting one. These cells do not divide, are terminal members of a cell "family" and are relatively short lived. Hepatocytes are long-lived, a significant fraction of them divide *in situ*, but as it has been shown recently, they contain a protease with a high specificity for insulin (21). However, we decided to attempt to target our liposomes towards hepatocytes and then check whether there is any significant uptake of the exogenous DNA by these cells. Ghosh *et al*. (22) had shown that the asialo-glycolipid GM_1, incorporated in the lipid bilayer of phospholipid liposomes, would target the liposomes to hepatocytes, apparently because of the asialoglycoprotein receptor present on hepatocytes plasma membranes (23). A very similar lectin has been reported on the Kupffer cells (24) and lectin-mediated endocytosis has been described for these cells (25). The existence of a specific sugar-binding site on the Langerhans islets-membrane would possibly permit the association of correspondingly targeted liposomes with these cells. In a recent *in vitro* study we demonstrated that rat Langerhans islets and renal cortex cells have on their plasma membrane a glycoconjugate receptor with specificity for lactose (26). The results of the *in vitro* work indicate that whereas PC:PS:Chol liposomes do not associate with isolated Langerhans islets the liposomes containing glycolipids (lactosylceramide) associate to a considerable extent ($\sim$ 35 % of the radioactivity of the liposomes is found associated with the cells, when the label is ^{14}C-oleyl-palmitoyl phosphatidyl choline (OPPC)).

In order to check whether the homing of liposomes was also associated with enhanced delivery of their contents to the target cells, we prepared liposomes containing phosphatidyl choline : phosphatidyl serine : cholesterol : asialo GM_1 (or lactosyl ceramide) in the molar ratios 8:2: 10:1. We encapsulated in those liposomes the rat preproinsulin I gene and injected them i.v. to rats and to mice. Prior to injection the liposomes were tested for the presence of the glycolipid and its binding properties. For this a trace of ^{14}C-OPPC was added to the liposomes, and after the encapsulation of the DNA according to (14) they were treated with DNase I and gel-filtered through a Sepharose 2-B column in order to separate encapsulated from non-encapsulated DNA (1).

The DNA-loaded liposomes were then passed through a Pea Nut agglutinin (PNA)-Ultrogel affinity column and the radioactivity of the non-selective eluate measured. The results are shown in Table 2.

Table 2 - Affinity of the GL-liposomes for Pea Nut agglutinin.

System	% of the liposomes adhering to the column
Asialo GM_1-liposomes	52.8
Lactosylceramide-liposomes	43.5
PC-PS-Chol liposomes	$\sim$ 0

3) Detection of exogenous plasmid DNAs in liver and spleen cells (27)

A subclone of p(gRl 9.4), designated p 007, was constructed by in-serting the central 2.5 Kb Bam HI-Hind III segment containing the insulin coding sequences into pBR 322. Swiss mice were injected with 150 ng of p 007, prior or following encapsulation in large PC:PS:Chol liposomes prepared by the ether-injection method (28) ; liposomes containing a 1/10 molar ratio of lactoceramide (LC) to phospholipids were used as "targeted" liposomes. Total DNA was extracted from the liver and spleen of these mice 5 hours following intravenous inoculation and subjected to Southern blot analysis following digestion with Pst I (Fig. 2).

In the case of the liver DNAs, the samples recovered from mice injec-ted with phospholipid and LC liposomes revealed the presence of the two bands, 4.7 Kb and 1.8 Kb, expected from Pst I cleavage of p 007, in addi-tion to several other bands also present in the controls and presumably corresponding to endogenous cross-hybridizing insulin gene sequences. In the case of the spleen DNAs, these two bands could only be detected in those samples originating from mice injected with LC liposomes.

The influence of the presence or the absence of lactoceramide in li-posomes containing the plasmid DNA, as detected with spleen DNAs, suggests that different type of cells in the spleen -and perhaps in the liver- might react with both types of liposomes. Since it is well documented that phospholipid liposomes are essentially taken up by the Kupffer cells in the liver (11), and since a galactose binding lectin, which might also react with lactoceramide, has been observed on the surface of hepatocytes (23), it seems possible that the LC liposomes would be taken up, at least in part, by hepatocytes. In order to check this possibility, rats were injected with 3 µg of free or LC-liposome entraped p(gRl 9.4) plasmid, livers were removed 5 hours following intravenous inoculation, and hepa-tocytes were purified following digestion of Kupffer and endothelial

1 2 3 4 5 1 2 3 4 5

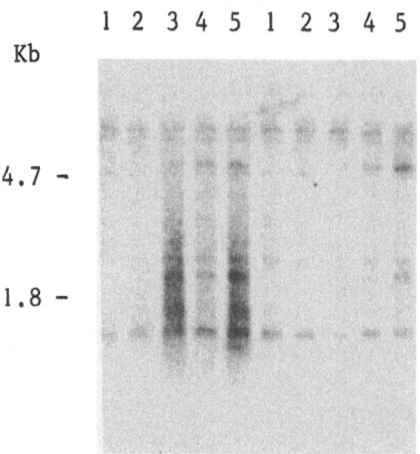

Kb

4.7 -

1.8 -

Liver   Spleen

Figure 2 - Detection of liposome transferred plasmid DNA in liver and
spleen from mice injected with p 007. 20 µg of DNA from mice 1) non injected, 2) injected with the free p 007, 3) injected with liposome-entraped p 007, 4) and 5) injected with LC-liposome-entraped p 007 (two different animals) were digested with Pst I, fractionated on a 0.6 % agarose
gel, transferred to nitrocellulose filter and hybridized with nick translated p 007 as described in Ref. 29. Plasmid bands are indicated.

cells with collagenase followed by metrizamide density gradient centrifugation (30). DNA was extracted from the purified hepatocytes, digested
with Pst I and subjected to Southern blot analysis by hybridization with
p 007 (p(gRl 9.4) contains a repeated sequence which does not allow it
to be used as a probe). Hepatocyte DNA recovered from rats injected with
LC liposomes shows three bands, 8.7, 2.1 and 1.1 Kb, expected from Pst I
cleavage of p(gRl 9.4) and homologous to sequences in p 007, as well as
endogenous bands, also present in the controls, and corresponding to the
endogenous rat insulin I and II gene sequences (Fig. 3). Since these
genes are present as single copy in the rat genome, the relative intensity of the exogenous plasmid bands suggests that the foreign DNA is present, on the average, in multiple copies per cell.

     Detection of foreign DNA in both liver and spleen cells raises in
turn additional questions which are currently under investigation. The
first question concerns the intracellular distribution of the plasmid
DNAs ; it seems unlikely that the liposomes containing the DNA would
remain adsorbed on the surface of the cells, since these were recovered
five hours after injection and in any case they would have been removed
from the cell by collagenase treatment. The plasmid DNA may therefore
either be located in the nucleus, where it could lead to some degree of
expression, or in the cytoplasm, entraped or not in liposomes. If however
specific gene transcripts are found, this will imply that at least part

of the plasmid DNA which enters the cell finds its way to the nucleus
by liposome transport to the nuclear membrane ; liposome release in the
cytoplasm would then be excluded since direct microinjection of cloned
genes into the cytoplasm does not lead in any case to the expression of
the gene (31). The Southern blot data does not allow to distinguish between these possibilities ; nuclear fractionation and detection of RNA
transcripts should allow us to better understand these processes.

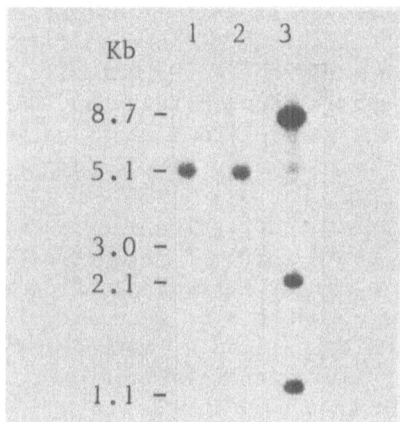

Figure 3 - Detection of LC liposome transferred plasmid in hepatocytes
from rats injected with p(gRl 9.4). 5 µg of hepatocyte DNA from rats
1) non injected, 2) injected with the free p(gRl 9.4) and 3) injected
with LC-liposome entraped p(gRl 9.4), were digested with Pst I, fractionated on a 0.6 % agarose gel, transferred to nitrocellulose filters and
hybridized with nick translated p 007 as described in ref. 29. Plasmid
bands are 8.7, 2.1 and 1.1 Kb ; bands corresponding to endogenous rat
preproinsulin I and II genes are 5.1 and 3.0 Kb, respectively.

        The second point of interest concerns the state of the plasmid in
the cell : is the plasmid maintained in a free form or is it integrated
in the genome ? A partial answer to this question is obtained by perfor-
ming Southern blots on the undigested DNAs ; this analysis reveals that
a certain proportion of the DNA is not integrated in the genome since a
hybridization band of higher mobility than the non digested DNA can be
detected. This result does not preclude integration of a small amount of
the plasmid DNA. The results obtained in vitro with the calcium phosphate
technique shows that independent transformants are transformed differently (32). This suggests that different cells within a target organ would
also contain plasmid DNA in different states ; the Southern blot analysis performed cannot provide an answer since the expected -and observed-
plasmid band pattern will be an average of what is found in all transformed cells, e.g. stoechiometric plasmid bands ; along the same line,

further restriction analysis would only reveal the restriction map of the plasmid. An answer to this question might be obtained by transfecting animals with another gene -a dominant selectable gene- and, by culturing clones (e.g. from hepatocytes subsequently transformed by SV40 or isolated from hepatoma rats) on the appropriate selection media, of the cells transformed *in vivo*. Analysis of the DNA from a sufficient number of clones might help with providing an answer.

A third point raised by these studies has to do with the short time during which expression is observed. Is this transient expression due to ejection of the gene or to a "quenching" of some sort ? The study of the presence of the foreign DNA in the transfected animals at different time points after the inoculation should show whether the short time expression is due to "ejection" of the gene or to a modification such as methylation (32) or to a change in chromatin structure.

Lastly, the use of appropriate expression vectors might circumvent many problems. Long terminal repeats (LTR) of retroviruses might enhance integration of the genes in the target cells (34). Promoter sequences adapted to the target cells may produce a selected regulation of the gene. The long term aim of these studies will be a lasting and controllable expression of foreign genes in living animals.

III - CONCLUSIONS

The work reported here presents the expression of a gene in adult animals after i.v. inoculation of the liposome-encapsulated gene. The liposome distribution in the organism is reported, as well as the physiological effects of the gene expression. Southern blot analysis indicate the presence of the exogeneous DNA in the liver and spleen cells. Use of "targeted" liposomes, containing the glycolipid GM_1 or lactosyl ceramide permitted the transfer of the insulin gene to hepatocytes, where it was found by Southern blot analysis.

The data presented raise a number of questions concerning the targeting of the liposomes, the transient nature of the gene expression, the possible secretion of the expressed insulin by splenic macrophages and Kupffer cells, the mechanisms of the termination of expression, etc... These aspects and others was well are currently under investigation.

ACKNOWLEDGEMENTS

This report presents work performed in the Group de Biophysique Cellulaire at the Centre de Biophysique Moléculaire in Orléans. The following have contributed to the presented results : Françoise Fargette, Danielle Gagliardi, Elisabeth Grosse, Marie-Françoise Juhel, Alain Legrand and Alain Le Pape.

REFERENCES

1. Wong, T.K., Nicolau, C. and Hofschneider, P.H.: 1980, Gene 10, pp. 87-94.
2. Fraley, R., Subramani, S., Berg, P. and Papahadjopoulos, D.: 1980, J. Biol. Chem. 255, pp. 10431-10435.
3. Schäffer-Ridder,M., Wang, Y.C. and Hofschneider, P.H.: 1982, Science 215, pp. 166-168.
4. Nicolau, C. and Rottem, S.: 1982, Biochem. Biophys. Res. Commun. (in the press).
5. Fraley, R., Straubinger, R.M., Rule, G., Springer, E.L. and Papahadjopoulos, D.: 1981, Biochemistry 20, pp. 6978-6987.
6. Sené, C. and Nicolau, C. : in "Liposomes, Drugs and Immunocompetent Cell Functions", C. Nicolau and A. Paraf eds., Academic Press, London 1981, pp. 67-79.
7. Nicolau, C. and Sené, C.: 1982, Biochim. Biophys. Acta (in the press).
8. Jonah, M.M., Cerny, A. and Rahman, Y.E.: 1975, Biochim. Biophys. Acta, 401, pp. 336-348.
9. Osborne, M.P., Richardson, V.J., Jeyasingh, K. and Rahman, Y.E.: 1979, Int. J. Nucl. Med. 6, pp. 75-83.
10. Gregoriadis, G. and Senior, J.: 1980, FEBS Letters 119, pp. 43-46.
11. Roerdink, F., Dijkstra, J., Hartman, G., Bolscher, B. and Scherphof, G.: 1981, Biochim. Biophys. Acta 677, pp. 79-89.
12. Kirby, C., Clarke, J. and Gregoriadis, G.: 1980, Biochem. J. 186, pp. 591-598.
13. Cordell, B., Bell, G., Tischer, E., De Noto, F.M., Ullrich, A., Pictet, R., Rutter, W.J. and Goodman, H.M.: 1979, Cell, pp. 533-543.
14. Nicolau, C., Le Pape, A., Soriano, P., Fargette, F. and Juhel, M.F.: 1982, Proc. Natl. Acad. Sci. USA (in the Press).
15. Hinkle, G.H., Born, S.G., Kessler, W.V. and Shaw, M.S.: 1978, J. Pharm. Sic. 67, pp. 795-798.
16. Nicolau, C., Le Pape, A., Soriano, P., Fargette, F., Juhel, M.F., Muh, J.P.: 1982, Proc. FEBS Spec. Meeting, Athens, A. Liss, ed. (in the press).
17. Okajima, S. and Ui, M.: 1978, Amer. J. Physiol. 234, pp. 106-111.
18. Kahn, C.R. and Roth, J.: 1975, J. Amer. J. Clin. Pathol. 63, pp. 656-668.
19. Freychet, P., Roth, J., Neville, D.M.: 1971, Proc. Natl. Acad. Sci. U.S.A. 68, pp. 1833-1837.
20. Chan, S.J. and Steiner, D.F.: 1977, Trends Biochem. Sci. 2, pp. 254-258.
21. Freychet, P., Khan, R., Roth, J. and Neville Jr. D.M.: 1972, J. Biol. Chem. 247, pp. 3953-3961.
22. Ghosh, P., Das, P.K., Bachhawat, B.K.: 1982, Arch. Biochem. Biophys. 213, pp. 266-270.
23. Ashwell, G. and Morell, A.G.: 1974, Adv. Enzymol. 41, pp. 99-128.
24. Kolb, H., Vogt, D., Herbertz, L., Cerfield, A., Schauer, R. and Schlepper-Schäfer, J.:1980, Hoppe-Seyler's Z. Physiol. Chem. 361, pp. 1747-1750.
25. Kolb-Dachofen, V., Schlepper-Schäfer, J. and Vogell, W.: 1982, Cell 29, pp. 859-866.

26. Nicolau, C., Monsigny, M., Le Pape, A., Roche, A.C. and Grosse, E.:
    1982 (submitted for publication).
27. Soriano, P., Londos-Gagliardi, D., Legrand A. and Nicolau, C.: 1982,
    manuscript in preparation.
28. Deamer, D. and Bangham, A.D.: 1977, Biochim. Biophys. Acta 443, pp.
    629-634.
29. Soriano, P., Szabo, P., Bernardi, G.: 1982, Embo J. 1, pp. 579-583.
30. Munthe Kaas, A.C. and Seglen, P.O.: 1974, FEBS Letters 43, pp. 252-
    256.
31. Cappechi, M.R.: 1980, Cell 22, pp. 479-488.
32. Pellicer, A., Wigler, M., Axel, R. and Silverstein, S.: 1978, Cell
    14, pp. 133-141.
33. Weintraub, H., Larsen, A. and Groudine, M.: 1981, Cell 24, pp. 333-
    344.
34. Temin, H.M.: 1981, Cell 27, pp. 1-3.

II - PROTEIN STRUCTURE AND DYNAMICS

THEORETICAL AND EXPERIMENTAL CONFORMATIONAL ANALYSIS OF SOME SYNTHETIC AND NATURAL PEPTIDES

Marius PTAK, Françoise VOVELLE and Monique GENEST
Centre de Biophysique Moléculaire (CNRS) et Université
d'Orléans, 1A avenue de la Recherche Scientifique,
45045 ORLEANS Cedex (France)

Abstract.- Several selected examples of theoretical and experimental
conformational studies are cited to show how intra- and intermolecular
interactions determine the conformations of small peptides. A combination of empirical energy computations and of solute-solvent interactions
modelling enable us to display the connections existing between calculated and experimental conformations of a series of cyclic dipeptides
containing the Asp or Glu, His, Ser or Thr residues. Starting from this
study, we have calculated the stable configurations of a model of the
active site of serine proteases. Remarkable similarities are found
between some calculated configurations and the experimental ones for
α-chymotrypsin and subtilisin Novo. All the results disagree with the
existence of a charge relay system. In a last part, a NMR study of natural lipopeptides is reported, which allows to show how sequences and
conformations are determined by combining several spectroscopic parameters. The relationships between conformations, self-association properties and the channel forming capacity of such compounds are briefly
discussed.

INTRODUCTION.

Natural peptides form a broad family of intermediate molecular
weight compounds exhibiting a large diversity of composition, conformation and biological activity. With the twenty L aminoacids usually found
in proteins and some others (several aminoacids from bacterial origin
are D), one can form a very high number of sequences containing from
two to some dozen of residues. The backbone of these sequences contains
the rigid stereochemical unit -CONH- which is allowed to rotate (ϕ,ψ
angles) around the C^{α} joint (fig.1). The conformational specificity and
the specificity of interaction is given by side chains which are characterized by their hydrophobicity or hydrophilicity, by their polarity,
their steric hindrance, their flexibility (χ_i angles), etc... The biological activity of these peptides is based on their capacity to specifically interact with macromolecular receptors (nucleic acids, proteins),

C. Hélène (ed.), Structure, Dynamics, Interactions and Evolution of Biological Macromolecules, 209–225.
Copyright © 1983 by D. Reidel Publishing Company.

with membranes or, on the other hand, with ions and small molecules. There are close connections between composition, conformation and activity.

The conformational analysis is a first obligatory step in the understanding of what is generally called "structure-activity relationships". In the past decade, a spectacular renewal of peptides studies came with the increasing recognition that various hormones are peptides. More than two dozen of peptides are neurotransmitters candidates [1] and there is a very intense activity in studying their biosynthesis and their receptors and in synthetizing analogues and derivatives capable of having pharmacological activities. Another considerable interest is in peptides and depsipeptides having antibiotic activity among which are found ionophores which selectively complex cations and mediate their translocation through membranes. Natural peptides also include different toxins, numerous substrates, inhibitors and more generally effectors.

Synthetic peptides are generally investigated as models and analogues of natural peptides. The conformational analysis of such peptides has played a fundamental role in the understanding of conformational properties of proteins inasmuch as the secondary structure of a polypeptide chain originates in short intramolecular interactions [2]. The stable conformations of short peptide sequences define a stereochemical code [3] which is at present one of the basis of the structural analysis of proteins.

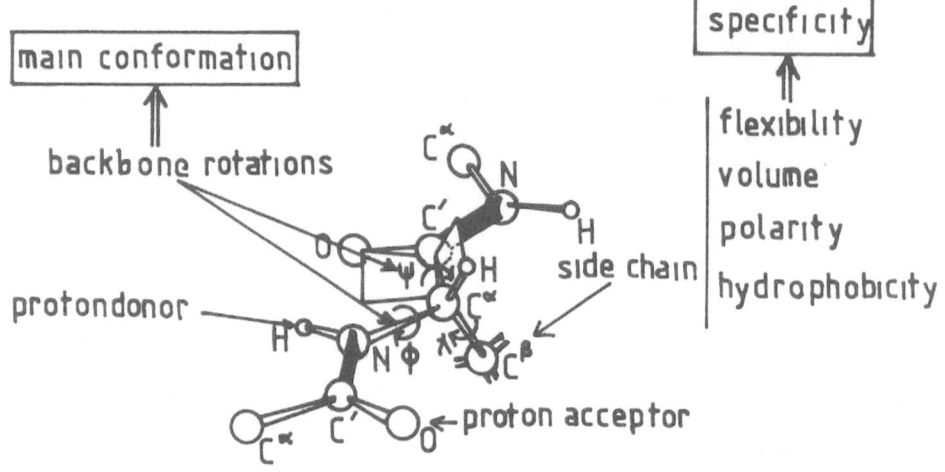

Figure 1. A peptide fragment labelled according to IUPAC-IUB conventions (see : Biochemistry (1970), 9, 3471-3479).

Interesting enough is to recall that the theoretical analysis of the conformations of biomolecules in its present state, has its origin in the works of Ramanchandran et al. [4] (hard spheres method) and of Liquori [5] (first empirical potentials) on peptides. In parallel with

empirical and semi-empirical methods of computation of the intramolecular energy, sophisticated quantum mechanical methods have been developed,
such as *ab initio* methods currently used at present for small molecules
[6]. The more recent developments deal with molecular systems of bigger
size and with environmental effects on conformations [7].

Among the experimental methods currently used for determining conformations in solution, Nuclear Magnetic Resonance (NMR) is certainly
one of the more efficient. Peptides having few hundred to few thousand
molecular weights define an ideal range of applications for this spectroscopic method [8] which has made spectacular progress for last years.
Direct informations on the three dimensional structure of small proteins have been recently obtained [9] which are complementary of crystallographic data.

In our contributions to developments of theoretical and experimental conformational analysis, we have selected several results of computations and of NMR studies proving to be effective in elucidating conformations of different kinds of peptides. In a first part we will describe a combined analysis of small synthetic peptides in which we will
demonstrate how intra- and intermolecular interactions dictate stable
conformations. Rather clear examples will prove the validity of classical empirical computations as compared with experimental data. In a
second step, we will show how such computations can help us to approach
the more complex problem of inter-residues interactions within a model
of the active site of serine proteases. In a third step, we will report
briefly some results of a NMR study of a new class of natural peptides,
activity of which is certainly closely related to their conformations
and self-association modes.

CONFORMATIONAL ANALYSIS OF SMALL SYNTHETIC PEPTIDES.

Conformations in the solid state are currently determined by X-rays
crystallography whereas those existing in solution are analyzed by
using NMR and other spectroscopic methods. In both cases these conformations result from a combination of intra- and intermolecular interactions. Classical energy computations lead to stable conformations corresponding to minima of intramolecular energy of an isolated molecule
immersed in an homogeneous medium. It is a permanent care to compare
these two approaches that explain several recent attempts to include
environmental effects in energy computations [7].

In the field of methodology, we have focussed our efforts on such
a problem for the last few years by investigating a series of small
synthetic peptides containing Asp (or Glu), His and Ser (or Thr) residues. Asp, His and Ser residues are present in the active site of serine
proteases in which they interact with the substrate and participate in
the catalytic reaction. Since the resolution of the three dimensional
structure of α-chymotrypsin by Birktoft and Blow [10] a famous model of
"charge relay" and "proton transfert" was proposed to account for the
special reactivity of the Ser residue and for the mechanism of the
amide bond cleavage. This model is now more and more challenged especially with regard to the existence of an hydrogen bond connecting the

His and Ser residues and to the protonation state of the Asp residue. That stimulated us to undertake a theoretical and experimental investigation of (Asp-His) and (His-Ser) interactions in small peptides by following two guiding lines : a) the relationship between the own conformations of these residues and the intermolecular interactions in which they are engaged ; b) the relationship between the interactions within a peptide and within a model of the active site.

We have synthetized for that a series of cyclic dipeptides : c(Thr-His), c(Ser-His), c(Glu-His) and c(Asp-His) in which the side-chains could establish specific interactions (all the results reported here concern *L-residues*. D-residues have been studied elsewhere [14]). These peptides have been investigated experimentally in the solid state and in solution and theoretically by using empirical methods to calculate intra- and intermolecular interactions energy.

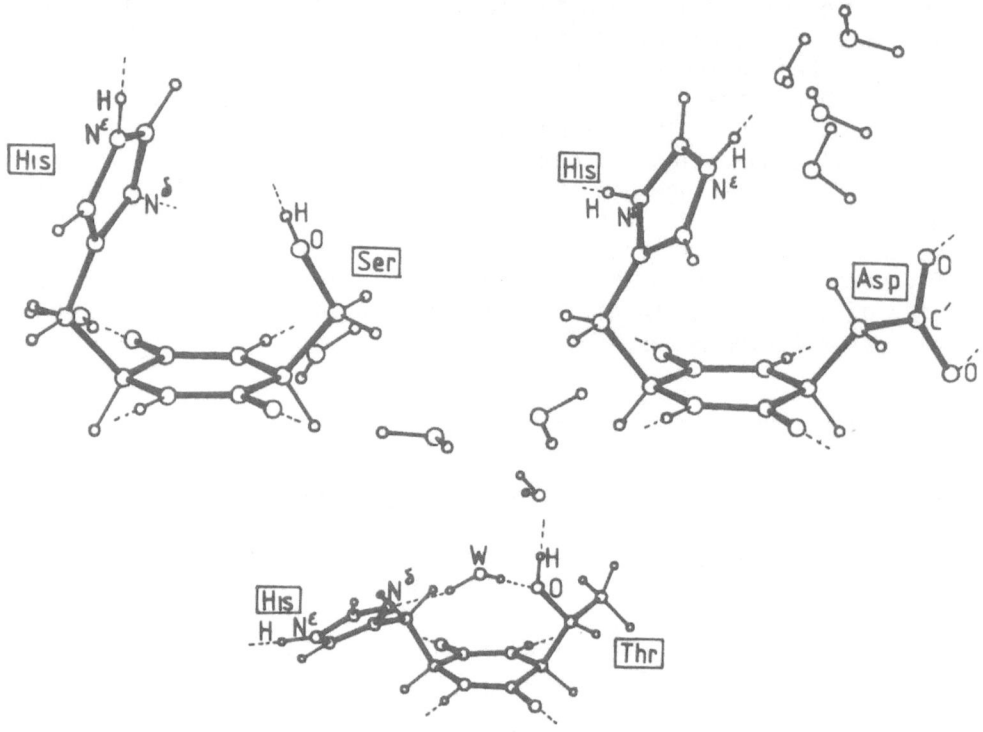

Figure 2. Conformations in the solid state of c(L-His-L-Ser), c(L-His-L-Thr) and of c(L-His-L-Asp) (after Ramani et al. [13]).

Conformations in the solid state.

Conformations in the solid state of c(Thr-His) [11], c(Ser-His) [12] and of c(Asp-His) (established by Ramani et al. [13]) are represented in the figure 2.

The conformation of c(Ser-His) is characterized by a folding of both side-chains above the central diketopiperazine ring without direct interaction between the Ser hydroxyl group and the His imidazole ring. There is only one H_2O molecule per peptide molecule which fills a cavity in the crystal lattice and connects different peptide molecules. An interesting difference is found in the conformation of c(Thr-His) : the Thr side chain is folded whereas the His one is an open position and interacts with a H_2O molecule which is hydrogen bonded on the other side to the Thr OH group. The peptide is dihydrated ; water molecules establishing intra- and intermolecular bridges as discussed later. Another situation occurs for c(Asp-His) in which the His side chain is folded whereas the ionized Asp side chain is in an open position and strongly interacts with neighbouring water and peptide molecules. There are now three water molecules per peptide molecule which form a rather complicated intermolecular lattice. A first simple idea emerges from the comparison between these three structures : the Thr, Ser, Asp and His side chains seem to have privileged conformations which are open and folded forms depending on intermolecular interactions including hydration. The validity of such an assumption will be now tested by determining conformations in solution and calculated conformations of the same molecules.

Conformations in solution.

The conformations in aqueous solution have been analyzed by using ^1H and ^{13}C NMR [14]. Averaged conformations are related to the $^3J = HC^\alpha - C^\alpha H$ coupling constants. For the Thr residue, we have taken advantage of the possibility to incorporate a ^{13}C enriched ($\simeq 85\%$) residue in the peptide, for selecting unambiguously the predominant conformations. For this residue, there is a large predominance of the folded form $(\chi_1 \simeq 60°)$ in H_2O and DMSO independently from the protonation state of the imidazole ring. The folding is less complete for the Ser side chain and an open form $(\chi_1 \simeq -60°)$ predominates for Asp.
For the His chain, the equilibrium between folded and open forms depends on the protonation state of the imidazole ring and on the structure of the second side chain. A completely extended form is never found, in spite of the very strong solvation that one could expect to find for imidazole or imadazolium ion. Environmental effects seem to be very weak for Thr and, to a lesser extent, for Ser side chains which remain folded in aqueous solution. It is quite clear that conformations existing in solution are averaged conformations because of solvation effects and thermal motions. Nevertheless, the existence of folded and open forms seems to be rather well confirmed, though their geometry cannot be accurately determined by NMR.

Minima of intramolecular energy.

The conformational behaviour of these peptides and some aspects of their interactions with water should be more easily understood by considering energy computations. Classical empirical methods have been used first to calculate intramolecular energy. The limits of such methods

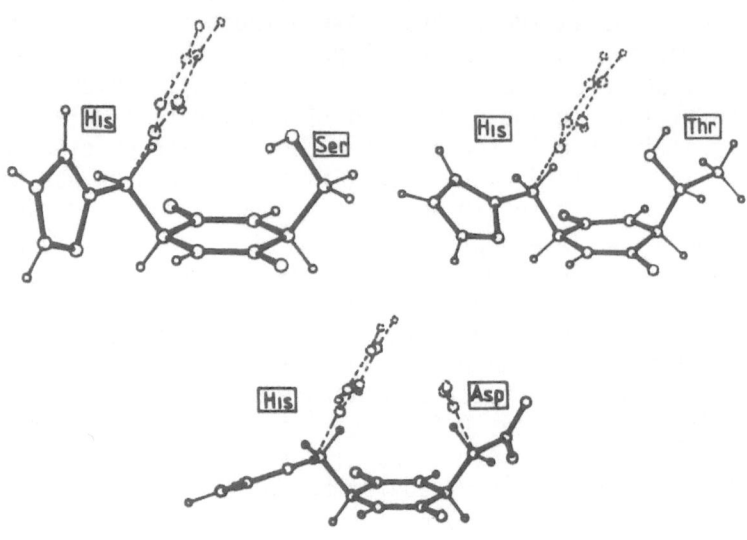

Figure 3. Averaged conformations in aqueous solution of c(L-His-L-Ser),
c(L-His-L-Thr) and of c(L-His-L-Asp). The dotted line indicates that
there is an equilibrium between folded and open forms.

have been often discussed especially with regard to additivity, trans-
ferability and accuracy [2,6]. Let us consider an isolated molecule
surrounded by an homogeneous dielectric medium, permittivity of which
is $1 < \varepsilon_r < 4$. The minima of intramolecular energy of side chain - side
chain and side chain - peptide backbone interactions define a set of
stable forms which are depicted in the figure 4 for c(Thr-His). More
generally, for all investigated peptides [15,16], two classes of conformations have been found for each category of side chains. Folded
forms ($x_1 \simeq 60°$) for which the side chain faces the diketopiperazine
ring. Such a position corresponds to a minimum of the energy of rotation around the $C^\alpha-C^\beta$ bond which is strengthened by 1) interactions of
the polar group with the diketopiperazine ring ; 2) interactions
between the two side chains. For open forms, one finds a secondary minimum ($x_1 \simeq 180°$) corresponding to   completely extended conformations
and more stable positions for which the polar group strongly interacts
with the NH or CO groups of the peptide backbone. A remarkable analogy
appears between the bifolded form found in computations and the conformation of c(Ser-His) in the crystal (fig.2). For this peptide the minimum of intramolecular energy coincides with the global minimum of the
three dimensional lattice. In other words, intermolecular interactions
are without appreciable effects on the positions of side chains determined by short internal interactions. For the other two peptides, these
intermolecular interactions have more discriminating effects : a) in
c(Thr-His), the His side chain takes an open position ($x_1 \simeq -60°$) principally because of interactions with water ; b) in c(Asp-His) strong
ionic interactions and hydration force the Asp side chain to take an

open position.

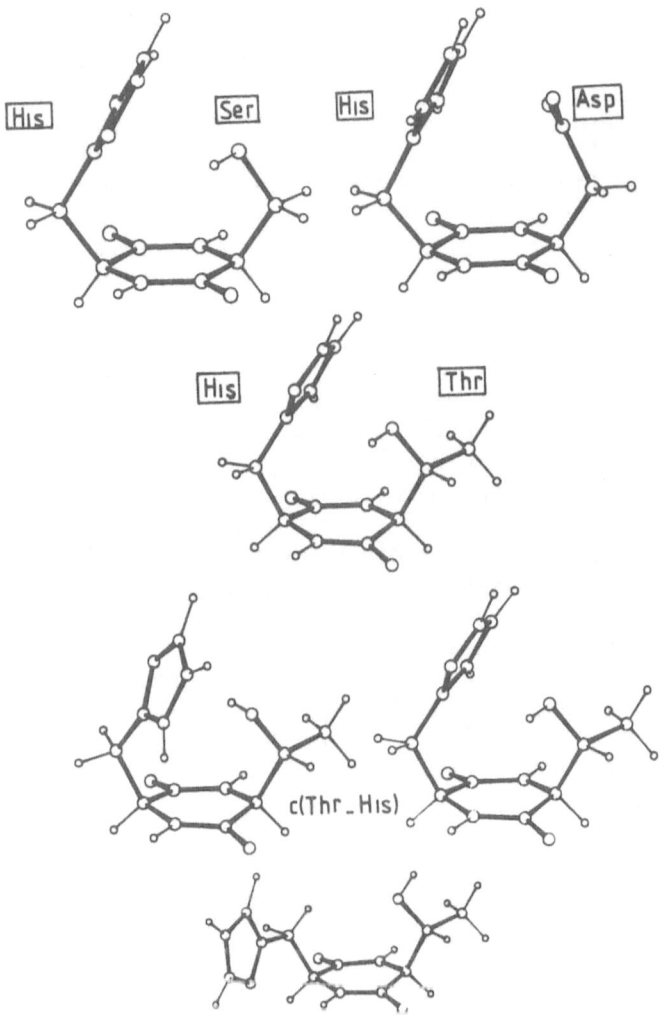

Figure 4. Above : the stablest calculated conformations of the three
cyclic dipeptides are bifolded forms. Below : three main stable conformations of c(L-His-L-Thr) corresponding to minima of the intramolecular
energy.

     A fundamental concept seems to emerge from these very simple con-
siderations : minima of intramolecular energy define a set of stable
conformations which are then "selected" by environmental effects in
order to be compatible with the organization of a crystal lattice or of
a solvent.

Solute-solvent interactions.

     Molecular dynamic methods [17] and Monte-Carlo methods [18] have
been proposed to describe the dynamics of a peptide-water systems. Such

methods seem to be specially suitable to analyze the fluctuations of
the organization of the solvent around the solute and also to define
the concept of "statistical hydration site" [19]. They have the great
disadvantage to be very expensive and their use seems to be reserved
– at the moment – for some special small systems. More simple approaches are then required, especially when structural features are wanted
for comparison with experimental results issued from crystallography
and spectroscopy. That is the reason why we have developed two empirical approaches based firstly on the concept of accessible surface area
and secondly on the concept of supermolecule.

Accessible surface areas of proteins have been first investigated
by Lee and Richards [20] in an attempt to describe the interactions of
these globular macromolecules with water. Though this method seems
especially suitable for describing the morphology of macromolecules,
Ponnuswamy et al. [21] have explored some possibilities of application
to small peptides. Following a very similar line we have investigated
our cyclic peptides with a special emphasis on c(Thr-His), conformations
of which have been previously carefully analyzed. We have calculated
accessible surface areas of the more stable conformations including the
experimental one [22]. Particularly interesting is to examine how accessible surface areas of polar atoms depend on the conformation. In
aqueous solution, these atoms should have nearly optimized interactions
with water, i.e. should have a high accessibility. Superposition of all
accessibility maps shows immediately that one cannot reach simultaneously the maximum accessibility for all polar atoms. One of the best com-
promise is found for a half folded form (fig.5) for which accessible
surface areas of N^δ(His) and $O^\gamma H$(Thr) have intermediate values around
16 Å^2 (for an hydrogen bond contact with one water molecule, this area
should be $\simeq$ 12 Å^2).

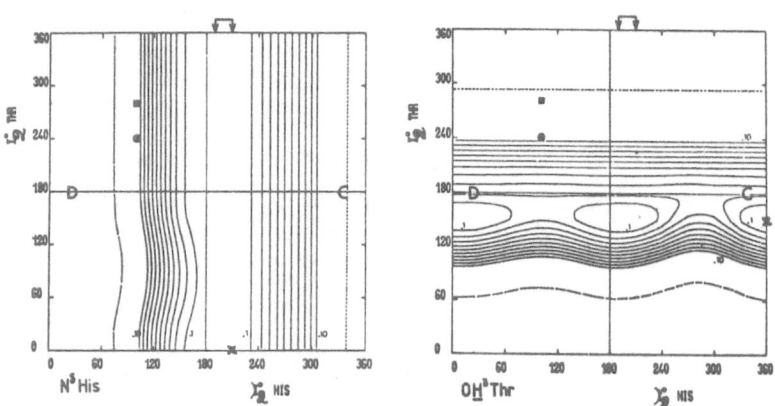

Figure 5. (χ_2His,χ_2Thr) accessibility maps of N^δ and $O^\gamma H$ atoms for a
semi-folded form : χ_1His = 300°, χ_1Thr = 60°. ●: conformation in the solid state, X = maximum of accessibility, ⬇⬇ = low stability.

An oversimplified assumption could associate hydration and maximum
accessibility. Clearly, the correlations existing between the accessible areas of neighbouring polar atoms determine conformations which are
not totally open forms. One can compare such a qualitative conclusion

to the striking feature previously noted according to which the Thr
side chain is nearly totally folded in water. Another parameter has
been considered which is the ratio of hydrophilic to hydrophobic accessible surface areas. This ratio has been found to be minimum for a bifolded form very similar to those found for c(Ser-His) in the crystal.
This peptide is the least hydrated one, since only one water molecule
is present in the lattice. Under these conditions and in spite of the
existence of several intermolecular hydrogen bonds, the partial hydrophobicity of the molecule contributes more significantly to determine
its conformation in a crystal lattice.

From the previous brief discussion , it clearly appears that the
accessible surface area method is a very qualitative one inasmuch as it
does not give any information on energy of hydration and on the structure of hydration sites. By taking one's inspiration from classical
energy computations on peptides, we have developed an empirical method
which enables us to describe more quantitatively peptide-water interactions.

One of the more popular models of water molecule is the ST2 model
introduced by Rahman and Stillinger [23]. The H_2O molecule is represented by four point charges localized on H atoms (positive charges) and
on the axis of oxygen lone pairs orbitals (negative charges). By analogy with this model we have localized on the axis of lone pairs orbitals
of O and N atoms in peptides fractional point charges, global value of
which is equal to the atomic charge usually considered in empirical
calculations [24]. That allows to confer some directivity to intermolecular hydrogen bonds. Empirical potentials have been then ajusted by
refering to the structure of the water dimer, to experimental data on
hydration of crystallized peptides and by comparing with several quantum mechanical calculations [25]. The general form of these potentials
is :

$$U = \frac{1}{4\pi\epsilon_0} \sum_i \sum_j \frac{q_i q_j}{R_{ij}} + \sum_i \sum_j \frac{A}{r_{ij}^{12}} - \frac{C}{r_{ij}^6} + \sum_i \sum_j \frac{A'}{r_{ij}'^{12}} - \frac{C'}{r_{ij}'^6} \quad_{HB}$$

R_{ij} = distances between q_i (peptide) and q_j (water) point charges.
r_{ij} = distances between i (peptide) and j (water) atoms non engaged in
       H-bonds.
r_{ij}' = distances between i (peptide) and j (water) atoms engaged in H-
       bonds.
A set of *ad hoc* (A,C) and (A',C') coefficients [25] allows to calculate
peptide-water interactions in a supermolecule. For shell computations,
a polarization term was added in order to account for the non pair
additivity of H-bonds in water.

In spite of the empirism of such a model, interesting conclusions have
been reached about structural aspects of peptide hydration [26]. In a
first step, an hydration site has been defined as the position of the
water molecule corresponding to a minimum of the solute-solvent energy
as it is classicaly done in the supermolecule method [27]. Such a static definition of an hydration site could be compared to the dynamical
picture emerging from dynamics computation [19]. Normally a deep minimum
of the solute-solvent interaction energy should correspond to a maximum

of probability of configuration of the system. That will be well illustrated below ; the existence of static hydration site in the solid state
will be also well proved.

From the exploration of the different kinds of interactions of one
water molecule with model molecules and peptides [26], two categories
of static hydration sites have been defined : a) individual sites in
which one H_2O molecule interacts with only one polar group of the solute
through one (or two for carboxyclic groups for instance) hydrogen bond.
b) water bridges in which one (or in special cases two and even three)
molecule of water interacts simultaneously with several polar groups of
the solute. A very remarkable example of one water bridge is found with
c(Thr-His) for which one H_2O molecule can be inserted between the two
side chains as shown in the figure 6.

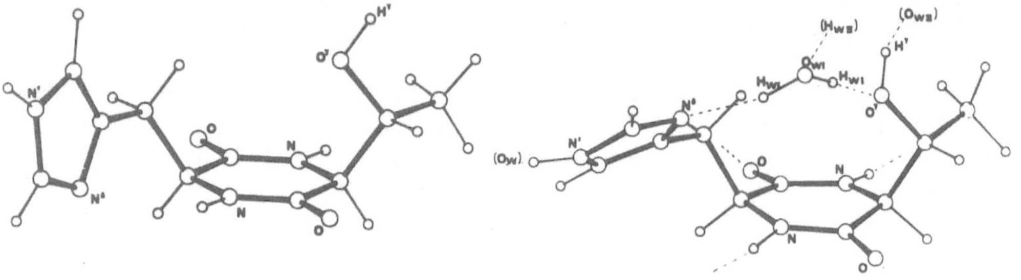

Figure 6. Left : a calculated stable conformation of c(L-His-L-Thr).
Right : a calculated supermolecule including a water bridge is quasi-
identical to the hydrated molecule existing in the crystal.

To get this very specific mode of hydration, the Thr side chain
remains folded in a position corresponding to a minimum of intramolecular energy, whereas the His side chain needs to be slightly rotated
from one of its stable position in order to interact with the water
bridge. The formation of such an hydration site requires only a small
loss of intramolecular energy which is compensated by a gain of the
global energy, i.e. there is a stabilizing effect of hydration. This
hydrated peptide can be inserted into a crystal lattice where a
second water molecule fills the remaining gaps.

The next step deals with the organization of the first hydration
shells. As previously said, the global organization of the solute-
solvent system is dynamical and involves a very fluctuating hydrogen
bonds network linking water molecules. Nevertheless, for water which is
in close contact with the peptide, strong interactions with proton
donor or acceptor groups should considerably reduce these fluctuations.
It seems then relevant to use static methods to establish some basic
structural features in this hydration process. By covering a peptide
molecule blocked in a given conformation with water molecules and after
minimizing the total energy, one gets possible configurations of the
first hydration shell. The addition of a second shell allows to have a
first idea about the effect on the surrounding "bulk" solvent on the

organization of these water molecules interacting with the solute. In
spite of obvious deficiencies of such a method, due to the difficulty
to define realistic boundary conditions, several important features are
worth discussing. Though the minimum of energy of the peptide - first
hydration shell system is broad, several constants are formed among a
lot of different allowed configurations. Proton donor and acceptor
groups act as anchoring points for water which organizes itself from
these points. A "radial" organization of water around polar groups [28]
contrasts with a "shell" organization around hydrophobic groups. Such a
particular arrangement is not maintened in the successive hydration
shells. That is in relatively good agreement with one of the results
of dynamic computations [17] : the influence of the solute on the solvent decreases rapidly and does not exceed the second or the third
hydration shell.

      Dynamic computations are not available for our cyclic dipeptides,
therefore we cannot compare static hydration sites defined by minimization methods and statistical sites. Such comparison is quite possible
for a linear protected dipeptide : Ac-LAla-LAla-Me that we have investigated elsewhere [26]. In this case, the most stable static sites coincide with several of the more probable sites found in a Monte-Carlo simulation [18]. That supports the validity of static methods for establishing the basic mechanisms of small peptides hydration.

Interactions within a model of the active site of serine protease.

      Another aim of the previous studies is to clarify the question of
interactions between His and Asp and Ser and His residues within the
active site of serine proteases. Let us recall that if strong interactions have been observed in dipeptides between His and Asp (neutral and charged residues), specific bonding has never been detected between His and
Ser, though such an interaction is sterically allowed. A simplified model of the active site of α-chymotrypsin and subtilisin Novo has been
elaborated [29] by starting from crystallographic coordinates of Asp,
His and Ser residues in the enzymes. Empirical methods have been used
to determine the equilibrium configurations of the three side chains
for different protonation states of the carboxylic and imidazole groups.
In all cases the Asp-His interactions greatly prevail upon the His-Ser
ones. The more striking result is the remarkable similitude between
several calculated configurations and the experimental one for the
α-chymotrypsin site (fig.7). That strongly suggests a weak influence of
the environment defined by neighbouring residues on the stability of
the catalytic triad. For subtilisin Novo, a "crystal-like" configuration
has been also determined which still supposes a more pronounced effect
of the environment. In the resting enzyme , the configuration of the
site mainly depends on the Asp-His interactions and in both cases, any
hydrogen bond connects the His and Ser residues. Let us recall that a
careful reexamination of recent crystallographic data leads Matthews
et al. [30] to question the existence of such an hydrogen bond. In the
three dimensional structures of several proteases established more recently [31,32], no more interactions have been detected between these
two residues. Consequently, any change in the organization of the

active site requires a modification of the interactions between His and
Asp residues and possibly some variations in the polypeptide backbone
conformation.

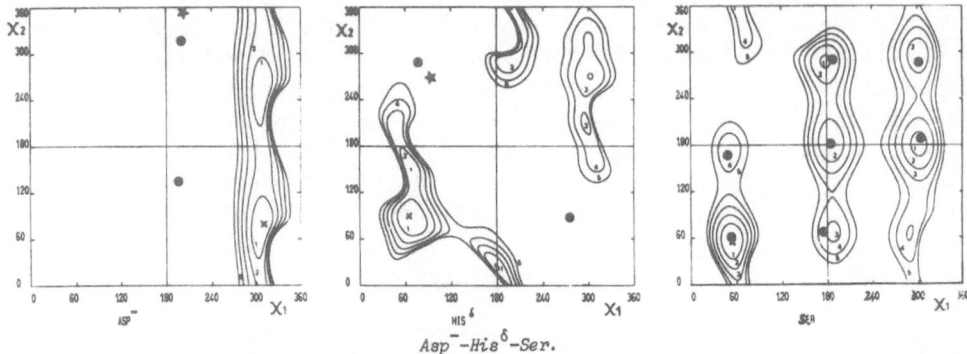

$Asp^--His^\delta-Ser.$

Figure 7. Conformational maps for the catalytic triad in the Asp^-His^δ
Ser state. The isoenergetic curves define energy minima of the isolated
residues. (X : global minimum). ● : minima of intermolecular energy.
★: conformations in the crystal.

All these facts provide evidence that a "charge relay" should be not
involved in the hydrolysis of substrates. A major role should be attributed to the fixation of the substrate in the oxyanion hole which initiates a reorganization of the site in which the dynamics of the protein contribute. The current progress in molecular dynamics computations give indication of the possibility to analyze in a very next future the global system including the protein, a substrate and some
water molecules.

NMR STUDIES OF CONFORMATIONS AND SELF-ASSOCIATION OF NATURAL
LIPOPEPTIDES.

Let us turn now to another category of problems dealing with the
use of NMR in conformational analysis of peptides in solution. Recently
we have undertaken an investigation of natural lipopeptides of microbial
origin, some of which have antibiotic activity. These compounds contain
a cyclic or a linear peptide moiety linked to a long hydrophobic alkyl
chain. A class of lipocyclopeptides has been extensively studied by
Michel et al. who have established the interactions between such compounds and cytoplasmic membranes [33]. In connection with these studies,
we have especially studied peptidolipin NA extracted from *Nocardia aste-*
roides [34] and having the formula :

$$
\begin{array}{ccccccc}
1 & 2 & 3 & 4 & 5 & 6 & 7
\end{array}
$$

$$CH_3(CH_2)_{16}CH-CH_2-CO-L-Thr{\rightarrow}L-Val{\rightarrow}D-Ala{\rightarrow}L-Pro{\rightarrow}D-aIle{\rightarrow}L-Ala{\rightarrow}L-Thr$$

$$\underset{O}{\overset{\mathsf{L}}{\rule{0pt}{0pt}}}\rule{7cm}{0.4pt}\mathsf{\rfloor}$$

A 400 MHz ¹H NMR study of peptidolipin NA and of its L-Val(6) analogue has been recently reported [35] in which we have selected some
demonstrative features. In spite of a considerable improvement of resolution due to an increase of working frequencies, difficulties often
remain in the assignments of resonances to specific positions in a
peptide sequence, when several identical residues are present. In addition to a lot of   classical methods, a very efficient mode of establishing peptide sequences is now available which is based on intramolecular nuclear Overhauser effects (nOe).

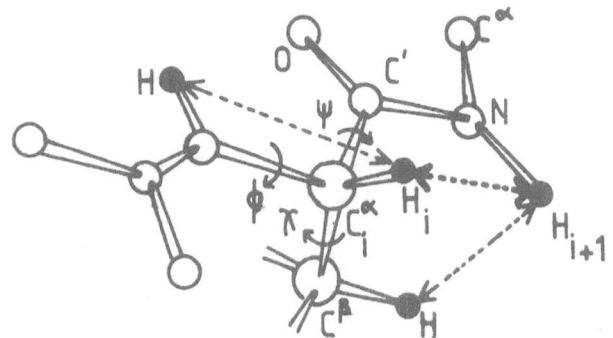

Figure 8. Through space interactions between a N$\underline{H}$ proton and neighbouring C$^\alpha$$\underline{H}$ and C$^\beta$$\underline{H}$ protons.

In many peptides, the nOe enhancements which are observed for
interacting C$^\alpha$H$_i$ and NH$_{i+1}$ protons depend on the dihedral angle ψ_i, i.e.
are relatively small for helical conformations and larger for extended
ones [36]. For many peptides containing no more than a dozen of residues, the only detection of nOe for these protons is often enough for
removing most of ambiguities,the assignments of one NH$_{i+1}$ proton allowing
to assign successively C$^\alpha$H$_i$, NH$_i$, C$^\alpha$H$_{i-1}$, etc... by a recurrent process.
In a second step, one attempts to relate the values of these nOe's to
interprotons distances by taking into account − if necessary − different relaxation mechanisms and conformational averaging. A combination
of all different techniques has enabled us to determine the conformations and self-association properties of peptidolipin NA and of its
L-Val(6) analogue in three solvents of different polarities [35]. A
typical 400 MHz ¹H spectrum of the L-Val(6) compound is shown in the
figure 9.
^{3}J$_{NH-C^\alpha H}$ coupling constants, $\Delta\delta_{NH}/\Delta T$ temperature coefficients, nOe's,
¹H ⇄ ²H rates of exchange, free radical NH accessibilities, a model of
conformation can be proposed for both lipopeptides. There are no drama-
tic differences between the conformations of peptidolipin NA and its
L-Val(6) analogue in different solvents. The peptide cycle is folded
into a γ-turn around the L-Pro residue and is stabilized by several
intracyclic hydrogen bonds. Nevertheless, this cycle is somewhat
flexible since its conformation slightly depends on the polarity of
the environment, on the peptide sequence and also on intermolecular
interactions. One of the most important property of these lipopeptides
is to self-associate into solvents of low polarity such as chloroform.

Dimers (and probably oligomers at high concentrations) are formed, which involves two intercyclic hydrogen bonds.

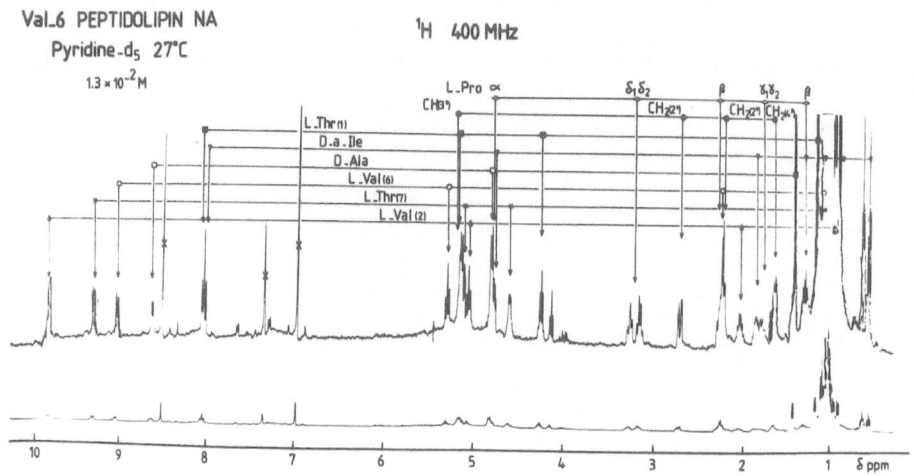

Figure 9. 400 MHz ^{1}H NMR spectrum of Val(6) peptidolipin NA in pyridine-d_5.

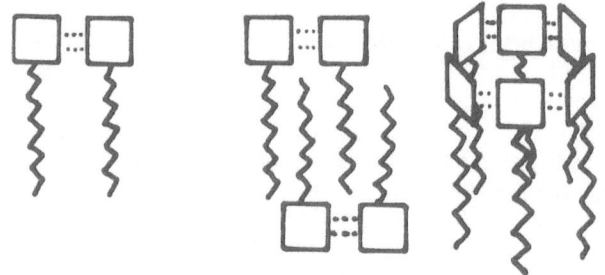

Figure 10. Different possibilities of peptidolipin NA self association in non polar solvents. The existence of dimers has been proved in CDCℓ_3.

In very polar solvents (DMSO - water mixtures), other types of agregates exist which can form gels, structure of which will be investigated elsewhere.

Associations formed in organic solvents are of special interest when considering a possible mode of action of lipopeptides on membranes. Recent preliminary experiments [37] show that peptidolipine NA induces the formation of channels in monooléate BLM (bimolecular lipid membranes). Irregular current fluctuations appear in such a membrane exposed to low concentration of lipopeptide (<< 10^{-6}M). The lifetimes of these channels range from some second tenths to some dozen of seconds and their conductance is around $\sim$ 10$^{-11}\Omega^{-1}$. For long exposures, larger channels can be formed which have a lifetime of several minutes.

Qualitatively, there is a striking similarity between such behaviour
and those of melittin interacting with lipid bilayers [38]. One could
suggest that lipopeptides oligomers are needed to form a channel through
which K$^+$ ions could be displaced by the external electrical field.

Other experiments with different kinds of lipopeptides reveal similar channel-forming properties. It is then possible that such properties may account for some of the biological effects of the compounds.
The low concentrations which are required to form channels could suggest
that channel-forming properties could be more important for the antibiotic activity than fusion or lysis phenomena which can be induced by
lipophilic compounds such as lipopeptides.

CONCLUDING REMARKS.

In this short survey of our theoretical and experimental investigations of conformations of peptides, we have tried to show that a combination of experimental techniques and of simple energy calculations
can be effective in elucidating how intra- and intermolecular interactions dictate stable conformations of small peptides. We have focussed
more on the methodology since these peptides have not biological acti-
vity.From our discussions, it clearly appears that empirical calculations provide valuable information about the manner in which interatomic
interactions influence the conformations of peptides. Intermolecular
interactions modulate these effects in determining the stable conformation which is immobilized in a crystal lattice or which can be in equilibrium with others in solution. Various aspects of hydration have been
discussed, which show that water can be directly involved in the selection of conformations especially by forming internal bridges between
donor or acceptor groups. The necessity to undertake dynamical calculations appears very clearly when more complex intermolecular interactions
are considered as in an active site of enzymes or when larger peptides
are studied. A linear peptide containing a dozen of residues can be
highly flexible and requires very special environmental influences to
take on well defined conformations. Cyclic peptides have more restricted
possibilities and are more easily stabilized by intramolecular hydrogen
bonds. Nevertheless, in this range of molecular weights, a statistical
approach including solute-solvent interactions must be added to a static
study based on energy computations. More generally let us emphasize on
the fundamental differences that can exist between the conformational
behaviour of a peptide in homoegenous solution and its behaviour in the
presence of a biological receptor. That the conformations existing in
aqueous solution are the biologically active forms has been dogmatic
for a long time. Several corrections must be proposed that take into
account the particular mode of interaction with the receptor, the conformational properties of this receptor, etc... That is the reason why
a static sequential approach [39] can also provide usefull information**s**
especially concerning the accessibility, the very short interactions
between neighbouring residues and the relative stabilities of different
conformations.

In another part we have briefly reported a NMR study of natural peptides which demonstrate the power of this spectroscopic method to determine conformations in solution. In addition to this contribution to a development of methodology, we have shown that for lipopeptides there are close connections between conformations, self-association properties and biological activity. The proving of such connections is the major aim at present of conformational analysis of natural peptides.

Acknowledgments.- This article is specially dedicated to Professor Charles Sadron on his 80th birthday. We are extremely grateful to him for having introduced us in the field of molecular biophysics and for his help for many years. We acknowledge A. Heitz and M. Cotrait who have taken an important part in the NMR and crystallography works and G. Michel and N. Guinand for a fruitful collaboration on lipopeptides.

References

1. SNYDER, S.H., 1980, Science, 209, pp.976-983.
2. NEMETHY, G., and SCHERAGA,H.A.,1977, Quart. Rev. Biophys., 10, pp. 239-252.
3. LIQUORI, A.M., 1969, Quart. Rev. Biophys., 2, pp.65-92.
4. RAMACHANDRAN, G.N., RAMAKRISHNAN, C., and SASISEKHARAN, V., 1963, J. Mol. Biol., 7, pp.95.
5. DESANTIS, P., GIGLIO, E., LIQUORI, A.M., and RIPAMONTI, A., 1965, Nature, 206, pp.456-458.
6. PULLMAN, B., and PULLMAN, A., 1974, in "Advances in Protein Chemistry", ANFINSEN, C.B., EDSALL, J.T., and RICHARDS, F.M., ed., Academic Press, New York, vol.28, pp.347-526.
7. See "Environmental effects on molecular structure and properties", 1975, PULLMAN, B., ed., D. Reidel Pub. Comp.
8. BYSTROV, V.F., 1976, in "Progress in NMR spectroscopy", EMSLEY, J.W., FEENEY, J., and STUCLIFFE, H.L., ed., Pergamon Press, Oxford, England, vol.10, pp.41-81.
9. NAGAYAMA, K., and WÜTHRICH, K., 1981, Eur. J. Biochem., 114, pp. 365-374, and WAGNER, G., KUMAR, A., and WÜTHRICH, K., 1981, Eur. J. Biochem., 114, pp.357-384.
10. BIRKTOFT, J.J., and BLOW, D.M., 1972, J. Mol. Biol., 68, pp.187-240.
11. COTRAIT, M., PTAK, M.,BUSETTA, B., and HEITZ, A., 1976, J. Am. Chem. Soc., 98, pp.1073-1076.
12. COTRAIT, M., and PTAK, M., 1978, Acta Cryst., B34, pp.528-532.
13. RAMANI, R., VENKATESAN, K., and MARSH, R.E., 1978, J. Am. Chem. Soc., 100, pp.949-953.
14. PTAK, M., HEITZ, A., and DREUX, M., 1978, Biopolymers, 17, pp.1129-1148.
15. GENEST, M., and PTAK, M., 1978, Int. J. Peptide Protein Res., 11, pp.194-208.
16. GENEST, M., and PTAK, M., 1980, Int. J. Peptide Protein Res., 15, pp.5-19.

17. ROSSKY, P.J., and KARPLUS, M., 1979, J. Am. Chem. Soc., 101, pp. 1913-1937.
18. HAGLER, A.T., OSGUTHORPE, D.J., and ROBSON, B., 1980, Science, 208, pp.599-601.
19. MEHROTRA, P.K., MARCHESE, F.T., and BEVERIDGE, D.L., 1981, J. Am. Chem. Soc., 103, pp.672-673.
20. LEE, B., and RICHARDS, F.M., 1971, J. Mol. Biol., 55, pp.379-400.
21. PONNUSWAMY, P.K., and MANAVALAN, P., 1976, J. Theor. Biol., 60, pp. 481-486 and MANAVALAN, P., and PONNUSWAMY, P.K., 1977, Biochem. J., 167, pp.171-182.
22. GENEST, M., VOVELLE, F., and PTAK, M., 1980, J. Theor. Biol., 87, pp.71-84.
23. STILLINGER, F.H., and RAHMAN, A., 1974, J. Chem. Phys., 60, pp. 1545-1557.
24. ZIMMERMAN, S.S., POTTLE, M.S., NEMETHY, G., and SCHERAGA, H.A., 1977, Macromolecules, 10, pp.1-9 and references cited herein.
25. VOVELLE, F., and PTAK, M., 1979, Int. J. Peptide Protein Res., 13, pp.435-446.
26. VOVELLE, F., GENEST, M., PTAK, M., and MAIGRET, B., 1981, in 14th Jerusalem Symposium on "Intermolecular Forces", PULLMAN, B., ed., D. Reidel Pub. Comp., pp.299-315.
27. PULLMAN, A., and PULLMAN, B., 1975, Quart. Rev. Biophys., 7, pp. 505-566.
28. LANGLET, J., CLAVERIE, P., PULLMAN, B., and PIAZZOLA, D., 1979, Int. J. Quantum Chem., Quantum Biology Symposium, 6, pp.409-437.
29. GENEST, M., and PTAK, M., 1982, Int. J. Peptide Protein Res., 19, pp.420-431.
30. MATTHEWS, D.A., ALDEN, R.A., BIRKTOFT, J.J., FREER, S.T., and KRAUT, 1977, J. Biol. Chem., 24, pp.8875-8883.
31. SIELECKI, A.R., HENDRICKSON, W.A., BROUGHTON, C.G., DEBAERE, L.T.J., BRAYER, G.D., and JAMES, M.N.G., 1979, J. Mol. Biol., 134, pp.781-804.
32. BRAYER, G.D., DELBAERE, L.T.J., and JAMES, M.N.G., 1979, J. Mol. Biol., 131, pp.743-775.
33. BESSON, F., PEYPOUX, F., MICHEL, G., and DELCAMBE, L., 1978, Biochem. Biophys. Res. Comm., 81, pp.297-304.
34. GUINAND, M., and MICHEL, G., 1966, Biochim. Biophys. Acta, 125, pp. 75-91.
35. PTAK, M., HEITZ, A., GUINAND, M., and MICHEL, G., 1982, Int. J. Biol. Macromol., 4, pp.79-90.
36. For a review see : ROQUES, B.P., RAO, R., and MARION, D., 1980, Biochimie, 62, pp.753-773.
37. HEITZ, F., personal communication.
38. TOSTESON, M., and TOSTESON, D.C., 1981, Biophys. J., 36, pp.109-116.
39. COTRAIT, M., and PTAK, M., 1981, J. Comp. Chem., 2, pp.460-469.

LIGAND BINDING AS A PROBE OF PROTEIN DYNAMICS

Hans Frauenfelder
Department of Physics, University of Illinois at Urbana-
Champaign, 1110 West Green Street, Urbana, IL 61801, USA

The binding of small ligands (O_2,CO) to heme proteins can serve as a
probe of protein dynamics. Investigations over a wide range in temperature show that protein motion is important in the migration of
the ligand through the protein matrix, motion within the pocket, and
binding at the heme. The results of the binding studies, together
with complementary work using X-ray diffraction, Mössbauer effect,
and infrared spectroscopy begin to yield detailed information on
protein motions.

1. INTRODUCTION

     Proteins are dynamic systems; they wiggle and breathe. This fact
follows from their general structure and the principles of statistical
mechanics and has been verified and observed by many techniques (1-8).
The question is no longer if motion is relevant for many biological
functions, but how function and motion are connected. Here I consider one of the simplest biological processes, the binding of a
small molecule to a heme protein, to show where and how dynamic
features are involved.

     The binding of oxygen and carbon monoxide to hemoglobin and myo-
globin has been studied extensively for many years and by many
different techniques. Earlier work is reviewed in the book by
Antonini and Brunori (9); I will concentrate here on newer results.
The general features of the binding process are shown in Figure 1.
Ligand binding occurs in a number of successive steps: The ligand
(e.g. CO) diffuses through the solvent to the protein, enters the
protein, migrates through the protein matrix M to the pocket B, and
finally binds at the heme iron. In thermal dissociation, the Fe-CO
bond is broken spontaneously, the ligand in the pocket either rebinds
or enters the protein matrix and migrates through the matrix into the
solvent. In photodissociation, the Fe-CO bond is broken by light.
Dissociation and association can be represented schematically by the
following sequence:

C. Hélène (ed.), Structure, Dynamics, Interactions and Evolution of Biological Macromolecules, 227–236.
Copyright © 1983 by D. Reidel Publishing Company.

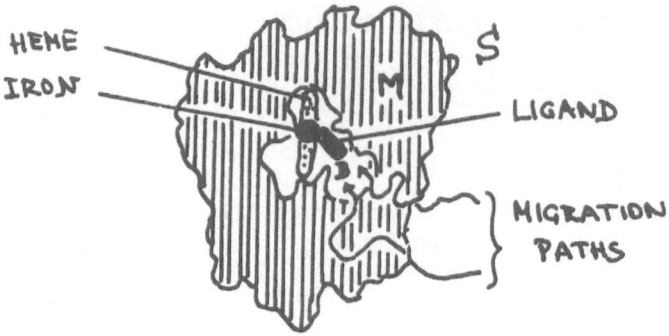

Fig. 1 Schematic cross section through myoglobin, with two possible
        pathways for the motion of the ligand.

$$
\begin{array}{ccccccccc}
 & \xrightarrow{k_{SM}} & & & \xrightarrow{k_{MB}} & & & \xrightarrow{k_{BA}} & \\
S & \rightleftharpoons & M & & \rightleftharpoons & B & & \rightleftharpoons & A \quad\quad (1)\\
\text{solvent} & \xleftarrow{k_{MS}} & \text{protein} & & \xleftarrow{k_{BM}} & \text{heme} & & \xleftarrow{k_{AB}} & \text{heme}\\
 & & \text{matrix} & & & \text{pocket} & & & \text{iron}
\end{array}
$$

Experimentally, the coefficients in Eq. (1) can be determined by
flash photolysis experiments performed over a wide range of temperatures, say from 40 to beyond 300 K (10-14). The system to be studied,
for instance MbCO, is placed in a cryostat, the Fe-CO bond is broken
by a laser flash, and the subsequent rebinding followed optically or
with an infrared detector. Such experiments imply that protein
motion is important in the following steps in the sequence Eq. (1):
   (i)   entrance and exit at the protein-solvent interface,
   (ii)   migration through the protein matrix,
  (iii)   motion within the heme pocket,
   (iv)   covalent binding at the heme iron.

2. BINDING AT PHYSIOLOGICAL TEMPERATURES

In general, all the rate coefficients k_{ij} shown in Eq. (1) are
needed to characterize binding and dissociation. At the physio-
logically important temperatures around 300 K, however, the situation
simplifies and binding can be described as follows: The ligands in
the solvent establish a preequilibrium with the pocket B and the
rate-limiting step is the transition B → A within the heme pocket
(14). Consider first the extreme case where $k_{BA} = k_{AB} = 0$. An
equilibrium will then be established between the ligands in the
solvent and in the pocket and the probability of finding a ligand in
the pocket of a single Mb molecule is given by

$$P_B(c,T) = k_{SM}k_{MB}/k_{BM}k_{MS} \ . \tag{2}$$

$P_B(c,T)$ is a function of the temperature T and is proportional to the ligand concentration c in the solvent; we call it the pocket occupation factor. At easily obtainable concentrations c, $P_B(c,T)$ is much smaller than unity.

We next consider the case, realized for MbCO at 300 K, where the direct rebinding $B \rightarrow A$ is small, but not zero. After photodissociation, essentially all the ligands move into the solvent. A preequilibrium is established between the ligands in the solvent and in the heme pocket and the fraction of protein molecules in state B is approximately equal to the equilibrium value $P_B(c,T)$. The association rate from the solvent, λ_{on}, is given by the pocket occupation factor P_B times the rate coefficient k_{BA} for the final step:

$$\lambda_{on}(c,T) = k_{BA}(T)\ P_B(c,T) . \tag{3}$$

If k_{BA} is not small, but if binding from the solvent is much slower than all internal processes, Eq. (3) can be generalized to read

$$\lambda_{on}(c,T) = k_{BA}(T)\ P_B(c,T)\ N^{out} , \tag{4}$$

where N^{out} is the fraction of photodissociated protein molecules in which the ligand moves into the solvent (14).

The advantage of the approach given here is the separation of λ_{on} into three factors which can be determined individually. The separation permits a comparison of various systems and makes it clearer where control is exerted.

The pseudo-first order rate coefficient λ_{on} and the factor N^{out} can be measured directly at the temperatures of interest, between 270 and 320 K. The factor k_{BA} can be found by extrapolation from experiments below 180 K as we will discuss in the following section. From these three numbers, the pocket occupation factor $P_B(c,T)$ can be found.

3. BINDING AT THE HEME

Studies of the photodissociation of MbCO and HbCO demonstrate that the photodissociated ligand does not leave the heme pocket B below about 180 K but rebinds from the pocket (10). The step $B \rightarrow A$ can consequently be investigated in detail by measuring the rebinding after a laser flash between about 60 and 180 K. Below about 60 K, the transition $B \rightarrow A$ is dominated by quantum mechanical tunneling (15-17). Above 60 K, however, k_{BA} satisfies an Arrhenius equation; if the parameters are determined accurately between 60 and 160 K, extrapolation to 300 K is possible. With k_{BA} in Eq. (4) known, $P_B(c,T)$ can be determined. Table I gives the relevant data for a number of systems (14-18).

Table I.   Parameters describing ligand binding at 300 K,
           ligand pressure 1 bar, pH 7.

System	$\lambda_{on}/10^3 s^{-1}$	$k_{BA}/10^6 s^{-1}$	N^{out}	$P_B/10^{-4}$
MbCO	0.72	6.6	1	1.1
MbO_2	28.	41	0.8	8
$\beta^A CO$	8.8	66	1	1.3
$\beta^A O_2$	41	540	0.45	1.5
LbCO	20	200	1	1
LbO_2	167	200	0.7	10

β^A is the normal beta chain from adult human hemoglobin.
Lb is soybean leghemoglobin.

The values of P_B and k_{BA} in Table I provide some new insights
into the process of ligand binding. The values of P_B are all
remarkably similar, they vary by about a factor of 10. The rate
coefficients k_{BA}, in contrast, vary by about a factor 100. Ligand
binding is mainly governed by the last barrier at the heme. This
fact also implies that these reactions are not diffusion controlled
at 300 K.

4.   CONFORMATIONAL SUBSTATES AND PROTEIN MOTION

Eq. (4) describes ligand binding phenomenologically, but does not
indicate where protein motion is important. Information concerning
the connection between protein dynamics and k_{BA} comes, however, from
a closer look at the low-temperature rebinding after photodissociation (19,10). The rebinding curve is not exponential in time, as
naively expected, but is closer to a power law, $N(t) \approx (1 + t/t_0)^{-n}$,
where $N(t)$ is the fraction of photodissociated protein molecules that
have not rebound a ligand at time t after the flash, and t_0 and n are
two temperature-dependent parameters. The observed behavior can be
explained by assuming that proteins can exist in a large number of
conformational substates. Proteins in different substates perform
the same biological function, albeit with different rates. Below
about 200 K, a particular protein remains frozen in a given substate;
at physiological temperatures, a protein changes rapidly from one
substate to another. Different substates have different barrier
heights, H_{ij}, and consequently different rates k_{ij}. The distribution
of barrier heights can be characterized by a probability density
$g(H_{ij})$, where $g(H_{ij})dH_{ij}$ gives the probability of finding a barrier

with height between H_{ij} and $H_{ij}+dH_{ij}$. The distribution function $g(H_{ij})$ has been determined experimentally for the transition $B \rightarrow A$ for various protein-ligand combinations from the observed nonexponential time dependence (10,11). If $g(H_{BA})$ is known, the average barrier height $\bar{H}_{BA}$ can be determined easily.

Evidence for conformational substates comes from many different experiments and from molecular dynamics calculations (1-8). Particularly powerful is the combination of X-ray diffraction and Mössbauer spectroscopy. The Debye-Waller factor in X-ray diffraction yields the mean-square displacement (msd) for every atom in a protein crystal. If this information is combined with measurements of the Lamb-Mössbauer factor of a particular atom (usually Fe), the msd due to conformational substates can be found (20).

We can now ask if a connection between the msd of the various atoms and residues and the protein function can be established. To find a tentative answer, we first describe the crucial features of the final binding step $B \rightarrow A$ in ligand binding to Mb. In state B, the ligand is essentially free in the heme pocket (13), the heme group is domed towards the proximal side, the iron is about 0.5 Å out of the mean heme plane (21) and has spin 2. In the bound state A, the CO has established a covalent bond with the heme iron, the iron has changed its spin to 0 and has moved into the heme plane, and the heme is more planar. The transition $B \rightarrow A$ consequently involves a motion of the heme, the iron, and the ligand. Where is the origin of the distributed activation barrier? A speculative answer can be found by considering the msd of the atoms near the heme as obtained by X-ray diffraction at 80 K (22). The data show that the residues on the distal side, where the ligand binds, have a small msd at 80 K. In contrast, many of the atoms of the F helix on the proximal side show a larger msd. It is therefore possible that "tension" on the F helix determines how easily the iron moves into the heme plane and thus governs the activation barrier. The situation is sketched in Figure 2 which shows the msd of the backbone atoms of the main residues in the F helix. Also shown as dashed line is the average value of the msd for the iron atom and the four nitrogens to which it is bound. To explain the activation energy spectrum, we can postulate the following mechanism: The F helix can for instance be in substates that are close to the heme group: it is then bent in such a way that the His F8-Fe bond distance is small and the iron can move easily into the mean heme plane on binding the ligand. In contrast, if the F helix assumes a position as far away from the heme group as allowed by the msd, the His F8-Fe bond will be stretched and it will be more difficult for the Fe to move into the heme plane -- the activation barrier will be large. At physiological temperatures, the F helix will rapidly move from one substate to another and the ligand sees an average barrier.

5. MOTION WITHIN THE POCKET

In Figure 1 and in Eq. (1) it is assumed that the transition
B → A in the pocket, the final step in ligand binding, consists of

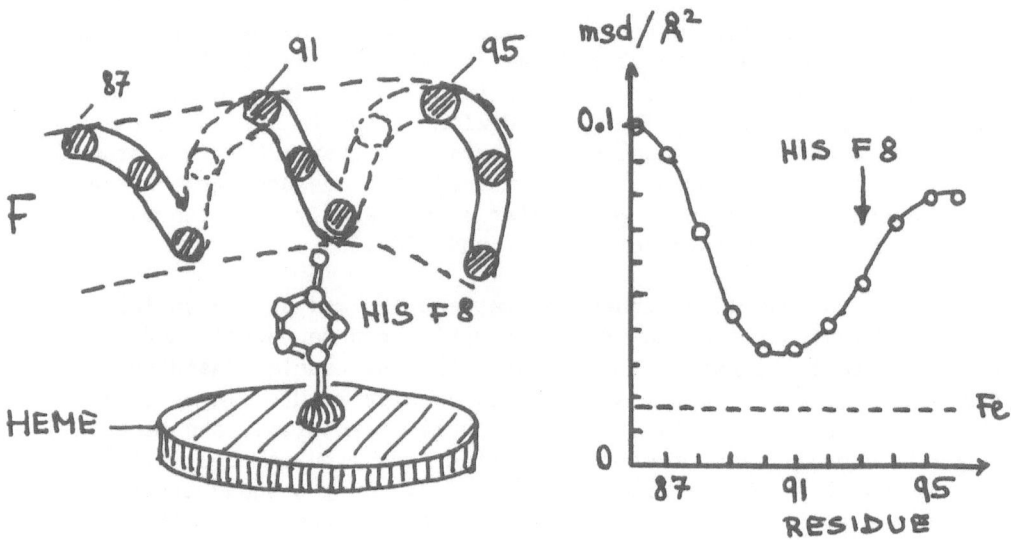

Fig. 2   Schematic arrangement of the heme and the F helix in the
         deoxy state. At right: Mean square displacement (msd) for
         the backbone atoms of the residues 86-96, for metMb (22).

just one maneuver. It is well known, however, that the infrared
spectrum in the bound state A reveals more than one CO stretching
vibration and thus points to additional complexity (23). A
measurement of the CO stretching vibrations in the bound and the
photodissociated states at low temperatures shows 3 lines in the
bound state A and three lines in the photodissociated state B (13).
The wavenumbers of the states B are close to that of the free CO and
imply that the ligand in the pocket is nearly free. The temperature
and time dependences of the intensities of the A and B states
demonstrate that the actual binding process B → A is more complex
than previously assumed. Rebinding B → A is, for instance, preceeded
by transitions between B states. Moreover it is known that the
distal histidine affects the binding of CO and O_2 to Mb and Hb
(14). Dynamic features of the residues forming the pocket therefore
may be important for the binding process; the distal histidine may,
for instance, move during binding. More data from fast IR systems,
Raman spectroscopy, and low-temperature X-ray diffraction will be
required to elucidate the details of the crucial transition B → A and
to extract the role of protein motion.

6. THE ROLE OF THE PROTEIN MATRIX

The discussion in Sections 2 and 3 shows that the binding at physiological temperatures occurs from a preequilibrium state in the heme pocket. This observation raises the question as to the role of the protein matrix in binding. Do the properties of the globin matrix affect the association rate? A clue to an answer comes from the rate coefficients k_{BM}, k_{MB}, k_{MS}, and k_{SM} for various systems, determined by flash photolysis experiments at temperatures between 180 and 300 K. These rate coefficients provide some insight into the motion of the ligand through the protein matrix and consequently give information about transport properties of the protein matrix.

The values of the coefficients k_{BM}, k_{MB}, and k_{MS} depend on the model that is chosen for the evaluation of the experimental data. In the simplest model, shown in Figure 1 and described by Eq. (1), the coefficients have clear meanings. We have explored many other models to find the features that are common to all and can be assumed to represent proteins realistically. The following remarks describe such features.

Comparison of different models shows that motion through the protein matrix is best described by the rate coefficient k_{MB}. In a very crude approximation we assume that the distance from the pocket to the edge of the protein is given by R = 1 nm. The time for this motion is given by $1/k_{MB}$. We can define a coefficient for the diffusion through the protein matrix from the relation $R^2 = 2 D_M t$ as

$$D_M = R^2 k_{MB}/2 . \qquad\qquad (5)$$

The values of k_{MB} determined from the experimental data can be extrapolated to 300 K and D_M is found from Eq. (5). Values of k_{MB} and D_M for four systems are collected in Table II. Also shown for comparison is the diffusion coefficient for O_2 through water.

Even though the values for D_M in Table II are model-dependent and crude, they give valuable additional information about the protein matrix. Motion through the protein matrix is very different for different proteins. Lb is the most fluid of the three proteins listed, Mb the most viscous. The diffusion coefficient through Mb is about a factor 1000 slower than through water.

We now return to the role of protein motion. X-ray data show that a static protein would not permit diffusion of the ligand through the matrix: O_2 or CO could not enter or leave the pocket of a rigid protein (24). The protein matrix must be flexible. Calculations by Case and Karplus also show that barriers in a rigid protein are too high to permit the necessary ligand motion (25). These remarks hold not only for migration through the matrix, but also for the exit and entry at the protein-solvent interface.

Table II.   Values of the rate coefficient k_{MB} and the
            diffusion coefficient for motion through
            the matrix at 300 K.

System	$k_{MB}/10^6 s^{-1}$	$D_M/10^{-8} cm^2 s^{-1}$
MbO_2	2	1
$\beta^A O_2$	20	10
LbO_2	200	100
$O_2 in H_2O$	--	2000

A possible picture for the motion through the protein matrix
emerges from computer graphics studies of the protein interior
(26).   These studies, based on X-ray diffraction data, show the
presence of cavities. It is possible that the migration through the
protein matrix takes place from cavity to cavity. The transition
from one cavity to another could involve large fluctuations that open
gates between cavities. Experiments in which the binding of CO and
O_2 to heme proteins were investigated as a function of the viscosity
of the solvent lend support to such an idea (12). These experiments,
moreover, show that standard transition state theory cannot be used
to describe protein reactions. The theoretical approach pioneered by
Kramers (27), however, provides a satisfactory description.

The picture that emerges from these tentative data is a protein
that is moving and in which motion and function are intimately
connected.

7.   ACKNOWLEDGMENTS

The work was supported in part by grants PHS GM 18051 from the
Department of Health and Human Services and by grant PCM 79-05072
from the National Science Foundation. Much of the work discussed
here was performed at the University of Illinois and I should like to
thank all my collaborators for spirited discussions and many valuable
and incisive comments.

REFERENCES

1. Careri, G., Fasella, P., and Gratton, E.: 1979, Ann. Rev.
    Biophys. Bioeng. 8, pp. 69-97.
2. Gurd, F.R.N. and Rothgeb, T. M.: 1979, Adv. Prot. Chem. 33, pp.
    74-165.
3. Woodward, C. K. and Hilton, D. B.: 1979, Ann. Rev. Biophys.
    Bioeng. 8, pp. 99-127.
4. Cooper, A.: 1980, Sci. Prog. Oxf. 66, pp. 473-497.
5. Karplus, M. and McCammon, J. A.: 1981, CRC Crit. Rev. Biochem.
    9, pp. 293-349.
6. Debrunner, P. G. and Frauenfelder, H.: 1982, in "Experiences in
    Biochemical Perception" (Ornstein, L. N. and Sligar, S. G.,
    Eds.) Academic Press, New York, pp. 327-345.
7. McCammon, J. A. and Karplus, M.: 1982, Acct. Chem. Res. in
    press.
8. Debrunner, P. G. and Frauenfelder, H.: 1982, Ann. Rev. Phys.
    Chem. in press.
9. Antonini, E. and Brunori, M.: 1971, "Hemoglobin and Myoglobin in
    Their Reactions with Ligands", North Holland.
10. Austin, R. H., Beeson, K. W., Eisenstein, L., Frauenfelder, H.,
    and Gunsalus, I. C.: 1975, Biochemistry 14, pp. 5355-5373.
11. Alberding, N., Chan, S. S., Eisenstein, L., Frauenfelder, H.,
    Good, D., Gunsalus, I. C., Nordlund, T. M., Perutz, M. F.,
    Reynolds, A. H., and Sorensen, L. B.: 1978, Biochemistry 17, pp.
    43-51.
12. Beece, D., Eisenstein, L., Frauenfelder, H., Good, D., Marden,
    M. C., Reinisch, L., Reynolds, A. H., Sorensen, L. B., and Yue,
    K. T.: 1980, Biochemistry 19, pp. 5147-5157.
13. Alben, J. O., Beece, D., Bowne, S. F., Doster, W., Eisenstein,
    L., Frauenfelder, H., Good, D., McDonald, J. D., Marden, M. C.,
    Moh, P. P., Reinisch, L., Reynolds, A. H., Shyamsunder, E., and
    Yue, K. T.: 1982, Proc. Natl. Acad. Sci. USA, 79, pp. 3744-3748.
14. Doster, W., Beece, D., Bowne, S. F., DiIorio, E. E., Eisenstein,
    L., Frauenfelder, H., Reinisch, L., Shyamsunder, E.,
    Winterhalter, K. H., and Yue, K. T.: 1982, Biochemistry in
    press.
15. Alberding, N., Austin, R. H., Beeson, K. W., Chan, S. S.,
    Eisenstein, L., Frauenfelder, H., and Nordlund, T. M.: 1976,
    Science 192, pp. 1002-1004.
16. Frauenfelder, H.: 1979, in "Tunneling in Biological Systems" (B.
    Chance et al., Eds.) Academic Press, pp. 627-649.
17. Alben, J. O., Beece, D., Bowne, S. F., Eisenstein, L.,
    Frauenfelder, H., Good, D., Marden, M. C., Moh, P. P., Reinisch,
    L., Reynolds, A. H., and Yue, K. T.: 1980, Phys. Rev. Letters
    44, pp. 1157-1160.
18. Stetzkowski, F. et al., to be published.
19. Austin, R. H., Beeson, K., Eisenstein, L., Frauenfelder, H.,
    Gunsalus, I. C., and Marshall, V. P.:1974, Phys. Rev. Letters
    32, pp. 403-405.

20. Frauenfelder, H., Petsko, G. A., and Tsernoglou, D.: 1979,
     Nature 280, pp. 558-563.
21. Perutz, M. F., Hasnain, S. S., Duke, P. J., Sessler, J. L., and
     Hahn, J. E.: 1982, Nature 295, pp. 535-538.
22. Hartmann, H., Parak, F., Steigemann, W., Petsko, G. A., Ponzi,
     D. R., and Frauenfelder, H.: 1982, Proc. Natl. Acad. Sci. USA in
     press.
23. Makinen, M. W., Houtchens, R. A., and Caughey, W. S.: 1979,
     Proc. Natl. Acad. Sci. USA 76, pp. 6042-6046.
24. Perutz, M. F. and Matthews, F. S.: 1966, J. Mol. Biol. 21, pp.
     199-202.
25. Case, D. A. and Karplus, M.: 1979, J. Mol. Biol. 132, pp. 343-
     368.
26. Connolly, M. L., Kuntz, I. D., Ferrin, T. E., and Langridge, R.:
     1982, J. Mol. Biol. in press.
27. Kramers, H. A.: 1940, Physica 7, pp. 284-304.

DYNAMICS OF PROTEINS AND OF PROTEIN ASSEMBLIES.

Roger CERF
Laboratoire d'Acoustique Moléculaire, Université Louis
Pasteur, Strasbourg, France.

In part 1 is proposed an interpretation of the viscosity scaling law that governs the kinetics of ligand rebinding to proteins (experiments of Frauenfelder's group). This interpretation, it is hoped, discloses the trick by which proteins achieved a form of adaptability, and the ability to fulfil a function under various conditions in the medium.

In part 2, earlier and recent results of this laboratory on the ultrasonic absorption in protein self-assemblies are reviewed and discussed. Their interpretation in terms of a dynamic effect specific to the assembly, and their possible biological significance, are discussed on the basis of several models.

1. DYNAMICS OF PROTEINS : HOW DID THEY CIRCUMVENT STOKES' LAW ?

1.1. Frauenfelder's scaling law.

The results of Frauenfelder's group at Urbana ((1) ; see also Professor Frauenfelder's article in this book) showed that in heme proteins: 1) a ligand encounters several barriers in succession on its way from the solvent to the binding site at the ferrous heme iron ; 2) some of the rate constants depend on the solvent viscosity, demonstrating that interior groups must move in order to let the ligand find its way to the binding site ; 3) the corresponding transition probabilities P obey the following "strange" scaling law :

$$P = (\frac{\alpha}{\eta^\kappa} + \alpha^0) \exp(-\Delta H/RT) \qquad 0 < \kappa < 1, \qquad (I,1)$$

in which η is the solvent viscosity and α, α^0, and ΔH (the activation enthalpy) are constants.

This experimental law, which was found to hold over five decades in viscosity, has two remarkable, even unexpected, features : first, one would expect instead a constant to be added in the denominator of the

C. Hélène (ed.), Structure, Dynamics, Interactions and Evolution of Biological Macromolecules, 237–251.

η-dependent term ; second, one would expect the power of η in the denominator of the first term to be equal to 1, whereas the experimental exponent κ is less than 1 and in some cases is as small as 0.4. The reason why the experimental findings are contrary to expectation will become clear in section 1.2 (see Eq.(I,4)).

As a starting point, let us accept Frauenfelder's model for the rebinding kinetics, based on gates that are controlled by conformational changes of the polypeptide chain. Let us assume further that these conformational changes are of sufficiently long range for the true viscosity of the outer medium to come into play. From this, it will be possible to propose one way in which the scaling law, Eq.(I,1), may be derived from known principles of the dynamics of polymeric chains (see also Ref. (2)).

1.2. Kramers' theory. Examples of application : rate of enzyme catalysis; oligopeptide dynamics.

In his study of the escape over a potential barrier of a particle that is embedded in a viscous medium, Kramers (3) defined a reaction rate constant in terms of a steady-state flow of particles from one well to another over the barrier. The parameters are the frequencies of oscillations ν_B and $i\nu_T$ ($i = \sqrt{-1}$) at the bottom and the top of the potential curve, the barrier height Δw (which is assumed to be large compared to kT), the mass m, and the friction coefficient ζ of the particle.

In the case of "high friction", which is defined by the inequality $\zeta/m \gtrsim \nu_B$ (kT/Δw), the effect of the Brownian forces on the velocity of the particle is much larger than that of the external forces, and the rate of escape is determined by a diffusion-process in phase-space. Note that m/ζ is the time constant for the decay of the velocity autocorrelation function of the Brownian particle. In the case of "low friction", which is defined by the inequality $\zeta/m \lesssim \nu_B$ (kT/Δw), the Brownian forces cause only a small variation of the particle energy during the time of one of its oscillations in the well, and their main effect is the gradual change of the distribution of the ensemble over the different energy values.

The results are shown in Fig.1, in which the reciprocal transition probability P^{-1} is plotted as a function of the viscosity-proportional parameter ζ/m, for $\nu_B = \nu_T = \nu$. From Kramers' original theory, the "low-friction" range of values of ζ/m can be ignored in the applications to follow, because even the viscosity of water is too high for the low-friction regime to be reached. Furthermore, the "high-friction" domain is advantageously subdivised in "high-viscosity" and "low-viscosity" ranges (see Ref.(4)), in which the behavior is *diffusive*, with

$$P^{-1} \simeq 2 \, A\eta \, , \qquad\qquad\qquad (I,2)$$

and *non-diffusive*, with

$$P^{-1} \simeq A\eta + B , \tag{I,3}$$

respectively ; A and B are functions of the parameters enumerated above. The P^{-1}-versus-η curve given by Eq.(I,3) shows a non-zero intercept B, that in Kramers' theory nearly equals the transition-state-theory value (TST in Fig.1).

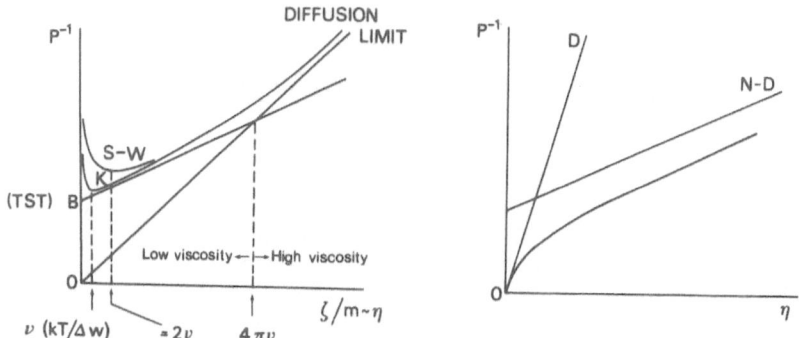

Fig.1(left). Reciprocal transition probability P^{-1} as a function of the "viscosity" ζ/m, according to the theories of Kramers (K; (3)) and of Skinner and Wolynes (S-W;(7)). The Kramers results are represented in the particular case $\nu_B = \nu_T = \nu$. The vertical arrow $\zeta/m = \nu(kT/\Delta w)$ separates the ranges of high friction (right to arrow) and low friction (left to arrow), respectively.

Fig.2(right). Author's interpretation of the high-frequency dynamic properties of certain polymeric chains (5) and of oligopeptides (6). Note the almost parabolic shape of the P^{-1}-versus-η curve. It is proposed in the present paper that this diagram also gives the clue to Frauenfelder's scaling law and to a form of protein adaptability.

Equations (I,2) and (I,3) were extensively used by the present author ((5),(6)), for instance in interpreting fluorescence decay experiments in oligopeptides. The recent results of Skinner and Wolynes (7) differ from those of Kramers in that their "low-friction" regime extends to higher values of η (see Fig.1 ; compare, in particular, the values of ζ/m at which P^{-1} is minimal in each theory).

It remains to be seen, however, whether either theory can be acceptably applied to local motions in chain molecules. The application of Kramers' theory to these molecules has received attention recently, although only at the diffusion limit, and with no essential modification of the rates of conformational changes of a segment resulting from interaction of this segment with other parts of the chain. Here Kramers' theory will be used, and it will be assumed simply that the preceding viscosity dependencies of P^{-1} remain valid for a local conformational change of a chain molecule. This approach is based on the two following arguments : 1) no "low-friction" behavior has ever been found, to the author's knowledge, in polymeric chains ; 2) for the oligopeptides already mentioned, Eqs.(I,2) and (I,3) are found valid, as in Kramers' original theory, over a wide range of values of η (see below and the figure in Ref.(6)).

For a potential barrier in a protein, I shall consider in Section 1.3 a fairly large number of processes, each contributing to the overall rate in a relatively narrow range of viscosity values. It will then be justifiable (see Fig.1) to represent the rate constant for each process by the linear law :

$$P = (a\ \eta + b)^{-1}\ \exp(-\Delta H/RT). \tag{I,4}$$

Note the difference from the form of P found experimentally by Frauenfelder and collaborators (Eq.(I,1)).

The use that may be made of Kramers' theory for biopolymers will be illustrated first, with the aid of two examples.

Rate of enzyme catalysis. Gavish and Werber (8) have recently studied the viscosity dependence of the rate of catalysis of carboxypeptidase A. They suggested that at saturating substrate concentration, the rate-limiting step may be a conformational change preceding the formation of the product. When the rate constants were "corrected" for the viscosity of the medium and the dielectric constant of the solvent was taken into account, a universal Arrhenius plot was obtained, but with one exception : the rate of catalysis was abnormally low in the aqueous solution (by a factor of about 2.5). In constructing their reduced Arrhenius plot, Gavish and Werber invoked the limiting diffusive law (I,2) of Kramers' theory. I suggest that, on the contrary, the entire P^{-1}-versus-η curve in Kramers "high-friction" range be considered. This obviously offers a possible explanation of the "too-low" rate observed in aqueous solution, since at the lower viscosity values the above term B of Eq.(I,3) comes into play and lowers the value of P.

Oligopeptide dynamics. A slightly more sophisticated use of Kramers' theory is required in the interpretation that I proposed (6) of the fluorescence decay in oligopeptides, as measured by E.Haas, E.Katchalsky-Katzir, and I.Steinberg (9). The repeat unit G in these peptides was N^5-(2-hydroxyethyl)-L-glutamine, and the chromophores used as donor and acceptor were naphthalene and dansyl, respectively ; these DG_1N oligomers were studied with $l = 4,5,8$ and 9. The fluorescence decay was determined by the change with time in the number of excited molecules, after an almost infinitely short excitation light-pulse had been applied at time $t = 0$. The experiment resulted in the measurement of an effective diffusion coefficient D of the ends of the molecules relative to one another. The value of D was obtained in different solvents, of viscosities ranging from 1 cP up to at least 20 cP, and in one case up to 120cP, depending on the oligomer studied.

A simple description of the molecular dynamics emerged when the original plot of D versus η (9) was transformed into the plot of D^{-1} against η (6). In particular, the curve thus obtained for DG_5N followed the trend of that shown in Fig.2. In the low-viscosity range, for $\eta < 4$ cp, the dependence of D^{-1} on η is linear and the intercept at $\eta = 0$ is almost zero. At higher solvent viscosities a linear law also shows up, the intercept being however, non-zero, and the slope much smaller than at low viscosities. These results were interpreted as follows (6). We have here an example of the competition between diffusive (D) and non-diffusive (N-D) processes. The solvent friction is smallest for the non-diffusive process, giving a smaller slope for the corresponding straight line (see Fig.2). As the renewal of conformations is controlled for each value of the solvent viscosity by the fastest movements, the actual curve, always follows the lower line. Thus, the behavior is diffusive at low values of η. At high viscosities however, the behavior is non-diffusive, resulting in a non-zero intercept of the curve, even though the processes are considerably faster than would follow from the diffusion law valid at lower viscosity values. In addition to offering an interpretation of the results for DG_5N, the D^{-1}-versus-η plot of Ref.(6) shows, for the other oligopeptides of Ref.(9) as well, the usefulness of considering both the linear laws (I,2) and (I,3).

1.3. A possible explanation of Frauenfelder's scaling law and of how proteins circumvented Stokes' law.

The shape of the curve shown in Fig.2, which reproduces that obtained for DG_5N (see Section 1.2), contains the essential features of Frauenfelder's scaling law. The reciprocal of a diffusion constant, indeed, behaves like the reciprocal of a transition probability, so that using arguments of the kind just presented, we may expect to find :

$$P \sim \eta^{-\kappa} ; \qquad 0 < \kappa < 1 . \qquad (I,5)$$

Of course, the curve in Fig.2 is not exactly a parabola, contrary to that described by Eq.(I,5). However, in order to obtain a parabola, in a prescribed range of values of η (2), and the constant term in Eq.(I,1) as well, we only need to consider a few more processes, rather than just two. Accordingly, the η-dependent term in the transition probability is now written :

$$P_1(\eta) = \sum_x n_x (a_x \eta + b_x)^{-1} \exp(-\Delta H_x/RT) . \qquad (I,6)$$

It is thus assumed that several conformational transitions, each identified by the value of the index x, may lead to the local event of interest, say, the opening of a gate ; n_x is the probability of each conformational state. It may be necessary also to account for several possible paths for the ligand's approach to its binding site ; each path would then contribute to this sum ; a_x and b_x are chosen so as to reflect the fact that, when η increases, the whole range of processes is spanned, from diffusive ones to strongly non-diffusive ones.

The summation in Eq.(I,6) may be performed by going over to an integral. It turns out that Eq.(I,5) is obtained, and that the exponent κ is, indeed, smaller than 1, *provided that the non-diffusive processes are heavily weighted* (see Ref.(2)). This may be due both to hindrance of conformational changes and, in our compact molecules, to hindrance of the motion of parts of the polypeptide chain with respect to one another. The second, constant term in Eq(I,1) may be interpreted as being produced, in exactly the same way, by what we may call "totally non-diffusive processes", i.e. processes in which the term B (see Section 1.2) is overhelmingly large relative to the term Aη.

This description, which is equivalent to saying that protein molecules exhibit high *inner viscosity*, of the η-independent type, undoubtedly emphasizes non-diffusive processes more than the diffusion picture favored by McCammon, Gelin and Karplus (10), but it offers one way to understand the experimentally determined viscosity scaling. On the other hand, the preceding interpretation offers no reason why both the terms in P should have the same activation enthalpy. It is not certain, however, that there is at present clear-cut experimental evidence in favor either of a unique value of the activation enthalpy, or of two different values, which would each appear in one of the terms of P (see Eq.(I,1)).

Inner viscosity must here be understood in a generalized sense (see Ref.(2)). For statistical coils, inner viscosity is currently associated with their well-known normal modes of deformation, whereas for proteins a detailed description of their motions is still lacking. Nevertheless, the present interpretation of the ligand-binding experiments suggests that movements in proteins are greatly hindered by inner viscosity. Similarly, my proposed interpretation of Gavish and Werber's measurements of the rate of catalysis of carboxypeptidase in aqueous solution amounts to suggesting that inner viscosity may play a role in enzyme catalysis.

Gavish (11) proposed that the viscosity scaling law (I,1) reflects the fact that variations in solvent viscosity are only partially transferred to the reaction site. There certainly is partial transfer of viscosity effects, but this need not produce the deviation from Stokes' law that is required to explain the data. This type of explanation also makes use of a local solvent viscosity in the interior of the protein molecule, which need not be meaningful.

One point of interest in the present interpretation of the ligand-rebinding experiments is that a fairly large number of conformational transitions is considered, each of which makes its own contribution in a restricted domain of values of the solvent viscosity, while covering altogether several decades in the viscosity scale. Furthermore, each process contributes most to the overall transition probability in the range of values of the solvent viscosity in which this process is the fastest one, as illustrated in Fig.2 in the case of only two processes.

This may be biologically significant. If the diffusion limit of Kramers'
theory were obeyed throughout the range of values of η, that is, if P
did obey the generalized Stokes' law

$$P \sim \eta^{-1} , \tag{I,7}$$

severe restrictions would be imposed on the transition rates in highly
viscous media such as membranes, in which η may reach several poise. In
the above interpretation of the experimentally found viscosity scaling,
proteins circumvented Stokes' law by the trick illustrated in Fig.2 :
as a result, the transition rates span a narrower range of values than
the viscosity. I suggest that we may have here a manifestation of the
adaptability of proteins, and of their ability to fulfil a function under various conditions in the medium.

2. DYNAMICS OF PROTEIN SELF-ASSEMBLIES : SPECIFIC STRUCTURAL FLUCTUATIONS.

     A few years ago we made some measurements of the absorption of ul-
trasound in viral capsids, and got a surprising result : the capsids turned out to absorb more than the dissociated protein did, and this led us
to suspect that spontaneous structural fluctuations must exist in protein assemblies. The information which may be obtained from such measurements will be outlined first.

2.1. What ultrasonic measurements can tell us.

     As an example of a molecular process producing absorption, consider
the monomolecular reaction

$$A \rightleftharpoons B \tag{II,1}$$

where A and B may simply be two conformers. The sound wave perturbs the
equilibrium populations of A and B, either by the excess pressure produced, which is periodic in time, or by the excess temperature, or both.
A molecular relaxation is associated with this reaction, and the corresponding relaxation time is

$$\tau = \frac{1}{k_{AB} + k_{BA}} \tag{II,2}$$

where k_{AB} and k_{BA} are the rate constants for the forward and backward
reactions. The reaction (II,1) may consist in a molecular conformational
change ; the rate constants then are the transition probabilities considered in Part 1, and they depend on the barrier's free energies of activation.

     When only a pressure change can occur, as in aqueous solutions,
the absorption per wavelength, αλ, is proportional to the product of
several factors :

$$\alpha\lambda \sim \frac{\alpha}{\omega} \sim c_r(\Delta V)^2 \frac{K}{(1+K)^2} \frac{\omega\tau}{1+\omega^2\tau^2} \qquad\qquad (II,3)$$

where $\sim$ means "proportional to", c_r is the molar concentration of the relaxing systems , ΔV is the molar volume change associated with the reaction (II,1), K is the equilibrium constant, and ω is the circular frequency. The factor $(1+\omega^2\tau^2)^{-1}$ in the right-hand side of Eq.(II,3) is the power spectrum of the fluctuations (by virtue of the fluctuation-dissipation theorem). The results will be presented in $\Delta\alpha/N^2$-versus-N plots, (wherein $\Delta\alpha$ is the difference between the measured absorption and that of the solvent, and $N = \omega/2\pi$). From Eq.(II,3), $\Delta\alpha/N^2$ is seen to tend to zero as N tends to infinity. For full exploitation of a $\Delta\alpha/N^2$-plot the low-frequency plateau must be obtained, or at least a downward curvature, as for instance in Fig.4. If, furthermore, the spectrum obeys Eq.(II,3), values of both the relaxation time τ and of the molar volume change ΔV may be obtained.

2.2. A dynamic effect in capsids of icosahedral plant viruses.

Icosahedral plant viruses, like bromegrass mosaic virus (BMV) and tomato bushy stunt virus (TBSV), were studied in collaboration with the Laboratory of Virology of the Institute of Molecular and Cellular Biology at Strasbourg ((12),(13)). An Eggers cell was used for the absorption measurements.

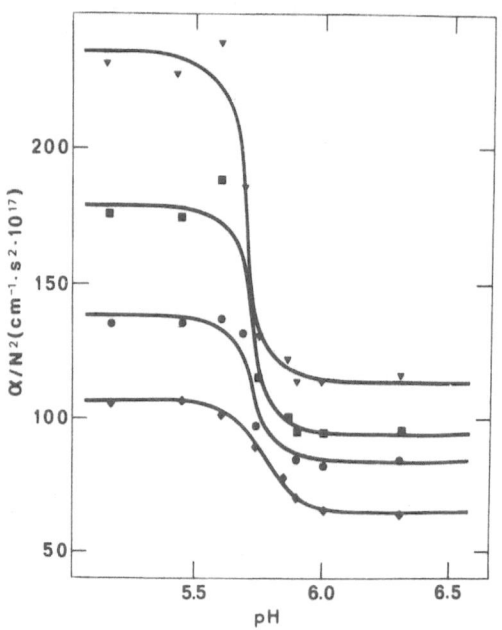

Fig.3. The first results for bromegrass mosaic virus (13) that pointed to a dynamic effect specific to certain protein assemblies. The ultrasonic absorption is higher in BMV capsids (low pH values) than in the dissociated protein (high pH values) ; α/N^2 is highest at the lowest frequencies.
(With permission from Ref. (13)

Figure 3 shows the ultrasonic absorption in a solution of 11 mg of BMV protein per ml in 1 molar sodium chloride at 23°C at the frequencies: 0.6, 0.9, 1.4, and 2.6 MHz, plotted as a function of the solution pH.

The mid-point of the dimer-capsid transition was in the range of pH values 5.7 to 5.8. At lower pH values, the protein is fully associated in the form of capsids, each consisting of 180 protein molecules ; at higher pH values, the protein is fully dissociated in the form of dimers.

The result is just the opposite of what would be expected for the perturbation by the sound wave. There is a volume change associated with each kind of group, whether polar, electrically charged, or non-polar, when the group is in contact with water. Because each specific protein-water contact contributes to the sound absorption by a term proportional to the corresponding $(\Delta V)^2$, the total absorption should decrease when protein-water contacts are replaced by protein-protein contacts, as the capsid assembles. Since we found the contrary, there must have been a relaxation process that existed only in the assembled particle. Furthermore, the ultrasonic absorption in BMV virions was found to be no greater than the sum of that of their component capsids and RNA (13) ; therefore we concluded that RNA-protein interactions cannot contribute much to the absorption. Thus, the observed relaxation was tentatively attributed to structural changes within the protein assembly, and subsequent work was aimed at testing this proposal.

The next experiment showed that, if we were truly detecting structural fluctuations, these fluctuations must have been characteristic of the highly symmetrical, compact form of the virion. The experiment consisted in swelling the virion, according to known procedures, by removing bound divalent cations. Swollen virions were found to absorb ultrasound less than compact virions (13).

The question may be raised whether a contribution to the ultrasonic absorption is expected that would reflect the interconversion between the compact and swollen forms of a virion. No such contribution is expected in icosahedral viruses like TBSV and SBMV (southern bean mosaic virus), in which the interconversion is known to be very cooperative, and therefore slow with regard to our time scale. Although protein-protein and protein-RNA interactions are weaker in BMV than in the other two viruses, there again no dynamic contribution associated with the interconversion is expected.

Our interpretation of the excess absorption observed in virions in terms of structural fluctuations is also supported by the reduction of the excess absorption when virions are treated with a crosslinking agent and the rigidity of the shell is thus increased (13). In BMV, crosslinking adjacent protein subunits inside particles with the difunctional reagent glutaraldehyde proved not to be feasible at the particle concentration used for these experiments. However, the same reagent did cross-. link TBSV. We checked that crosslinking had occurred between protein molecules of a given virion, not between different virus particles.

2.3. How general is the dynamic effect in protein self-assemblies ?
What are the molecular processes involved ?

From the results summarized in Section 2.2 the dynamic effect found
in capsids of icosahedral plant viruses seems to be *specific* to compact,
highly symmetrical structures, in which the intrinsically flexible subunit is locked into "quasi-equivalently related conformations" (14). The
next questions are : how general in protein self-assemblies is the effect,
and what are the molecular processes involved ?

Dynamic effects were found also in other highly symmetrical assemblies, for instance in tobacco mosaic virus (TMV ; see Section 2.4), as
well as in microtubules and hemocyanins (15). The evidence was more conflicting in less structured nucleo-protein assemblies ; thus, the core
particle of the bacterial virus FV3 (see Section 2.4) showed a large effect, whereas none was observed in ribosomes.

In most of the $\Delta\alpha/N^2$ ultrasonic spectra obtained for icosahedral
plant viruses, the relaxation time was too great (i.e. $\tau > 0.2$ μs) to be
measured. However, for BMV it could be shifted into the range of observation of our equipment by increasing the temperature. The result is
shown in Fig.4 : at 50°C, the relaxation time was about 60 ns. The relaxation amplitude could also be estimated, and we shall assume that the
total amplitude measured reflects our dynamic effect in the virion.

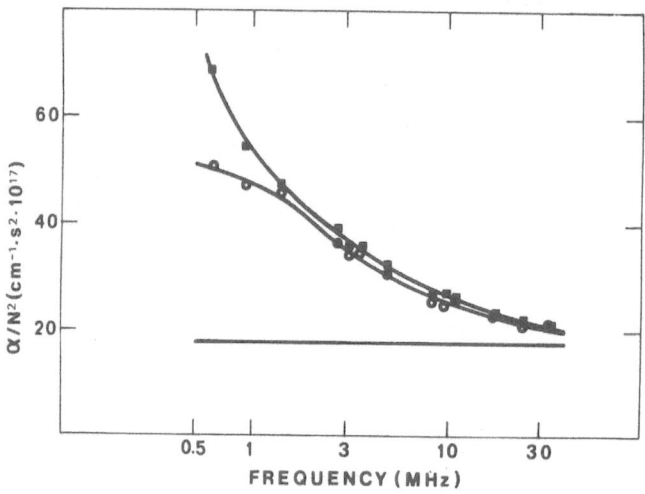

Fig.4. Ultrasonic $\Delta\alpha/N^2$-versus-N plots for TBSV and BMV at 50°C
(13). The spectrum obtained for BMV shows downward curvature at low frequency, and may be interpreted in terms of the two-state model, Eq.(II,
1). The values of the relaxation time τ and of the molar volume change
ΔV thus obtained are discussed in the text in terms of models (see next
figure).(With permission from Ref. (13)

In those protein-assemblies we may expect *a priori* that dimers,

pentamers or hexamers do exhibit an excited state giving rise to relaxation. For n-mer excitation in general, assuming that on the average
one n-mer per virion is excited, we have : $K = n/180 \ll 1$ and
$K(1+K)^{-2} = K \sim n$. On the other hand, if c is the protein concentration
in g/cc, the molar concentration of relaxing systems is : $c_r = c(nM)^{-1}$;
M is the molecular weight of one protein molecule. Hence, from Eq.(II,3),
we have :

$$\Delta V \sim n^o . \qquad\qquad\qquad (II,4)$$

When Eq.(II,3) is written explicitly (i.e. including a trivial factor
not written here), the value of ΔV can be obtained. Here we have the
additional condition (II,4), and the value of ΔV therefore is independent of that of n. For BMV at 50°C, using the results of Fig.4 :

$$\Delta V \simeq 200 \text{ cc/mol of relaxing n-mer.}$$

Since the molar volume per protein is about $1.4 \ 10^4$cc, $\Delta V/V$ is about
$(1.4/n) \ 10^{-2}$.

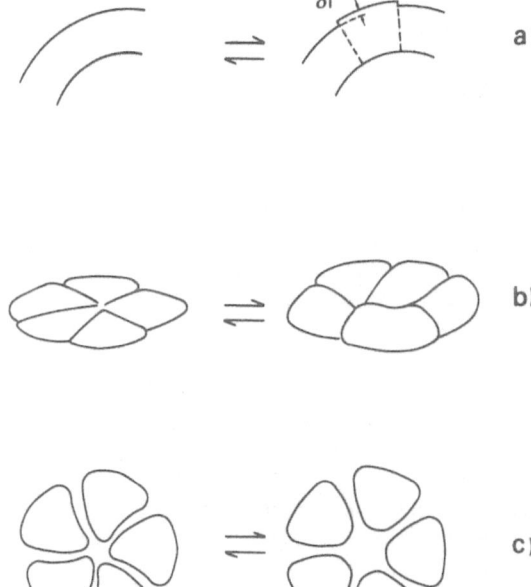

Fig.5. Three models for a
volume change in icosahedral
capsids : a) a subunit moves
radially outwards in the virion (or its outer surface
deforms); b) more detailed
picture of a possible collective deformation of the outer
surface of a pentameric subunit ; c) the deformation is
radial in the subunit (even
radially symmetric) ; see
text for origin of the volume change in each case.

        In the naive picture in which a conformational change of a dimer,
or a collective conformational change of a pentamer or of a hexamer
would result in the external surface of the excited subunit moving *radially in the virion* by δl (see Fig.5a), very small displacements δl are
required. Thus for dimer excitation, $\delta l \simeq 0.35$ Å, and for pentamer and
hexamer excitation $\delta l \simeq 0.13$ Å. The corresponding mean displacements per
excited n-mer, $\overline{\delta l} = \delta l(n/180)$, of course, are much smaller still. Such

fluctuations are well below those that can be detected in viruses with X-ray techniques.

In calculating the preceding values of δl it has been assumed that the true volume change of the particle is equal to the apparent volume change brought about by the outward motion of a surface that represents the external boundary of a subunit. In view of the very small values of δl thus obtained, this assumption represents a minor violation of close-packing. Radial translation of the outer surface of a subunit could result from the motion of the subunit as a whole, or from its deformation. Collective deformation of the outer surface, as schematically and pictorially represented in Fig.5b, seems quite as likely, however. Of course, neither the external surface of the virion as in Fig.5a, nor the ground state of a pentamer, as in Fig.5b, should be drawn as being flat.

Alternatively, let us assume that the central passage in a pentameric or a hexameric subunit may vary in size (one could also assume that intersubunit spaces may vary in size, or that parts of a subunit situated in the interior of the virion may move relative to each other), thus leading to a change of volume ΔV brought about mainly by changes of hydration. Assuming further that on the average half of the 32 pentameric and hexameric subunits of a virion are excited, therefore giving $K(1+K)^{-2}$ its maximum value of 1/4 and ΔV its minimum value, we now have :

$$\Delta V = 70 \text{ cc/mol of relaxing system.}$$

The volume change per protein in the assembly process is roughly this value. On the other hand, structural fluctuations in accordance with the latter process (see Fig.5c), in which the motion is *radial in the subunit*, involve changes of exposure to the solvent of the subunit ; these changes would easily be equivalent to the outer surface of one protein molecule, and the corresponding change of hydration could therefore explain the data.

It should be noted : 1) that a motion described here as "radial in the subunit" is not necessarily thought of as radially symmetric, although for simplicity, the case of a radially symmetric motion has been shown in Fig.5c ; 2) that mixed motions of the two types, radial in the virion and radial in the subunit may occur. The preceding calculations show, however, that motions radial in the virion may be quite efficient in producing ultrasonic absorption.

2.4. Possible biological significance.

The structural fluctuations considered in the preceding section, especially the collective excitation of a pentameric or a hexameric subunit resulting either in a minute deformation of the external surface, radially in the virion, or in an increase of the size of the central passage, radially in the subunit, may be of biological significance, since those movements may help release the RNA. It would be desirable to have better estimates of the amplitude of these motions and to com-

pare these estimates with data of high-resolution crystallographic measurements.

Small-angle X-ray diffraction measurements from solutions and mi-
crocrystals of swollen SBMV show that there is substantial variation
of the shape of the particles upon swelling. According to Caspar and
collaborators (14) , the positions of the subunits in the capsid vary
relative to the particle center, with a standard deviation of the order
of 6 Å. Furthermore, the swollen state shows some disorder, which is
also dynamic , as was found by comparing NMR spectra from the compact
and swollen forms (14).We have seen, however, that swollen BMV absorbs
ultrasound less than in the compact state (see Section 2.2). So, the
dynamic state of the swollen virion, which is likely to play an important role for the assembly and diassembly processes, must be different
from that we detect with ultrasound in compact structures. The structural fluctuations of the compact form could, however, act as nucleation centers for larger fluctuations. Clearly, it will be of interest
to have a better knowledge of the nature of the structural fluctuations
of the different states of a virion.

The study of another virus, tobacco mosaic virus (TMV), is of interest here, since the coat protein is capable of assembly in several
polymorphic states. The helical aggregate was found to exhibit quite
high ultrasonic absorption and was more absorbing than the virus. The
two-ring disk, on the other hand, absorbed scarcely more than the A-
protein (16).

The two-ring disk however, changes its structure slightly when it
becomes part of a finished assembly. From the work of Durham, Finch,
and Klug (17), the helical aggregate is thought to arise from the polymerization of double disks converted into some active structure, which
may be a double spiral. More generally, protein molecules that build
structures appear to be synthesized in a conformational state in which
they do not spontaneously assemble. Therefore, in some cases the structural fluctuations we find in the assembled state could be characteristic of active subunits, or of assemblies of such subunits. In TMV,fluc-
tuations of active subunits might result in torsional dislocation of
the particle around its axis.

Structural fluctuations may be important also for other processes ;
for instance,they may play a role when the virion interacts with a cell
membrane. With this in mind,we have studied another virus, frog virus 3
(FV3), a large DNA virus, in which fusion with the plasma membrane of
liver macrophages and of cultured cells was observed at the Virology Laboratory of the Institute of Bacteriology of Strasbourg University (18).

Neutron scattering studies had previously shown that this large
icosahedral DNA virus is composed of three concentric domains : a central spherical core consisting of the DNA and the proteins associated
with it ; an intermediate zone, probably a lipid bilayer that includes
a noticeable amount of proteins ; and an outer icosahedral protein

capsid (19). The measurements showed, among other results (20), that the virion absorbs more ultrasound than the capsidless particle does. The difference between the absorption of the virion and that of the capsidless particle was found to be a simple relaxation curve, with a characteristic time of about 25 ns. The capsid of FV3 is likely to be much less rigid than that of icosahedral plant viruses, and movements may be quite different in the two structures. The change in volume, ΔV, in FV3 may therefore reveal a conformational change of some proteins of the capsid, or fluctuations in their exposure to the solvent.

Either of these structural fluctuations may play a role in the fusion of FV3 with membranes. In itself, the possibility that an ultrasonically detected change in volume, ΔV, characterizes fluctuations of the exposure of proteins to the solvent, makes it possible to anticipate further applications of the ultrasonic techniques to biological systems.

2.5. With another supporting technique.

Application of the ultrasonic techniques is, at present, limited to frequencies higher than a few hundred KHz, i.e. the relaxation times must be shorter than about 0.2 μs, when test samples of a few cc of liquid are to be used. In our experiments on self-assemblies, with the exception of some results obtained at temperatures above ambient, the downward curvature of the $\Delta\alpha/N^2$-spectra that would tell us that low-frequency behavior sets in is not observed. Clearly, a technique that would permit the measurement of relaxation times longer than 0.2 μs would open up new possibilities in these studies.

Application of the classical temperature-jump and pressure-jump techniques is hampered, for the former, by the fact that temperature changes do not, in general, perceptibly affect the optical absorbancies of virions and of other self-assemblies, and for the latter, by the fact that relaxation times longer than 100 μs are usually required.

It seemed that the transient Kerr effect (21) (transient electric birefringence) was a promising technique. One could hope that the electric field would induce dipoles in the subunits, and that the orienting effect of the electric field on these dipoles would so strain a spherical protein assembly that it would acquire optical anisotropy. The measurements were made by Yves Dormoy at the Center of Research on Macromolecules,with the equipment that was set up by Henri Benoît 35 years ago.

A large Kerr effect was observed in turnip yellow mosaic virus, as well as in its capsids. In the virion, the rising time, which characterized the deformation of the assembly, was about 50 μs, and was thus in the gap between the times that can be measured with the P-jump and ultrasonic techniques. The decay time was shorter, of the order of a few μs, in accordance with the expectation that thermal disorientation of the deformed and optically anisotropic particles controls the disappearance of the birefringence when the electric field has been switched off. The Kerr constant was larger in capsids than in virions. The dissociated protein showed almost undetectable birefringence. The results of these experiments will be published shortly (22).

It is a pleasure, at the end of this presentation, to have been able to report results that were obtained with the transient Kerr effect technique. In suggesting that these experiments be attempted, I remembered the happy time when Henri Benoît and I were starting to work in the cellar of the Institute of Physics, under the kind guidance of Professor Charles Sadron, whom we all honor today.

I would like also to express my gratitude to Professor Léon Hirth. With his proverbial open-mindedness and courtesy, he made possible the work on protein self-assemblies.

References

(1) Beece, D., Eisenstein, L., Frauenfelder, H., Good, D., Marden, M.C., Reinisch, L., Reynolds, A.H., Sorensen, L.B., and Yue, K.T. : 1980, Biochem. 19, pp.5147-5157.
(2) Cerf, R. : 1982, C.R.Acad.Sci. (Paris), série III, to appear.
(3) Kramers, H.A. : 1940, Physica 7, pp.284-304.
(4) Cerf, R. : 1960, C.R.Acad.Sci. (Paris) 250, pp.3599-3601.
(5) Cerf, R. : 1973, Chem. Phys. Lett. 22, pp.613-615.
(6) Cerf, R. : 1979, Biopolymers 18, pp.731-734.
(7) Skinner, J.L., and Wolynes, P.G. : 1978, J. Chem. Phys. 69, pp. 2143-2150.
(8) Gavish, B., and Werber, M.M. : 1979, Biochemistry 18, pp.1269-1275.
(9) Haas, E., Katchalski-Katzir, E.,and Steinberg, I. : 1978, Biopolymers 17, pp.11-31.
(10) McCammon, J.A., Gelin, B.R., and Karplus, M. : 1977, Nature 267, pp.585-590.
(11) Gavish, B. : 1980, Phys. Rev. Lett. 44, pp.1160-1163.
(12) Cerf, R. : 1976, 16th Solvay Conference on Chemistry, Brussels ; in 1979, Advances in Chemical Physics 39, pp.242-243.
(13) Cerf, R., Michels, B., Schulz, J.A., Witz, J., and Pfeiffer, P. : 1979, Proc. Natl. Acad. Sci. USA 76, pp.1780-1782.
(14) Caspar, D.L.D. : 1980, Biophysical J. 10, pp.103-136.
(15) Dormoy, Y., Cerf, R., Pantaloni, D., and Lamy, J. : to be published.
(16) Michels, B., Dormoy, Y., Schulz, J.A., Cerf, R., and Witz, J. : to be published.
(17) Durham, A.C.H., Finch, J.T., and Klug, A. : 1971, Nature New Biol. 229, pp.37-41.
(18) Braunwald, J., Nonnenmacher, H., and Tripier-Darcy, F. : 1981, 5th Internat. Cong. Virol., Strasbourg, France, Abst. P 37/01.
(19) Cuillel, M., Tripier-Darcy, F., Braunwald, J., and Jacrot, B. : 1979, Virology 99, pp.277-285.
(20) Robach, Y., Michels, B., Cerf, R., Darcy, F., Braunwald, J., and Kirn, A. : to be published.
(21) Benoît, H. : 1951, Ann. Physique 6, pp.561-609.
(22) Dormoy, Y., Cerf, R., and Witz, J. : to be published.

DYNAMICS OF PROTEINS

Martin Karplus
Department of Chemistry
Harvard University
Cambridge, Massachusetts   02138

ABSTRACT. A brief outline of theoretical approaches to the internal
motions of protein is presented. The magnitudes and time-scale of the
calculated fluctuations around the average structure are given and their
biological importance indicated.

In this lecture I shall discuss how one can examine theoretically
the internal motions of proteins; we have begun recently to apply the
same techniques to nucleic acids. Why are the motions of interest?
There are two primary reasons - first, proteins are intrinsically interesting systems from the physical chemistry viewpoint. They are long-
chain polymers, but unlike most polymers they have a well-defined average structure. This structure is amorphous or aperiodic, in the sense
that it does not have regular repeats. Since the structure is determined by weak, non-covalent interactions among the elements of the
polypeptide chain, large fluctuations are expected. It is important,
therefore, to know the nature of the fluctuations, to determine how
they take place and to evaluate their magnitudes and time scales.
Second, the motions in proteins play an important role in their functions. They may be the primary element, as in enzyme catalysis, or
they may be secondary, though essential, as they are in the oxygen
binding by myoglobin.

Before describing the details, it is useful to indicate the range
of motions that occur. They vary widely in amplitude, in energy, and
in their time scale (see Table I). In terms of their amplitudes, the
motions at room temperature range from a 10^{-2} to 10^{2} Å or so; the energies involved vary from 10^{-1} kT to 10^{2} kT; and the time scales go from
10^{-14} to 10^{3} sec. Another way of looking at the motions is in terms of
their characteristics; they are the local motions, rigid body motions
and a last category of large-scale motions. The local motions involve
the displacements of individual atoms or small groups of atoms, which
can be side chains or parts of the backbone (e.g., loops or chain
termini). Rigid body motions are those in which one part of the protein moves relative to another in an approximately rigid fashion; this

253

C. Hélène (ed.), Structure, Dynamics, Interactions and Evolution of Biological Macromolecules, 253–269.
Copyright © 1983 by D. Reidel Publishing Company.

Table 1. Classification of Internal Motions of Globular Proteins

Scales of Motions (300°K)

Amplitude	0.01 to 100 Å
Energy	0.1 to 100 Kcal
Time	10^{-14} to 10^3 sec

Types of Motions

Local	atom fluctuations, side chain oscillations, loop and "arm" displacements
Rigid Body Large-Scale	helices, domains, subunits opening fluctuation, folding and unfolding
Collective	elastic-body modes, coupled atom fluctuations, soliton and other non-linear motional contributions

may involve two or more helices, two domains, or the subunits of an oligomeric protein. The large scale changes are those that are involved in the partial unfolding and the folding of proteins.

One might expect that, as the amplitude of the motion increases, the associated energy of the motion will be larger and the time scale will be longer. This is true in many cases, but not always. There are motions that are slow because they are complex, involving the correlated displacements of many atoms. An example might be partial to total unfolding transitions. In such a case, the correlation of amplitude, energy, and time scale is expected to hold. However, there are much more localized events, often involving small displacements of a few atoms, in which the motion is slow because there is a high activation barrier; an example is provided by the aromatic ring flips in certain proteins. It is important to note that in this case the macroscopic rate constant can be very slow ($k \sim 1$ sec^{-1} at $300°K$) not because an individual event is slow (a ring flip occurs in $\sim 10^{-12}$ sec), but because the probability is very small ($p \sim 10^{-12}$) that a ring has sufficient energy to get over an activation barrier that is on the order of 16 Kcal.

It is clear that to examine protein motions, which cover a very wide range, one has to use a variety of methods. In this lecture I shall try to indicate how for each type of motion one can develop methods that are effective for studying them theoretically, and secondly to give some hint as to what one thinks the role of these motions might be in the biological function. Obviously, I do not have time to discuss all types of motion that can occur, so I will pick a number of examples from those that we have studied in detail.

SIMULATION METHODS

What theoretical approach does one use to study motions in proteins? The answer is very simple – one does exactly the same thing as one does when one wants to look at the motions in small molecules.

It is essential to have a knowledge of the potential energy surface, the energy of the system as a function of the atomic coordinates. The potential energy can be used directly to determine the stabilities of the different possible structures of the system; the relative populations of such structures under conditions of thermal equilibrium are given in terms of the potential energy by the familiar Boltzmann distribution law. The mechanical forces acting on the atoms of the systems are simply related to the first derivatives of the potential with respect to the atom positions. These forces can be used to calculate dynamical properties of the system, e.g., by solving Newton's equations of motion to determine how the atomic positions change with time. From the second derivatives of the potential surface, the force constants for small displacements can be evaluated and these can be used to find the normal modes; this serves as the basis for an alternative approach to the dynamics in the harmonic limit.

Although quantum mechanical calculations can provide a potential surface for small molecules, empirical energy functions of the molecular mechanics type are the only possible source of such information for proteins and their solvent surroundings. Since most of the motions that occur at ordinary temperatures leave the bond lengths and bond angles of the polypeptide chains near their equilibrium values, which appear not to vary significantly throughout the protein (e.g., the standard dimensions of the peptide group first proposed by Pauling in 1951), the energy function representation of the bonding can be hoped to have an accuracy on the order of that achieved in the vibrational analysis of small molecules. Where globular proteins differ from small molecules is that the contacts among nonbonded atoms play an essential role in the potential energy of the folded or native structure. From the success of the pioneering conformational studies of Ramachandran and co-workers in 1963 that made use of hardsphere nonbonded radii, it is likely that relatively simple functions (Lennard-Jones nonbonded potentials supplemented by a special hydrogen-bonding term and electrostatic interactions) can adequately describe the interactions involved.

The energy function used for proteins are generally composed of terms representing bonds, bond angles, torsional angles, van der Waals interactions, electrostatic interactions, and hydrogen bonds. The resulting expression has the form

$$E(\vec{R}) = \frac{1}{2} \sum_{\text{bonds}} K_b (b-b_o)^2 + \frac{1}{2} \sum_{\text{bond}} K_\theta (\theta-\theta_o)^2 +$$

$$\frac{1}{2} \sum_{\text{torsional}} K_\phi [1 + \cos (n\phi-\delta)] + \tag{1}$$

$$\sum_{\text{nb pairs}} \left(\frac{A}{r^{12}} - \frac{C}{r^6} + \frac{q_1 q_2}{Dr} \right) + \sum_{\substack{H \\ \text{bonds}}} \left(\frac{A'}{r^{12}} - \frac{C'}{r^{10}} \right)$$

The energy is a function of the Cartesian coordinate set, $\vec{R}$, specifying the positions of all the atoms involved, but the calculation is carried out by first evaluating the internal coordinates for bonds (b), bond angles (θ), dihedral angles (ϕ), and interparticle distances (r) for any given geometry, $\vec{R}$, and using them to evaluate the contributions to Eq. 1, which depend on the bonding energy parameters K_b, K_θ, K_ϕ, Lennard-Jones parameters A and C, atomic charges q_i, dielectric constant D, hydrogen-bond parameters A' and C', and geometrical reference values b_o, θ_o, n, and δ. For most protein atoms, an extended atom representation is used; i.e., one extended atom replaces a nonhydrogen atom and any hydrogens bonded to it. However, although the earliest studies employed the extended atom representation for all hydrogens, present calculations treat hydrogen-bonding hydrogens explicitly and generally use a more accurate function to represent hydrogen bonding interactions (e.g., angular terms are included) than that given in Eq. 1.

Given a potential energy function, one may take any of a variety

of approaches to study protein dynamics. The most exact and detailed
information is provided by molecular dynamics simulations, in which one
uses a computer to solve the Newtonian equations of motion for the atoms
of the protein and any surrounding solvent. With currently available
computers, it is possible to simulate the dynamics of small proteins for
periods of up to a few hundred ps. Such periods are long enough to
characterize completely the librations of small groups in the protein
and to determine the dominant contributions to the atomic fluctuations.
To study slower and more complex processes in proteins, it is generally
necessary to use other than the straightforward molecular dynamics simulation method. A variety of dynamical approaches, such as stochastic
dynamics, harmonic dynamics, and activated dynamics, can be introduced
to study particular problems.

Since the molecular dynamics simulation has so far provided the
most detailed and interesting results on protein motions, we briefly
describe the methodology. To begin a dynamical simulation, one must
have an initial set of atomic coordinates and velocities. These are
obtained from the X-ray coordinates of the protein by a preliminary calculation that serves to equilibrate the system. The X-ray structure is
first refined using an energy minimization algorithm to relieve local
stresses due to nonbonded atomic overlaps, bond length distortions, etc.
The protein atoms are then assigned velocities at random from a
Maxwellian distribution corresponding to a low temperature, and a dynamical simulation is performed for a period of a few ps. The equilibration
is continued by alternating new velocity assignments, chosen from
Maxwellian distributions corresponding to successively increased temperatures, with similar intervals of dynamical relaxation. The temperature, T, for this microcanonical ensemble is measured in terms of the
mean kinetic energy for the system composed of N atoms as

$$\frac{1}{2} \sum_{i=1}^{N} m_i <v_i^2> = \frac{3}{2} Nk_B T \qquad (2)$$

where m_i and $<v_i^2>$ are the mass and average velocity squared of the ith
atom, and k_B is the Boltzmann constant. Any residual overall translational and rotational motion can be removed to simplify analysis of the
subsequent conformational fluctuations. The equilibration period is
considered finished when no systematic changes in the temperature are
evident over a time of about 10 ps (slow fluctuations could be confused
with continued relaxation over shorter intervals); it is necessary also
to check that the atomic momenta obey a Maxwellian distribution and that
different regions of the protein have the same average temperature. The
actual dynamical simulation results (coordinates and velocities for all
the atoms as a function of time) for determining the equilibrium properties of the protein are then obtained by continuing to integrate the
equations of motion for the desired length of time. Several different
algorithms for integrating the equations of motion in Cartesian coordinates are being used in protein molecular dynamics calculations. Most
common are the Gear predictor-corrector algorithm, familiar from small

molecule trajectory calculations and the Verlet algorithm, widely used
in statistical mechanical simulations.

EQUILIBRIUM FLUCTUATIONS

Fig. 1 gives a qualitative picture of the fluctuations observed in
the molecular dynamics simulation of the basic pancreatic trypsin inhibitor (PTI), a small protein with 58 amino acids and 454 atoms; only
the α carbon atoms plus the three disulfide bonds are shown. The left-
hand drawing represents the X-ray structure and the right-hand drawing
an instantaneous picture of the equilibrated structure after 3 ps. It
is evident that the two structures are very similar but that there are
small differences throughout. The largest displacements appear in the
C-terminal end, which interacts with a neighboring molecule in the
crystal, and in the loop in the lower left, which has rather weak interactions with the rest of the molecule. Corresponding behavior and deviations from X-ray structure would be observed in "snap-shots" taken
at any other time during the simulation.

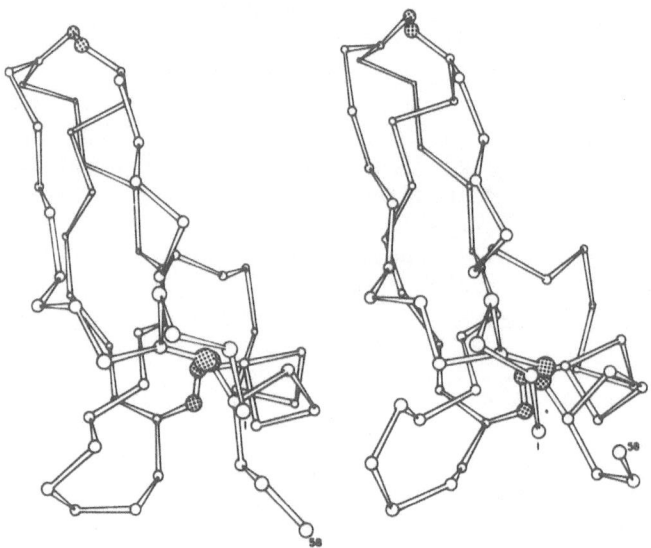

Fig. 1.   Drawing of α-carbon skeleton plus S-S bonds of PTI; left-hand
          drawing is the X-ray structure and right-hand drawing is a
          typical "snap-shot" during the simulation.

To obtain a more quantitative measure of the motions, we can calculate the mean square fluctuations of the atoms from their average positions. These can be related to the atomic temperature or Debye-Waller
factors, B, determined in an X-ray diffraction study of a protein crystal.
The mean-square positional fluctuation, $<\Delta r^2>_{dyn}$, with the assumption of
isotropic and harmonic motion, can be written

$$\langle\Delta r^2\rangle_{dyn} = \frac{3B}{8\pi^2} - \langle\Delta r^2\rangle_{dis} \tag{3}$$

where $\langle\Delta r^2\rangle_{dis}$ is the contribution to B from lattice disorder and other
effects that are difficult to evaluate experimentally. For a number of
proteins, the measured value of $(3B/8\pi^2)$ averaged over all of the non-
surface atoms of the protein is in the range $0.48 - 0.58$ $\AA^2$. Comparison
of this result with the mean value of $\langle\Delta r^2\rangle_{dyn}$ from the simulations
$(0.28 - 0.36$ $\AA^2)$ suggests that the non-motional contribution to the B
factor, $\langle\Delta r^2\rangle_{dis}$, is in the range $0.20 - 0.25$ $\AA^2$. The only experimental
estimate of $\langle\Delta r^2\rangle_{dis}$ is from Mössbauer data for the heme iron in myoglobin; for that one atom a somewhat smaller value $(0.14$ $\AA^2)$ was obtained.
Thus, in the cases examined, approximately half of the experimental B
factor is associated with thermal fluctuations in the atomic positions
and half with other sources. However, some protein and many nucleic
acid crystals, particularly those with a high percentage of water,
appear to have a larger disorder contribution (e.g., tortoise lysozyme).

For most proteins studied, there is an increase in the magnitude of
the experimental and theoretical fluctuations with distance from the center of the molecule. The magnitudes of the rms fluctuations range from
~0.4 $\AA$ for backbone atoms to ~1.5 $\AA$ for the ends of long sidechains.
The hydrogen-bonded secondary structural elements (α-helices, β-sheets)
tend to have smaller fluctuations than the random coil parts of the
protein. More generally, the magnitude of the fluctuations vary widely
throughout the protein interior, suggesting that the system is inhomogeneous and that some regions are considerably more flexible than others.
As to their time scale, the calculations show that the atomic fluctuations have relaxation times between 0.20 and 10 ps. This means that
they are associated with motions at "frequencies" in the range 5 to
300 cm^{-1}, which correspond to the low frequency coupled torsional modes
of polypeptide chains and proteins. If we analyze the motion of a given
atom in the protein, it is composed of two parts; one is a local libra-
tion which contributes about half of the mean-square amplitude, and the
other consists of more collective motions in which the atom moves with
parts of its surroundings. The latter can be a portion of the peptide
backbone, it can be a whole side chain, or even a larger region of the
protein.

Fig. 2 shows a comparison of the calculated and experimental rms
fluctuations on a residue by residue basis for reduced cytochrome c.
The experimental values have been corrected for an estimated disorder
contribution by subtracting from all of them $\langle\Delta r^2\rangle_{dis} = 0.25$ $\AA^2$, obtained
from the average calculated results for the protein interior. As is
evident, there is generally a very good correlation between the experimental and theoretical values. The most prominent differences involve
the residues that are calculated to have very large fluctuations; these
are all charged sidechains (particularly lysines) that protrude from
the protein and so are not correctly treated in the present vacuum

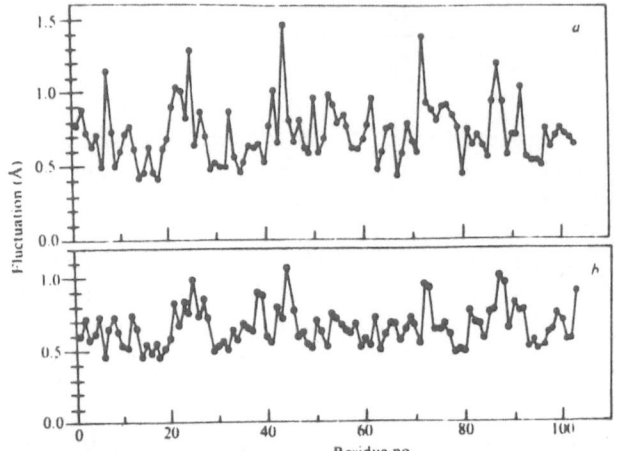

Fig. 2. Calculated and experimental rms fluctuations of ferrocytochrome
         c; residue averages are shown as a function of residue number:
         (a) molecular dynamics simulation; (b) X-ray temperature factor
         estimation corrected for mean disorder contribution.

simulation. A study of PTI in a Lennard-Jones solvent and in a crystal
environment shows that the motion of such outside residues is significantly perturbed by the surrounding medium; in particular, the interaction
between charged sidechains of a given protein and its crystal neighbors
can produce a reduction in the rms values. Such results for the external
residues contrast with those for the protein interior, where the medium
effects on the amplitude of fluctuations are found to be small.

     Of interest also are the results from the dynamic simulation concern-
ing deviations of the atomic motions from the isotropic, harmonic behavior
assumed in the X-ray analysis. The motions of many of the atoms in the
proteins examined are found to be highly anisotropic and somewhat anharmonic. For PTI in a 25 ps, room temperature simulation, the average
anisotropy obtained for all atoms, measured by the ratio of the largest to
the smallest eigenvalue of the mean-square fluctuation matrix for the
Cartesian coordinates, is found to be 2.5. Similar results have been obtained in an analysis of the anisotropic fluctuations in cytochrome c.
To explore the anharmonicity, we have done a detailed analysis of an
α-helix. A comparison was made between the full dynamics of the α-helix,
which includes harmonic and anharmonic terms in the potential, with the
results obtained from an harmonic analysis of the system done by determining the force constants, the vibrational modes and then calculating
from standard statistical mechanical formulas the expected mean square amplitudes of the atoms in the harmonic approximation. The study covered
the temperature range between 5° to 300°K. At very low temperatures
(5° to 50°K), the harmonic approximation is excellent and quantum corrections to the displacements are significant. Above 100°K, the quantum

corrections are negligible and anharmonic contributions become important.
It is between 100 and 200°K that the anharmonic part of the potential
comes to be sampled because the system has sufficient kinetic energy to
move out of the harmonic region.

One other point to mention is that the packing of the atoms in the
protein is so tight that it plays an important role in determining what
motions can occur. We have done a simulation in which we kept all of
the bond angles fixed and compared the results with one in which the bond
angles were allowed to vary. If the bond length in the protein are kept
fixed, it does not make any difference but if the bond angles are fixed
(even though they vary by only small amounts, 2-5°, in the normal dynamics of the protein), the protein is much more rigid and the mean-square
fluctuations are reduced by a factor of two. The essential point is
that a displacement of two atoms by a few hundredth of an Angstrom, which
can result from a small angular change, can greatly reduce the van der
Waals repulsion.

The motions that occur in a protein are very similar to those that
occur in a liquid, in the sense that the displacements are diffusive in
character and that the diffusion occurs basically by vacancies; i.e.,
there are correlated fluctuations of atoms that open vacancies into which
other atoms can move. Of course, in the protein (unlike a real liquid),
the atoms are not free to move but carry out their diffusive motion
around an average position that is fixed.

Although many of the individual atom fluctuations observed in the
simulations or obtained from temperature factors are probably not in
themselves important for protein function, they contain information that
may be of considerable significance. These motions serve to "lubricate"
the larger scale motions, such as those of the rigid body type. If the
atomic fluctuations were absent, the rigid body motions could not occur;
e.g., hinge bending motions require the local fluctuations to reduce the
effective potential to a range such that the required displacements are
possible at ordinary temperatures. In addition, it appears in a number
of enzymes (e.g., in ribonuclease A which we are studying in collaboration with G. Petsko) that the structures determined for the native enzyme
and for the enzyme with substrates, inhibitors, or transition state analogues bound differ in rather small displacements of the protein atoms in
and near the active site region. These displacements are on the order of
magnitude of the conformational subspace determined by the fluctuations
that occur in the dynamics simulation. We are now trying to understand
the specific role of these small changes. There is no question that
binding, say of the substrate, requires fluctuations and that for the
enzymatic reaction to take place motions must occur. Also, it is important to remember that the mean kinetic energy of a single atom is on the
order of 2 Kcal at 300°K (the exact value depending on the form of the
potential energy), so that fluctuations in the energy that are required
in an activated process, such as an enzymatic reaction, may involve a
relatively small number of atoms. There is then the possibility, though
there is no evidence for this so far, that there is some correlated

directional character present in the active site fluctuations as a result
of the remaining protein atoms and that this may play a role in catalysing the enzymatic reaction; e.g., fluctuations in enzymes may be directed along the reaction path by the structure of the surrounding protein.
The anisotropies that have been found in the atomic motions suggest that
this is a likely possibility. Further, the calculations show that there
are local correlations in the fluctuations and that some regions of the
protein are particularly flexible. This is found to be true not only in
comparing the inside and outside of a protein but one interior region
with another. Changes in the fluctuations induced by perturbations
(e.g., ligand binding), also can be important; e.g., the entropy differences may make a significant contribution to the binding free energy.

LIGAND-PROTEIN INTERACTION IN MYOGLOBIN

A biological problem where protein fluctuations are important concerns the manner in which ligands like carbon monoxide and oxygen are
able to get from the solution through the protein matrix to the heme
group in myoglobin and hemoglobin and then out again. These transport
proteins are systems in which the ligand is unchanged before and after
the interaction with the macromolecule, though a detailed description of
the heme iron-ligand bond would clearly require a quantum-mechanical
treatment. What makes this problem interesting is that examination of
the high-resolution X-ray structure of myoglobin does not reveal any
path by which ligands such as O_2 or CO can move between the heme binding
site and the outside of the protein. Since this holds true both for the
unliganded and liganded protein (i.e., myoglobin and oxymyoglobin),
structural fluctuations must be involved in the entrance and exit of the
ligands. Empirical energy function calculations have shown that the rigid protein would have barriers on the order of 100 kcal/mol; such high
barriers would make the transitions infinitely long on a biological time
scale. Fig. 3 shows a potential energy map for the shortest "path" from
the heme pocket to the exterior of the protein. The figure gives the
non-bonded potential contour lines seen by a test particle representing
on O_2 molecule in a plane (xy) parallel to the heme and displaced 3.2 Å
from it in the direction of the distal histidine; the coordinate system
in this and related figures has the iron at the origin and the z axis
normal to the heme plane. The low potential energy region in the center
is the so-called "heme pocket", with the energy minimum corresponding to
the observed position of the distal 0 atom of an O_2 molecule forming a
bent Fe-0-0 bond.

The shortest path for a ligand from the heme pocket to the exterior
(the low energy region in the upper left of the figure) is between His
E7 and Val E11. However, this path is not open in the X-ray geometry
because the energy barriers due to the surrounding residues indicated
in the figure are greater than 90 kcal/mol. Fig. 4 shows a possible path
to the exterior in a plan (xz plane) perpendicular to Fig. 3; again the
barriers in the X-ray structure are very large.

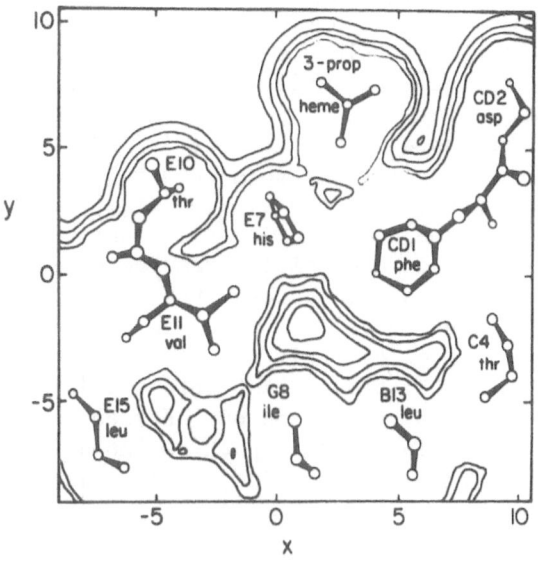

Fig. 3.   Myoglobin-ligand interaction contour map in the (x,y) plane
at z = 3.2 Å (see text).   Distances are in Å and contours in
kcal; the values shown correspond to 90, 45, 10, 0, and
−3 kcal/mol relative to the ligand at infinity.   The highest
contours are closest to the atoms whose projections onto the
plane of the figure are denoted by circles.

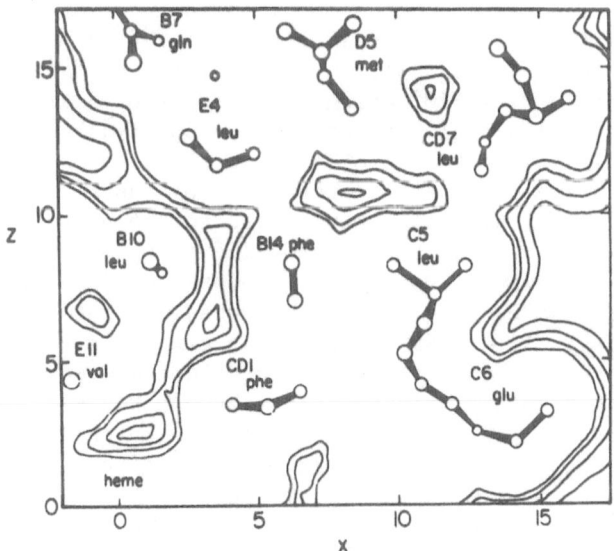

Fig. 4.   Myoglobin-ligand interaction contour map in the (x,z) plane
at y = 0.5 Å (see Fig. 3 legend).

To analyse pathways available in the thermally fluctuating protein, ligand trajectories were calculated using the static myoglobin X-ray structure together with a test molecule of reduced effective diameter to compensate for the absence of protein fluctuations. The trajectory was determined by releasing the test molecule with substantial kinetic energy (15 kcal/mol) in the heme pocket and followed its classical motion for a suitable length of time. A total of 80 such trajectories were computed; a given trajectory was terminated after 3.75 ps if the test molecule had not escaped from the protein. Slightly more than half the test molecules failed to escape from the protein in the allowed time; twenty-five molecules remained trapped near the heme binding site, while another twenty-one were trapped in two cavities accessible from the heme pocket. Most of the molecules which escaped did so between the distal histidine (E7) and the sidechains of Thr E10 and Val E11 (see Fig. 3). A secondary pathway was also found (see Fig. 4); this involves a more complicated motion along an extension of the heme pocket into a space between Leu B10, Leu E4, and Phe B14, followed by squeezing out between Leu E4 and Phe B14. A typical model trajectory following this path is shown in Fig. 5. Additional, more complicated pathways also exist, as is indicated by the range of motions observed in the trial trajectories.

In the rigid X-ray structure, the two major pathways described above have very high barriers for a thermalized ligand of normal size. Thus, it was necessary to study the energetics of barrier relaxation to determine whether either of the pathways had acceptable activation enthalpies. Local dihedral rotations of key sidechains were investigated and it was found that the bottleneck on the primary pathway could be relieved at the expense of modest strain in the protein by rigid rotations of the sidechains of His E7, Val E11, and Thr E10. The reorientation of these three sidechains and the resultant opening of the pathway to the exterior is illustrated schematically in Fig. 6; Panel I shows the X-ray structure (same as Fig. 3); in Panel II the distal histidine (E7) has been rotated to $\chi_1 = 220°$ at an energy cost of 3 kcal/mol; in Panel III, Val E11 has also been rotated to $\chi_1 = 60°$ (~5 kcal/mol); and Panel IV has the additional rotation of Thr E10 to $\chi_1 \cong 305°$ (<1 kcal/mol). In this manner a direct path to the exterior has been created with a barrier of ~5 kcal/mol at an energy cost to the protein of ~8.5 kcal/mol, as compared with X-ray structure value of nearly 100 kcal/mol. On the secondary path, however, no simple torsional motions reduced the barrier due to Leu E4 and Phe B14, since the necessary rotations led to larger strain energies. A test sphere with van der Waals radius of 3.2 Å was then fixed in the energy-refined structure at one of two positions in the bottleneck on the primary path (between His E7 and Val E11, or between His E7 and Thr E10) or in the bottleneck on the secondary path (between Leu E4 and Phe B14). The protein was allowed to relax by energy minimization (adiabatic limit) in the presence of the ligand and the resulting displacements in the polypeptide chains were monitored. There were local alterations in sidechain dihedral angles and bond angles. In addition, neighboring sidechains and the backbones of helices D and E participated in the globin response, mostly by small dihedral angle changes. Approximate values for the relaxed barrier heights were found

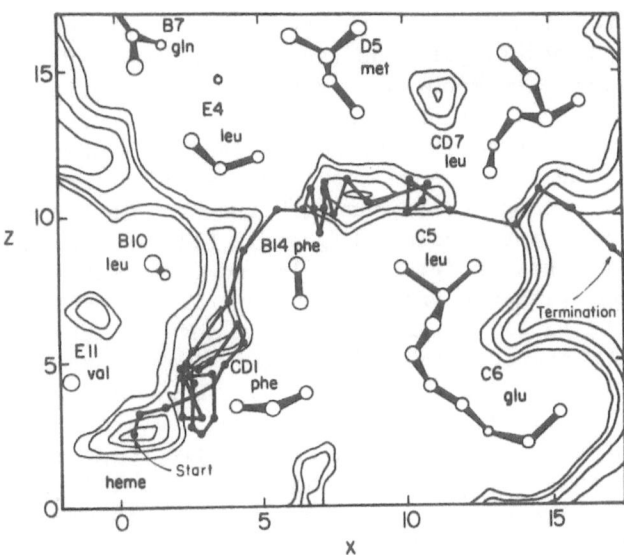

Fig. 5.   Diabatic ligand trajectory following the secondary pathway
          (see text); a projection on the plane of Fig. 4 is shown
          with the dots at 0.15 ps intervals. The start of the tra-
          jectory at the heme iron and the termination point exterior
          to the protein are indicated by arrows.

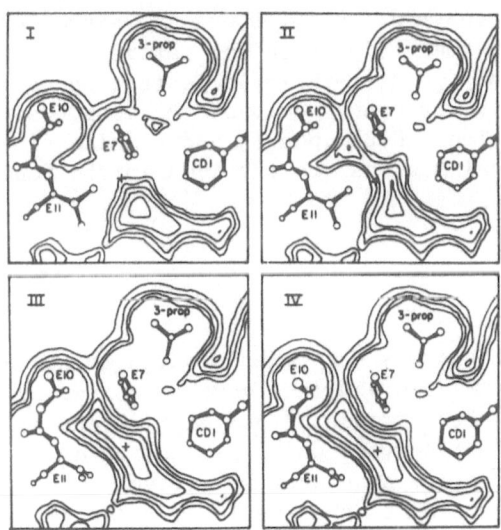

Fig. 6.   Myoglobin-ligand interaction contour maps in the (x,y)
          plane at z = 3.2 Å showing protein relaxation; a cross
          marks the iron atom projection onto the plane (see Fig. 3
          legend). The sidechain rotations relative to the X-ray
          structure (I) shown in Panels II-IV are discussed in the
          text.

to be 13 kcal/mol and 6 kcal/mol for the two primary path positions and 18 kcal/mol for the secondary path position. These barriers are on the order of those estimated in photolysis, rebinding studies for CO myoglobin. The type of ligand motion expected for such a several-barrier problem can be determined from the trajectory studies mentioned earlier. What happens is that the ligand spends a long time in a given well, moving around in and undergoing collisions with the protein walls of the well (see Fig. 5). When there occurs a protein fluctuation sufficient to significantly lower the barrier or the ligand gains sufficient excess energy from collisions with the protein, or most likely both at the same time, the ligand moves rapidly over the barrier and into the next well where the process is repeated. That the ligand spends most of the time in the low energy wells is evident from Fig. 5. However, it should be noted that in a completely realistic trajectory involving a fluctuating protein and ligand-protein energy exchange, the time spent in the wells would be much longer than that found in the diabatic model calculations. Further, from the complexity of the range of pathways in the protein interior, the motion of the ligand is expected to have a partly diffusive character.

From this analysis of myoglobin and more general considerations, it appears that in many cases the native structure of a protein is such that the small molecules which interact with proteins cannot enter or leave if the atoms are constrained to their average positions. Consequently, sidechain or other local fluctuations may be required for ligand binding by a variety of proteins and for the entrance of substrates and exit of products from enzymes. Also, there are situations in which the substrate pocket of enzymes is open in the native enzyme and sidechains (e.g., carboxypeptidase) or main-chain loops (triose phosphate isomerase) move after the substrate is bound to cover the active site and facilitate the catalytic event. Thus, both opening and closing fluctuations involving relatively local motions are involved in the function of many globular proteins.

RIGID-BODY MOTIONS

A large number of enzymes and other protein molecules (e.g., arabinose binding protein, immunoglobulins) consist of two or more distinct domains connected by a few strands of polypeptide chain which may be viewed as "hinges". In lysozyme, which will serve as an example, it was noted in the X-ray structure that when an active site inhibitor is bound, the cleft closes down somewhat as a result of relative displacements of the two globular domains that surround the cleft. Other classes proteins (kinases, dehydrogenases) have considerably larger displacements of the two lobes on substrate binding than does lysozyme.

In the study of lysozyme, the stiffness of the hinge was evaluated by the use of an empirical potential energy function. An angle bending potential was obtained by rigidly rotating one of the globular domains relative to a bending axis which passes through the hinge and calculating

the changes in the protein conformational energy. This procedure is expected to overestimate the bending potential, since no allowance is made
for the relaxation of unfavorable contacts between atoms that are generated by the rigid rotation. To take account of the relaxation, an adiabatic bending potential was calculated by holding the bending angle fixed
at various values and permitting the positions of atoms in the hinge and
adjacent regions of the two globular domains to adjust themselves so as
to minimize the total potential energy. As in a previous adiabatic ring
rotation calculation, only small (<0.3 Å) atomic displacements occurred
in the relaxation process; the differences between the rigid and adiabatic bending potential is largely due to small shifts in the relative
positions of a few atoms which have been forced too close together by
the rigid rotation model. The relief of these contacts can be effected
by localized motions (e.g., bond angle and local dihedral angle deformations). The frequencies associated with these deformations (>100 cm^{-1})
are expected to be much greater than the hinge bending frequency
(≈ 5 cm^{-1}), so that the use of the adiabatic bending potential appears to
be appropriate.

The rigid and adiabatic bending potentials were found to be approximately parabolic, with the restoring force constant for the adiabatic
potential about an order of magnitude smaller than that for the rigid
potential (see Fig. 7). However, even in the adiabatic case, the effective force constant is about twenty times as large as the bond-angle
bending force constant of an α carbon (i.e., N-C$_\alpha$-C); the dominant contributions to the force constant come from repulsive nonbonded interactions involving on the order of fifty contacts. If the adiabatic
potential is used and the relative motion is treated as an angular harmonic oscillator composed of two rigid spheres, a vibrational frequency
of about 5 cm^{-1} is obtained. This is a consequence of the fact that
although the force constant is large, the moments of inertia of the two
lobes are also large.

In considering the hinge bending motion it is essential to take
account of the fact that lysozyme is normally in solution. Although
fluctuations in the interior of the protein, such as those considered in
myoglobin, may be insensitive to the solvent (because the protein matrix
acts as its own solvent), the domain motion in lysozymes involves two
lobes that are surrounded by the solvent. To take account of the solvent
effect in the simplest possible way, we make use of the Langevin equation
for a damped harmonic oscillator. The friction coefficient for the solvent damping term was evaluated by modeling the two globular domains as
spheres and calculating the viscous frictional drag accompanying the
relative motion of these spheres by use of a modified Stokes law; the
internal friction of the protein was considered to be negligible compared to the hydrodynamic friction. From the adiabatic estimate of the
hinge potential and the magnitude of the solvent damping, it was found
that the relative motion of the two globular domains in lysozyme is over-
damped; i.e., in the absence of driving forces the domains would relax
to their equilibrium positions without oscillating. The decay time for
this relaxation was estimated to be about $2 \cdot 10^{-11}$s. Actually, the

lysozyme molecule will experience a randomly fluctuating driving force
due to collisions with the solvent molecules, so that the distance between the globular domains will fluctuate in a Brownian manner over a
range limited by the bending potential; the root-mean-square fluctuation
of the cleft width was estimated to be about 0.5 Å.

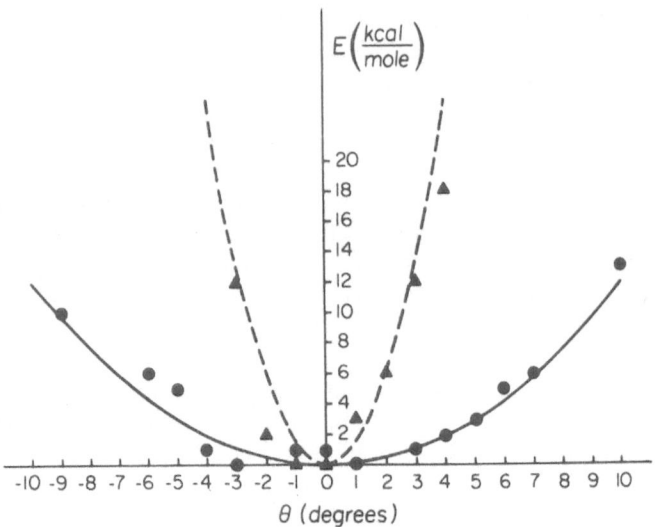

Fig. 7.   Change of conformational energy produced by opening ($\theta<0$) and
          closing ($\theta>0$) the lysozyme cleft; calculated values are for
          the rigid bending potential (triangles) and for the adiabatic
          bending potential (circles); the origins for the two calcula-
          tions are superposed.

     A system we have recently begun to study is liver alcohool dehydro-
genase, which is a dimer with each subunit composed of a coenzyme binding domain and a catalytic domain. The two domains move relative to
each other in an essentially rigid fashion by an angle of ~10° about a
fixed axis when the coenzyme is bound. Energy calculations of the open
enzyme structure (apoenzyme) indicate that the 10° motion required to
close the structure requires very little energy; i.e., one might expect
that the enzyme molecules without the coenzyme bound are fluctuating
between the open and closed structure at room temperature and become
fixed in the closed structure in the presence of coenzyme. Structurally,
there is the very interesting point that a "broken" helix connects the
two domains and that the rotation axis passes through the point at which
the helix changes direction. This suggests that nature has built a well-
defined flexible hinge into this enzyme. As we progress further in our
analysis of the system we hope to be able to show in a very clear way
how domain motions mediated by local atomic fluctuations play an important role in catalysis. In many such hinge bending enzymes, the product
release step is the rate determining step of the overall reaction. It
is interesting to suggest that it is in fact the opening of the molecule

by domain motion which is rate limiting. Since there is no experimental
evidence on this possibility, measurements involving, for example, fluorescence quenching or small angle X-ray scattering at low temperatures
are needed to provide the necessary data.

For a recent review with detailed references, see M. Karplus, J.A.
McCammon, 1983, Ann. Rev. Biochem. 52, to be published.

I am pleased to acknowledge the essential role played by my collaborators on various aspects of the work reviewed in this lecture. The
students, postdoctoral fellow, and colleagues who have contributed to
the work are B. Brooks, R. Bruccoleri, D. A. Case, F. Colonna, B. R.
Gelin, W. van Gunsteren, T. Ichiye, J. Kushick, C. T. Lee, R. M. Levy,
J. A. McCammon, S. H. Northrup, B. D. Olafson, M. R. Pear, D. Perahia,
D. States, S. Swaminathan, T. Takano, and P. G. Wolynes.

CONFORMATIONAL FLUCTUATIONS AND ENZYMATIC ACTIVITY

Jeannine M. YON
Laboratoire d'Enzymologie Physico-Chimique et Moléculaire,
Groupe de Recherche du C.N.R.S., Associé à l'Université de
Paris-Sud.
91405, F. ORSAY

S U M M A R Y

Protein conformation fluctuates around an equilibrium position in
a given environment. The free energy of stabilization of the native
structure is small and scarcely exceeds - 10 kcal/mole; thus, fluctuations of the environment can induce conformational fluctuations. Conformational oscillations and transmission of motions even at rather long
distance inside a protein play a role in determining and modulating the
functional properties of proteins. In this paper, we discuss the role
of conformational fluctuations on ligand binding and in enzyme catalysis
and we examine the particular aspect of the relative movement of two
structural domains or hinge bending mode in proteins.

I N T R O D U C T I O N

The remarkable progresses in biochemistry originate in a better
knowledge of structural properties of biological macromolecules, obtained
from X-ray three dimensional structures at atomic resolution. Crystallographic studies have presented protein conformation as a well compactly
packed structure. But folded proteins are not rigid bodies. Under thermal
motions, and fluctuations of the environment their conformation fluctuates
around an equilibrium position. The existence of sub-states more or less
populated according to the conditions is also suggested (1). Even
crystallographic data indirectly have provided information about flexibility of proteins. There are several examples of conformational changes
induced by ligand binding, revealed by X-ray crystallographic studies ;
hemoglobin (2) and carboxypeptidase (3) are well documented and classical examples. The recent improvement of crystallographic analysis allows
to detect motions in protein structure. Small motions have been observed
in human and hen egg white lysozyme (4,5) and in crystalline myoglobin
(1). A dynamic model of protein structures tends to supersede progressively the rigid view of crystal structure. After the determination of
space parameters, the time parameter is introduced in protein structure.
This new aspect of structural studies is very promising to understand

271

how is originated the biological activity of proteins, catalytic activity as well as regulatory properties which modulate the catalytic
function of enzymes.

   Reviews on motions in proteins have been published by Weber (6),
Careri et al (7,8), Gurd and Rothgeb (9). The importance of enzyme flexibility has been particularly emphasized by Citri (10).In this paper,
we discuss two particular aspects of conformational dynamics and enzymatic activity. The first refers to the conformational mobility of an
enzyme during catalysis; the second one is the relative motion of structural domains in two-domain proteins and its implication in enzyme activity.

1 - STABILITY OF THE NATIVE STRUCTURE.

   The native structure of a protein results from a balance between
a large number of stabilizing non covalent interactions (hydrophobic
interactions and hydrogen bonds) and conformational entropy which
tends to destabilize the structure. The resulting conformational free
energy scarcely exceeds -10 kcal/mole. Table 1 gives estimates of ΔG
for several proteins. These values have been obtained either from extrapolation procedure in studies of transition induced by guanidine hydrochloride (GuHCl) or urea, or by hydrogen exchange experiments (11-15).

T A B L E 1

Estimate of the free energy of conformational stabilization for several
proteins (a)

Protein		ΔG (kcal/mole)	
Ribonuclease pH 6.6		- 9.3 *	
pH 4.7		- 6.5°	
Hen egg white lysozyme pH 2.9		- 5.8*	
pH 5.4		- 7.4°	
Phage T4 lysozyme	pH 7.4	- 14	(b)
	pH 5.6	- 14	(c)
Chymotrypsinogen	pH 4.3	- 7.8*	
Lactoglobulin	pH 3.2	- 12.5*	
Myoglobin	pH 7	- 13°	
Horse muscle phosphoglycerate kinase			
	pH 7.5	- 3	(d)

   * linear extrapolation from GuHCl transition
   ° from hydrogen-deuterium exchange studies
   (a) from Pace (11)
   (b) from Desmadril et al. (13)
   (c) from Elwell and Schellman (14)
   (d) from Desmadril and Yon (15)

The stabilizing energy is small and fluctuations or weak variations of environment can induce either local or concerted motions in proteins.

2 - PACKING DENSITY AND DEFECTS IN PROTEINS

Globular proteins are characterized by a close packing of the atoms. However, packing defects giving rise to cavities inside the proteins have been reported (16,18). Packing density in the protein interior varies in different parts in the same protein. There are some parts with low densities, displaying packing defects (figure 1).

In myoglobin, the largest cavities have 2.85 A^3 and 2.29 A^3. The presence of cavities allows motion at the interior of a protein ; high density regions, which are not compressible , may transmit or correlate these motions over long distance through the molecule. Proteins contain flexible and rigid parts. The first ones are located near the surface of a protein and mainly at the active site of an enzyme. In sperm whale myoglobin, the packing of helix segments which are relatively rigid, allows a vectorial transmission of motions (19,20). Consequently to the various motions of the atoms, the internal cavities can be temporarily rearranged into channels allowing penetration of small molecules as oxygen. Such a mechanism has been proposed for the penetration of O_2 to the heme in myoglobin and hemoglobin (20,21). Penetration of oxygen inside several other proteins has been shown by fluorescence (22,23).

Fig. 1 - Superposition of van der Waal's contour and accessibility contour in a section of ribonuclease -S. The cavity inside the molecule is indicated by an arrow ; this cavity can accomodate a solvent molecule of radius 1.4 A°. (16).

3 - MOBILITY OF THE NATIVE STRUCTURE

3,1 - Time scale of dynamic events in proteins

There are various kind of motions in proteins ranging on a wide time scale, as indicated in table 2. Some of them are directly invol-ved in the catalytic process of an enzyme   reaction.

3,2 - Diversity of motions in proteins

Protein molecules are subjected to a great variety of motions. Local motions arise from external side-chains and also from internal side-chains located in cavities. Internal mobility of protein struc-ture results from vibrational and rotational motions, which have diffe-rent time scale. Vibrational motions range from 3.10^{-14} to 3.10^{-12} sec., whereas rotational motions occurring about either dihedral angles of the backbone or dihedral angles of side chains are ranging between 1-5 nanosecondes. Larger displacements are observed at the surface of proteins and at the active site. In lysozyme, the greatest displace-ments are observed in the cleft (4,5). In myoglobin, they occur in the region around the heme (1). The great mobility observed at the ac-tive site of proteins must be correlated with functional properties.

T A B L E   2

Time scale of dynamic events in proteins from Careri et al (7).

Determinants	Time (sec)
Protein surface	
Bound water relaxation	10^{-9}
Side-chains rotational correlation	10^{-10}
Proton transfer reaction of ionizable side chains	$10^{-7}-10^{-3}$
Protein conformation	
Local motion	$10^{-8}-10^{-9}$
Isomerization process	$10^{-2}-10^{-7}$
Folding-unfolding transition	$10^{+2}-1$
Enzyme substrate complex in solution	
Encounter rate	diffusion controlled
Estimated lifetime of the transition state in covalent reactions	10^{-10}
Change in metal ion coordination sphere in metalloenzymes	$10^{-6}-10^{-9}$
Enzyme-substrate local conformational motion	10^{-9}
Covalent enzyme-substrate intermediate lifetime	$10^{-2}-10^{-4}$
Enzyme-substrate complex conformational isomerization	$10^{-2}-10^{-4}$
Enzyme-substrate complex unfolding trantition	$10^{+2}-1$

There is evidence for concerted motions in proteins, such as breathing motions involving the entire molecule or limited regions, and the
relative movement of two structural domains called "hinge bending "mode
(24,25). Breathing motions have been shown by isotope exchange of internal hydrogens or by studies of the conformational equilibrium. These
approaches reveal the oscillation of protein molecules between a close
and an open conformation. The hinge bending mode was described for several proteins and studied by both theoretical and experimental methods.
These studies are reported in section 6.

4 - CONFORMATIONAL FLUCTUATIONS AND LIGAND BINDING

Conformational flexibility and motions in proteins determine the
expression of functional properties and their regulation. Several models
have been proposed to account for the efficiency of catalysis and for
regulatory properties of enzymes resulting from : substrate or ligand
binding. Conformational adaptability of a protein to a substrate is
illustrated by the induced-fit mechanism proposed by Koshland (26,27).
This mechanism provides a better respective orientation of the substrate and the catalytic groups which ensures the efficiency of an
enzymatic reaction. This model, so-called the instructive model (10)
was applied to allosteric enzymes built up of several subunits. It is
often presented as an alternative to the model introduced by Monod,
Wyman and Changeux (28), called the selective model. A generalized
model which includes both instructive and selective model as particular
cases, was proposed first by Weber (29,30) and by Wyman (31).

$$P_1 \rightleftharpoons P_2 \rightleftharpoons \cdots P_{j-1} \rightleftharpoons P_j \rightleftharpoons P_{j+1} \rightleftharpoons \cdots P_m$$

$$\Updownarrow \qquad \Updownarrow \qquad \Updownarrow \qquad \Updownarrow \qquad \Updownarrow \qquad \Updownarrow$$

$$P_1 X \rightleftharpoons P_2 X \rightleftharpoons \cdots P_{j-1} X \rightleftharpoons P_j X \rightleftharpoons P_{j+i} X \rightleftharpoons \cdots P_m X$$

$$P_1 X^{i-1} \rightleftharpoons P_2 X^{i-1} \rightleftharpoons \cdots P_{j-1} X^{i-1} \rightleftharpoons P_j X^{i-1} \rightleftharpoons P_{j+1} X^{i-1} \rightleftharpoons \cdots P_m X^{i-1}$$

$$\Updownarrow \qquad \Updownarrow \qquad \Updownarrow \qquad \Updownarrow \qquad \Updownarrow \qquad \Updownarrow$$

$$P_1 X^i \rightleftharpoons P_2 X^i \rightleftharpoons \cdots P_{j-1} X^i \rightleftharpoons P_j X^i \rightleftharpoons P_{j+1} X^i \rightleftharpoons \cdots P_m X^i$$

$$P_1 X^n \rightleftharpoons P_2 X^n \rightleftharpoons \cdots P_{j-1} X^n \rightleftharpoons P_j X^n \rightleftharpoons P_{j+1} X^n \rightleftharpoons \cdots P_m X^n$$

In this scheme, several sub-states of the enzyme are in equilibrium ; most of them are very little populated in the absence of ligand,
and may be stabilized preferentially by the binding of a given ligand.

In allosteric proteins concerted motions ensure transmission of
signals (for example that resulting from ligand binding) from one protomer to the other through the quaternary constraints. Such mechanism
allowing either amplification or attenuation of the catalytic efficiency of an enzyme is involved in regulation of cell metabolism. In collaboration with C. Ghélis (32,33) we have proposed an extension of the
concept of allostery to monomeric proteins organized in different structural domains, domains in monomeric enzymes being compared with subunits
in oligomeric proteins. We have emphasized the importance of a right
conformational coupling between domains allowing transmission of signals
from one domain to another for the optimization of the catalytic
efficiency.

5 - CONFORMATIONAL DYNAMICS AND CATALYSIS

Several researchers have emphasized recently the role of structural motions in enzymatic activity (6,10, 23,24). Catalytic efficiency
is conditionned by the substrate binding process. It is reasonnable
to assume that the most specific substrates are those which select and
stabilize one of the most active sub-states of an enzyme ; the binding
of allosteric activators might help in this selection.

During the catalytic pathway, an enzyme can transit by different
conformational sub-states. Feldhammer et al (34) have suggested that
chymotrypsinogen cannot reach its optimal conformation during catalysis
whereas chymotrypsin can.

With β-galactosidase from E. coli (wild type), we have shown that
a conformational step is the rate limiting process for the hydrolysis
of specific substrates. This step preceeds the formation of the galactosyl-enzyme. For different aryl galactosides with either electron
donating or electron withdrawn substituants, there is no correlation
between the enzymatic hydrolysis rate constant and the σ parameter
according to the Hammet equation (table 3). Thus k_2 does not correspond
to a chemical reaction but reflects some conformational modification
of the enzyme. By contrast for non specific substrates, k_2 reflects
the first chemical step i.e. the formation of a galactosyl-enzyme (35).
Our interpretation was supported by the analysis of the isotope effect
(36).

The positive effect of a conformational change is observed only
for the enzyme in the presence of Mg^{++}, which is an activator of the
enzyme (35, 37,38). With the Mg^{++} free enzyme no conformational step
was observed. It does not exist with a point mutant, the CZP protein,
which has a very weak activity (39). The remarkable efficiency of
β-galactosidase on specific substrates perhaps originates in this
conformational step.

Conformational rearrangements of β-galactosidase during catalysis
have been also evinced in the study of the action of various effectors
(glycosides or alcohols) on the reaction catalyzed by this enzyme (40).
Table 4 summarizes the results. Large effectors as tri-and tetrasaccha-

T A B L E   3

Kinetic parameters for the hydrolysis of several
glycosides by β-galactosidase (wild type) from E. coli
and in the presence of 1 mM $MgSO_4$ values for the
corresponding substrates.

Glycosides	k_{cat}	K_m	k_2	k'_3	σ
	s^{-1}	mM	s^{-1}	s^{-1}	
β-D-Galactosides :					
Phenyl	45	0.1	45		
o-Nitrophenyl	750	0.11	2100	1200	1.24
m-Nitrophenyl	800	0.15	1900	1400	0.71
p-Nitrophenyl	90	0.03	90		0.78
o-Aminophenyl	56	0.45	56		0.11
p-Aminophenyl	90	0.33	90		-0.66
α-L-Arabinoside :					
o-Nitrophenyl	44	4.3	44		

rides have no effect on the enzyme. Most of the β-galactosides produce
competitive inhibition. The other compounds behave either as non competitive or as uncompetitive inhibitors. The data indicate the existence
of two subsites : a galactose and a glucose subsite . This latter
site is in a more favorable conformation in the galactosyl-enzyme than
in the free enzyme. That indicates conformational rearrangements at
least at the active site between Michaelis complex and galactosyl enzyme.
The glucose subsite could be generated even by such a conformational
rearrangement induced by galactose binding.

T A B L E 4

Action of effectors of β-galactosidase as estimated
from kinetic parameters (40).

Compounds	K_1	K''_1	k_4
	mM	mM	s^{-1}
Competitive effectors			
Galactose	40		
Deoxygalactose	160		
Isopropyl-β-D-thiogalactoside	0.085		
Phenyl-β-D-thiogalactoside	0.19		
p-Aminophenyl-β-D-thiogalactoside	1		
o-Nitrophenyl-β-D-thiogalactoside	0.3		
Hydroxyethyl-β-D-thiogalactoside	0.32	2	0
Lactose	1		
Mesoerythritol	9.6		
Dulcitol	155		
Noncompetitive effectors			
Arabinose	400	180	0
Lyxose	105	78	0
Cellobiose	200	130	0
Maltose	260	135	0
L-(-)-Arabitol	105	150	0
D-(+)-Arabitol	70	20	2260
Uncompetitive effectors			
Fucose	300	25	110
Glucose	630	34	330
Melibiose	170	27	0
Saccharose	950	48	166
Mesoinositol	1060	50	420
Compounds without effect			
Trehalose			
Raffinose			
Melezitose			
Stachyose			

Conformational changes induced by ligand binding (isopropyl thiogalac-
toside or D-galactal) have been detected directly by difference spectroscopy (40). All these data suggest that at each step of the reaction
catalyzed by β-galactosidase, discrete conformational changes which
may be localized at the active site occur. One of these isomerizations
becomes the rate limiting step for specific substrates, with the wild
type enzyme in the presence of Mg^{++}. The dynamic conformation of the
active site seems related with the catalytic efficiency. Although we

have no arguments in this case to determine if the substrate has an instructive or a selective role, we can suggest that the more specific the substrate, the more favorable sub-state of the enzyme it selects and stabilizes.

6 - HINGE BENDING MODE AND ENZYMATIC ACTIVITY

Several proteins are built up of two lobes separated by a cleft which binds substrates. The binding of the substrates induces a conformational change by a hinge bending motion of the two domains and this closes the cleft between them. Hinge bending motion has been suggested for T4 lysozyme (41), hen egg white lysozyme (42), immunoglobulins (43) and for several kinases, such as hexokinase (44), pyruvate kinase (45), phosphoglycerate kinase (25,46). The possibility of such a hinge bending motion has been proposed from data obtained by X-ray crystallography of the enzymes in the presence and in the absence of substrates, or by X-ray scattering studies. The hinge bending mode has been theoretically analyzed by Karplus and coworkers using the method of molecular dynamics with hen egg white lysozyme (42) and immunoglobulins from IgG class (43).

Since hinge motions involves the active site cleft, they may have a role in enzymatic activity. It seems to be particularly involved in kinases. Reactions catalyzed by a monomeric kinase such as wheat germ hexokinase (47) and yeast phosphoglycerate kinase (48) deviate from Michaelis behavior. A careful kinetic analysis of wheat germ hexokinase-catalyzed reaction has led to the concept of enzyme memory (47). A correlation between the occurrence of a hinge bending motion and a mnemonic behavior may be assumed, at least as a working hypothesis.

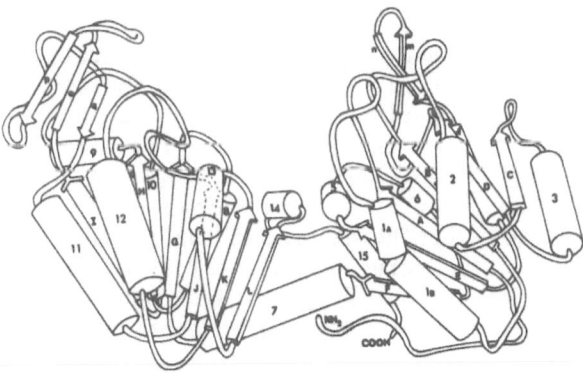

.Fig. 2 - Structure of horse phosphoglycerate kinase according to Banks and coworkers (25).

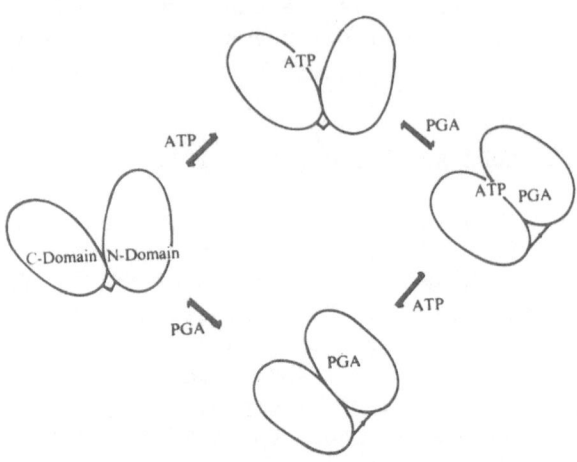

Fig. 3 - Mechanism proposed by Banks et al (25) for the
conformational change associated with substrate binding.

        A project in collaboration with M. Karplus involving theoretical
and experimental approaches is under study in our laboratory with horse
phosphoglycerate kinase. X-ray structure of this protein (molecular
weight 45.000) has been published by Banks and coworkers (25). This
enzyme catalyses the regeneration of ATP in the glycolytic pathway by
the following reaction :

$$1,3, \text{ diphosphoglycerate } + \text{ ADP } \underset{\longleftarrow}{\overset{Mg^{++}}{\longrightarrow}} 3, \text{ phosphoglycerate } + \text{ ATP}$$

        The molecule is organized into two structural domains, the N-
terminal domain binds specific substrates and the C-terminal domain binds
the nucleotides. A conformational change provoked by ligand binding has
been suggested as indicated in figure 3.
        The three dimensional structure is known at atomic resolution for
free enzyme, and more recently for the enzyme in the presence of nucleotide (49), but not yet for the enzyme in the presence of specific substrate. Examination of the three dimensional structure has suggested
that this enzyme is capable of hinge bending motion. A rotation of
10°-20° about a hinge located near the connection between helix 7 and
β-strand F has been assumed to be sufficient to close the cleft and
bring ATP and 3-phosphoglycerate into a contact sufficient for reaction.
        Several preliminary experimental results have been obtained with
this protein. The GuHCl induced transition has been studied ; the data
indicate the occurrence of intermediates in the folding pathway. The
overall structure of the molecule as probed by difference spectra and
fluorescence is regain for higher GuHCl concentration as the enzymatic
activity. Equilibrium studies and preliminary kinetic data are compatible with an independent refolding of domains. However, it is not yet

proved and further experimental informations are needed to propose a
plausible pathway of folding. A protection of the native structure by
phosphoglycerate was observed by following the reactivity of sulfhydril
groups and the susceptibility to proteolytic attack. X-ray scattering
of the enzyme in the presence and in the absence of ligands is under
study (50). For the moment we have only preliminary results, but many
signals can give information on motions in this protein. We want only
indicate the interest on such studies in the understanding of the mechanisms of enzyme action in terms of conformational dynamics.

B I B L I O G R A P H Y

1. Frauenfelder, H., Petsko, G.A. and Tsernoglou, D. 1979,Nature, 280
      pp. 588
2. Perutz et al, 1968, Nature, 219 , pp. 29 and 131.
3. Reeke, G.N., Hartsuck, J.A., Ludwig, M.L., Quiocho, F.A., Steitz, T.A.
      and Lipscomb, W.N., 1967, Proc. Natl. Acad. Sci., USA, 58, pp. 2220
4. Artymiuk, P.J., Blake, C.C.F., Grace, D.E.P., Oatley, J.J., Phillips,
      D.C.and Sternberg, M.J.E., 1979, Nature, 280, pp. 563
5. Sternberg, M.J.E., Grace, D.E.P. and Phillips, D.C., 1979, J. Mol.
      Biol. 130, pp 231
6. Weber, G., 1975, Adv. Prot. Chem. 29, pp. 1
7. Careri, G., Fasella, P. and Gratton, E., 1975, C.R.C. Crit. Rev.
      Biochem., 3, pp. 141.
8. Careri, G., Fasella, P. and Gratton, E., 1979, Ann. Rev. Biophys. Bioeng.
      8, pp. 69
9. Gurd, F.R.N. and Rothgeb, T.M., 1979, Adv. Prot. Chem., 33, pp. 74
10. Citri, N., 1973, Adv. Enzym., 37, pp. 397
11. Pace, C.N., 1975, C.R.C. Crit. Rev. Biochemistry, 3, pp. 1
12. Tanford, C., 1970, Adv. Prot. Chem., 24, pp. 1
13. Desmadril, M., Tempête-Gaillourdet, M., and Yon, J.M., 1982, submitted
      for publication.
14. Elwell, M. and Schellman, J., 1975, Biochem. Biophys. Acta, 386,
      pp. 309
15. Desmadril, M. and Yon, J.M., unpublished data
16. Lee, B. and Richards, F.M., 1971, J. Mol. Biol. 55, pp. 379
17. Richards, F.M., 1974, J. Mol. Biol. 82, pp. 1
18. Richards, F.M., 1977, Ann. Rev. Biophys. Bioeng., 6, pp. 151
19. Schoenborn, B.P., 1971, in Probes of structure and Function of Macro-
      molecules and Membranes (B. Chance, T. Yonetani and A.S. Mildvan
      ed.) Acad. Press,p.181, vol. 12
20. Richmond, T.J. and Richards, F.M., 1978, J. Mol. Biol. 119, pp. 537
21. Alberding, N., Chan, S.S., Eisenstein, L., Frauenfelder, H., Good, D.,
      Gunsalus, I.S., Nordlund, T.M., Perutz, M.F., Reynolds, A.H. and
      Sorensen, L.B., 1978, Biochemistry 17, pp. 43
22. Case, D.A. and Karplus, M., 1979, J. Mol. Biol. 132, pp. 343
23. Weber, G. and Laskowicz, J.R., 1973, Chem. Phys. Letters, 22, pp. 419
24. Karplus, M. and Mc Cammon, J.A., 1981, C.R.C. Critic. Rev. Biochem.
      9, pp. 293

25. Banks, R.D., Blake, C.C.F., Evans, P.R., Haser, R., Rice, D.W., Hardy, G.W., Merrett, M. and Phillips, A.W., 1979, Nature, London 279, pp. 773
26. Koshland, D.E., 1958, Proc. Natl. Acad. Sci., 44, pp. 98
27. Koshland, D.E., Némethy, G. and Filmen, D., 1966, Biochemistry, 5 pp. 365.
28. Monod, J., Wyman, J. and Changeux, J.P., 1965, J. Mol. Biol. 12, pp. 88
29. Weber, G. in Molecular Biophysics ed. by P. Pullman and M. Weissbluth Acad. Press, 1965, pp. 369
30. Weber, G. and Anderson, S., 1965, Biochemistry, 4, pp. 544
31. Wyman, J., 1975, Proc. Natl. Acad. Sci. USA, 79, pp. 3983
32. Ghélis, C. and Yon, J., 1979, C.R. Acad. Sciences, Ser. D, 289, pp. 197
33. Ghélis, C. and Yon, J., 1982, Protein Folding, Acad. Press, N.Y.
34. Feldhammer, H., Bode, W. and Huber, R., 1977, J. Mol. Biol.,111,pp.415
35. Viratelle, O.M. and Yon, J.M., 1973, Eur. J. Biochem. 33,pp.110
36. Sinnott, M. and Souchard, I.J.L., 1973, Biochem. J., 133, pp. 89
37. Tenu, J.P., Viratelle, O.M. and Yon, J.M., 1972, Eur. J. Biochem. 26, pp. 112
38. Viratelle, O.M. and Yon, J.M., 1980, Biochemistry, 19, pp. 41
39. Deschavannes, P., Viratelle, O.M. and Yon, J.M., 1978, Proc. Natl. Acad. Sci. USA, 75, pp. 1892
40. Deschavannes, P., Viratelle, O.M. and Yon, J.M., 1978, J. Biol. Chem. 253, pp. 833
41. Timchenko, A.A., Ptitsyn, O.B., Troitsky, A.V. and Denesyuk, A.I., 1978, FEBS Letters, 88, pp. 109
42. Karplus, M. and Wolynes, P.G., 1976, Nature, 262, pp. 325
43. Mc Cammon, J.A. and Karplus, M., 1977, Nature, 268, pp. 765
44. Bennett, W.S. and Steitz, T.A., 1980, J. Mol. Biol. 140, pp. 183 and 211
45. Muller, K., Kratky, O., Röschlan, P. and Hess, B., 1972, Hoppe Seyler's Z. Physiol. Chem., 353, pp. 803
46. Pickover, C.A., Mc Kay, D.B., Engelman, D.M. and Steitz, T.A., 1979, J. Biol. Chem., 254, pp. 11323
47. Ricard, J., Meunier, J.C. and Buc, J., 1974, Eur. J. Biochem., 49, pp. 195
48. Scopes, R.K., 1978, Eur. J. Biochem., 85, pp. 503
49. Blake, C.C.F., Personnal Communication
50. Yon, J.M. et al. unpublished data.

Fig. 1 is reprinted from Ref. (16), with permission from Academic Press Inc. (London) Ltd.
Figs. 2 and 3 are reprinted from Ref. (25), with permission from Macmillan Journals Limited.

X-RAY AND NEUTRON SCATTERING STUDY OF MEMBRANE PROTEINS IN SOLUTION

V. LUZZATI, A. TARDIEU, C. SARDET[+], M. le MAIRE
Centre de Génétique Moléculaire, CNRS, F91190 Gif-sur-Yvette

H.B. OSBORNE and M. CHABRE
Département de Recherche Fondamentale, CEN-G, F38041 Grenoble

[+] present address : Groupe de Biologie Marine du CEA, Station
Zoologique, FO6230 Villefranche-sur-Mer

Some chemical problems involved in the solution scattering study of
membrane proteins are discussed, as well as some theoretical and technical aspects of scattering experiments performed with systems containing
proteins and detergents. The results of the X-ray and neutron scattering
studies of rhodopsin are reported and compared : the conclusion is drawn
that X-rays are specifically suited for morphological analyses, whereas
neutrons are more sensitive to parameters related to internal structure.
The X-ray scattering study of Ca^{++} ATPase provides an example of how an
analysis of the complete scattering curve can lead to a fairly detailed
morphological description. Those examples illustrate three different
procedures for taking into account the influence of the detergent.

INTRODUCTION

One characteristic property of membrane proteins is their poor
solubility in aqueous solvents. In situ membrane proteins are indeed
anchored to the hydrocarbon core of membranes through hydrophobic interactions ; when these proteins are transferred to an aqueous medium their
hydrophobic surfaces associate to each other and aggregation occurs. On
the other hand, most of the techniques available for the study of proteins require homogeneous aqueous solutions ; if membrane proteins are
to be studied using those techniques then they must be solubilized. The
most successful procedure to achieve this goal is to use detergents.
Over the last decade membrane proteins-detergent interactions have been
thoroughly investigated and detergent solubilization has acquired a high
degree of sophistication (1). These investigations have paved the way to
the study of membrane proteins using a variety of physical and chemical
techniques (1).

The structure determination of membrane proteins by conventional
crystallographic techniques probably is one of the most urgent problems
in membrane research. Yet, progress in membrane proteins crystallization
has been extremely slow. Quite recently, two membrane proteins have

C. Hélène (ed.), Structure, Dynamics, Interactions and Evolution of Biological Macromolecules, 283–298.
Copyright © 1983 by D. Reidel Publishing Company.

been reported to crystallize, porin (2) and bacteriorhodopsin (3) ;
although the crystallographic analyses are still quite coarse, these
results may well be the forerunners of a major breakthrough. One excep-
tional case is bacteriorhodopsin which, <u>in situ</u>, is organized in highly
ordered two-dimensional arrays ; the electron microscope study of these
arrays (4) is the most sophisticated structure analysis performed so far
on membrane proteins. This exception, nevertheless, confirms the rule :
it has not been possible since to obtain nearly as detailed information
with any other membrane system.

For the time being, and in spite of these promising developments,
less ambitious approaches to the structure of membrane proteins are likely to yield valuable information. Among the techniques available, solution X-ray and neutron scattering are particularly powerful, especially
since the renewal that has taken place over the last decade, under the
impact of a variety of technical developments. For X-rays, the introduction of position sensitive detectors (5) and the use of synchrotron
radiation (6) had the overall effect of shortening exposure times by 4
to 6 orders of magnitude. On the other hand the construction of high
flux reactors and of improved facilities has transformed neutron scattering into a practical technique for the study of macromolecules in
solution (7).

We review in this paper the X-ray scattering study of two membrane
proteins, rhodopsin and Ca^{++}ATPase, and the neutron scattering study of
rhodopsin. These examples provide a clear illustration of structure analyses based upon solution scattering methods, and of the idiosyncrasy of
each of the two techniques.

CHEMICAL PROBLEMS

One severe requirement of solution scattering techniques, namely
chemical and physical homogeneity, is not easily fulfilled by detergent
solubilized membrane proteins. One cause of heterogeneity is the presence of free detergent micelles ; another frequent cause is the presence of polydisperse populations of protein-detergent particles (see
below the case of Ca^{++} ATPase). The problems raised by heterogeneity
can sometimes be circumvented if the chemical composition and the concentration of the different macromolecular components are kept under
control. In any event, several precautions must be taken :
a)- The choice of the detergent often is critical.The detergent must be
a good solubilizing agent and it must not upset the "native" structure
of the protein. Moreover, radioactive detergents are useful for chemical
analyses, deuterated ones for the neutron scattering study (see below).
b)- The purity of the protein preparation must be tested.
c)- The chemical composition of the protein-detergent particles must be
determined at the conditions of each scattering experiment - namely protein concentration, chemical and isotopic composition of the solvent -
and shown to be constant.
d)- The concentration of the free detergent micelles must be known, as

well as their contribution to the scattering curves.
e)- A variety of techniques must be used to test physical and chemical
homogeneity : velocity and equilibrium sedimentation are the most useful
ones, others (gel filtration, electron microscopy) can be of some help
(see review in (8)).

SOLUTION SCATTERING THEORY

This chapter is meant to provide an introduction to the experimental studies described below ; only the basic concepts and the essential
equations are presented. We adopt here the treatment and the notation
of a recent review (9). The results of that review - focussed on X-rays
- can easily be extended to neutrons by merely replacing "electron density" by "scattering density". It is worth to stress that whereas electron density is simply related to physical density, neutron "scattering
density" is dependent upon isotopic composition.

A most rewarding procedure in solution scattering studies is to
vary the scattering density of the solvent, and to perform the experiments as a function of that parameter. For X-rays, this can be achieved
by adding a dense component to the solvent (sucrose, ions, etc.), for
neutrons by varying the H_2O/D_2O ratio. When the solvent contains more
than one component it is also necessary to define, and to measure,
"solution", "solvent" and "concentrations" at constant chemical potential (10, 11).

The treatment below applies to ideal scattering curves, properly
corrected for collimation, chromatic and blanck scattering distortions.
We also assume that the solute - here the protein-detergent particles
or the detergent micelles - is chemically and physically homogeneous.

It is convenient to express the results of a solution scattering
experiment in the form of a normalized function $i_n(s)$ ($s=2\sin\theta/\lambda$, 2θ is
the scattering angle, λ the wavelength). For an ideal solution of identical particles in a solvent of electron density ρ_o the first two terms
of the expansion of $i_n(s)$ in powers of s takes the form :

$$\frac{i_n(s,\rho_o)}{c_e} = \{\frac{\bar{\rho}-\rho_o}{c_{ev}}\}^2 kM \{1 - \frac{4}{3} \pi^2 R^2(\rho_o) s^2 +....\} \qquad (eq. 1)$$

where c_e and c_{ev} express the concentration (solute/solution) respectively in electrons/electrons and electrons/A^3. M is the molar mass, k
the ratio (number of electrons)/(molar mass) of one particle of solute,
$\bar{\rho}$ and ρ_o are the average electron densities of solution and solvent,
$R(\rho_o)$ is the radius of gyration of the electron density contrast associated with one particle of solute (see eq. 5 below). Therefore, the
curvature of the scattering curve at s=0 yields the value of $R(\rho_o)$; if,
moreover, the intensities are properly calibrated, then the molar mass
of the solute can also be determined.

The first term of eq. 1 is as far as thermodynamic arguments can lead : it is hardly possible to retrieve additional information without reference to models. One model, particularly useful in solution X-ray and neutron scattering experiments, is specified by the <u>invariant volume hypothesis</u> which, in its more general formulation (9) is equivalent to assuming that, at all solvent density ρ_o, the electron density contrast associated with one particle of solute is linearly dependent upon ρ_o :

$$\Delta\rho(\underline{r},\rho_o) = \rho(\underline{r},\rho_o)-\rho_o = \Delta\rho(\underline{r})-(\rho_o-\bar{\rho}_1)v_1(\underline{r}) \qquad\qquad (eq.\ 2)$$

$\rho(\underline{r},\rho_o)$ is the electron density distribution inside one particle, $\bar{\rho}_1$ is the buoyant density of the solute :

$$\int\Delta\rho(\underline{r},\ \bar{\rho}_1)\ dv_{\underline{r}} = 0 \qquad\qquad (eq.\ 3)$$

$\Delta\rho(\underline{r})$ is the distribution of electron density contrast at buoyancy, $v_1(\underline{r})$ is the distribution of the volume fraction excluded to the compound used to alter the density of the solvent. Note that $-(\rho_o-\bar{\rho}_1)\ v_1(\underline{r})$ is the distribution of electron density contrast when $(\rho_o-\bar{\rho}_1)$ goes to infinity.

Under quite general conditions (11) the first term of eq. 1 takes the form

$$\frac{i_n(0,\ \rho_o)}{c_e} = \{\frac{\bar{\rho}-\rho_o}{c_{ev}}\}^2 kM = \frac{(\rho_o-\bar{\rho}_1)^2 v_1^2}{kM} \qquad\qquad (eq.\ 4)$$

Moreover, as a consequence of the invariant volume hypothesis the second term becomes :

$$R^2(\rho_o) = \int r^2 \Delta\rho(\underline{r},\rho_o)\ dv_{\underline{r}} = R_v^2 - \frac{a}{\rho_o-\bar{\rho}_1} - \frac{b}{(\rho_o-\bar{\rho}_1)^2} \qquad\qquad (eq.\ 5)$$

v_1 and $v_1 R_v^2$ are the zeroth and the second moments of the function $v_1(\underline{r})$ (the origin of $\underline{r}$ is taken at the centre of mass of $v(\underline{r})$) :

$$\int v_1(\underline{r})\ dv_{\underline{r}} = v_1 \qquad\qquad (eq.\ 6)$$

$$\int r\ v_1(\underline{r})\ dv_{\underline{r}} = 0 \qquad\qquad (eq.\ 7)$$

$$\int r^2\ v_1(\underline{r})\ dv_{\underline{r}} = v_1 R_v^2 \qquad\qquad (eq.\ 8)$$

a is a parameter which takes into account the relative distribution of the high and the low density regions of the particle

$$a = v_1^{-1} \int r \Delta\rho(\underline{r}) \, dv_{\underline{r}}$$ (eq. 9)

$b(\rho_o - \bar{\rho}_1)^{-2}$ is the distance between the centres of mass of $\Delta\rho(\underline{r}, \rho_o)$ and of $v_1(\underline{r})$.

The only mathematical restriction applied to the function $v_1(\underline{r})$ is to be independent of ρ_o. From the experimental standpoint the properties of this function are remarkably different in X-ray and in neutron scattering techniques. In X-ray contrast variation studies any point of the sample can be expected - and is indeed observed (9) - to be either totally impenetrable (inside the solute particles) or totally penetrable (outside the solute particles) to the dense component : in other words the function $v_1(\underline{r})$ takes only two values, 0 or 1. On the contrary deuterium, the "dense" component in neutron scattering studies, penetrates the interior of the solute molecules via the diffusible water molecules and the exchangeable protons ; partial penetration is equivalent to $v_1(\underline{r})$ taking values intermediate between 0 and 1. This difference has important consequence (see below).

If, in addition to the small s region (see eq. 1) the function i(s) is explored over a sufficiently extended range of s, then it is possible to calculate the autocorrelation function :

$$p(r) = \frac{2}{r} \int s \, i(s) \, \sin 2\Pi rs \, ds$$ (eq. 10)

which describes the spherically averaged distribution of the vectors joining two points of the particle, weighed by the product of the values of the electron density contrasts (eq. 2) at those points.

RHODOPSIN

Rhodopsin, the major protein component of rod outer segments, is the seat of the primary molecular events of vision. Rhodopsin is a glycoprotein of molar mass 39,000, which contains a chromophore, 11-cis retinal (see review in (12)).

The first steps in the solution scattering study - extraction of rhodopsin by detergents and chemical characterization of the protein-detergent complex - were carried out by Helenius, Osborne and Sardet (13). Some of the results of the chemical and hydrodynamic tests are of great relevance to the solution scattering studies :
a)- The chemical composition of the complex was shown to be the same at all the experimental conditions - namely sucrose concentration for X-rays, degree of deuteration of solvent and detergent for neutrons.
b)- Sedimentation velocity experiments are consistent with the complex being chemically and physically homogeneous.
c)- The molar mass of the complex was determined.

It is worth stressing that not all detergents are suitable for
the solution scattering studies : the binding ratio of Triton X100, for
example, was found to vary with sucrose concentration.

An important issue in this kind of work is to ascertain that the
native conformation of the protein is preserved in the protein-detergent
complex (14). In the case of rhodopsin several tests, performed on both
the intact membranes and the complex, provided a positive answer (see
references in (15, 16)).
a)- The spectral properties and the ability to bleach when exposed to
light remain unaltered. b)- Hydrogen isotope exchange displays similar
kinetic trends. c)- The same fragments are obtained by proteolytic degradation. On the other hand, the observation that rhodopsin, once bleached
in the complex cannot be regenerated by addition of the chromophore
suggests that, in the presence of detergent, irreversible denaturation
occurs upon bleaching.

X-Ray scattering study

The X-ray scattering study (15) was performed on both the detergent micelles and the protein-detergent complex, and in the presence of
variable amounts of sucrose, in the range 0-50%. The detergent used was
dodecyldimethylamine oxide (DDAO).

The plots $\{i_n(0,\rho_o)/c_e\}^{1/2}$ vs ρ_o and $R^2(\rho_o)$ vs $(\rho_o-\bar{\rho}_1)^{-1}$ turn
out to be linear, in keeping with the notions (see above) that the
protein/detergent ratio is the same in all the particles of the complex,
that the invariant volume hypothesis is fulfilled and that the term b
is very small (in other words, that the centres of mass of the low and
the high density regions coincide). The slope and intercept of the two
straight lines, the knowledge of the partial specific volume (measured
in the absence of sucrose) and of the binding ratio (measured by chemical analysis) lead to the determination of the five parameters M, a,
R_v, $\bar{\rho}_1$, v_1 (see eqs. 4, 5). This is all the information which can be
retrieved from an analysis of the scattering curves near the origin of
s (see table). The positive value of the parameter a shows (see eq.
9) that the outer region of the complex is predominantly occupied by the
denser component, in this case protein.

Two of the scattering curves, respectively at 0 and 49.5% sucrose,
were recorded over a sufficiently extended range of s to make the
calculation of the autocorrelation functions possible. Although the two
functions are not as accurate as in more recent experiments (see Ca^{++}
ATPase below) yet they display several revealing features (see below).

All these data can be exploited to gain some information on the
structure of the protein moiety in the complex. The most obvious result,
derived from the molar masses of the complex and of rhodopsin, and from
the binding ratio (see table), is that each particle of the complex
contains one molecule of rhodopsin. In order to determine some morphological parameters of rhodopsin from those of the complex one must

Table : Some morphological parameters

	Rhodopsin-detergent complex		Rhodopsin molecule in the complex		Ca^{++}ATPase-det. complex
	X-Ray (15)	neutron (17)	X-ray (15)	neutron (17)	X-ray (21)
detergent	DDAO	Cemusol LA90			DOC
binding ratio	0.92	2.20			0.20
M. (daltons)	74,800	120,000	39,000	38,000	132,000
hydration	0.40	-	0.44	-	0.26
$v_1(\text{Å}^3)$	162,900	180,000 (?)	74,600	47,000	229,000
$R_v(\text{Å})$	29.5($\pm$ 0.4)	24.5($\pm$ 0.4)	30($\pm$ 2.5)	19.2($\pm$ 1.2)	37($\pm$ 1)
R_v^3/v_1	0.157($\pm$ 0.009)	0.082 (?)	0.36($\pm$ 0.10)	0.15($\pm$ 0.03)	0.22($\pm$ 0.02)

Binding ratios are expressed in (g detergent)/(g protein), hydration in (g water)/(g dry particle). Note that the parameters v_1, R_v and R_v^3/v_1 refer to the hydrated particle in the case of X-rays, to the volume not accessible to diffusible and to exchangeable protons in the case of neutrons ; the neutron values are undelined to stress this difference. Standard errors, as estimated in the original papers, are given for a few critical parameters (see text). Remember that accuracy is a subtle problem in solution scattering studies.

eliminate or at least take into account, the contribution of DDAO. In
low resolution scattering studies any region with constant electron density can be "erased" by setting the electron density of the solvent at
that very level. In the case of the rhodopsin-DDAO complex one may
expect that the electron density be fairly constant over the protein
volume – this is indeed a general property of proteins in solution (9) –
but not in the detergent volume, since the density of the hydrocarbon
is quite different from that of the polar layer. Therefore it is not
possible to obliterate the detergent. Alternatively, one can rely upon
the results of the X-ray scattering study of DDAO micelles to evaluate
the contribution of the detergent to the volume and radius of gyration
of the complex. This analysis, discussed in (15), yields a reliable
value of the volume v_p of rhodopsin and a good estimate of the lower and
upper limits of the radius of gyration R_p (see table). It must be stressed that these parameters refer to the volume excluded to sucrose,
containing one molecule of rhodopsin and a fairly large amount of water.

The most striking conclusion of this analysis is that rhodopsin is
a highly anisometric object : indeed the ratio R_p^3/v_p is much larger that
for a sphere (0.36±0.10, see table ; 0.111 for a sphere).

The autocorrelation functions (fig. 1) show that the maximal chord
of the particle – defined by the value of r beyond which p(r) vanishes
– is much larger in 0 than in 49.5% sucrose. Since the electron density
of the latter solvent is very close to that of a hydrated protein, then
this result shows that the outermost regions of the complex are occupied
by protein. The shape of the function $r^2p(r)$ in 0 sucrose provides additional information. In this solvent the electron density contrasts of
the protein and of the detergent regions are of opposite signs ; therefore the contribution of the vectors joining protein and detergent is
negative. The conspicuous minimum centred around 31 Å can thus be explained by a high concentration of detergent-protein vectors, in keeping
with the model sketched in fig. 2. Besides, a strongly asymmetric model
formed by a compact protein with an excentric detergent cap (this model
was popular at the time of the X-ray study) is ruled out by the observation that, at the largest values of r the function p(r) is positive :
indeed p(r) should be negative for that model (see in(16) a case in which
this situation was observed).

The overall conclusions of the X-ray scattering study are summarized by the model of fig. 2 : rhodopsin is an elongated molecule which, in
the protein-DDAO complex spans a flat aggregate of DDAO molecules.

Neutron scattering study

An extensive neutron scattering study was performed on rhodopsin
solubilized in a variety of detergents (17). The use of specific deute-
rations offered the possibility of eliminating the effects of the scattering density fluctuations inside the detergent moiety. The detergent
used was polyethylene (9)-lauryl-myristoyl alcohol (Cemusol LA 90), in
three forms : totally hydrogenated, totally deuterated, deuterated in

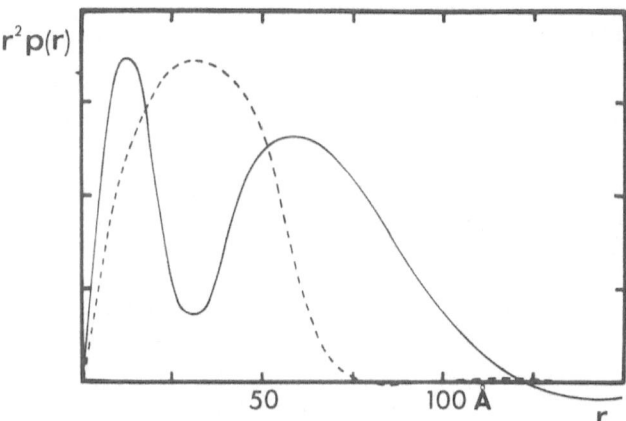

Figure 1. Autocorrelation functions (see eq. 10) of the
rhodopsin-DDAO complex, derived from the X-ray scattering
study (15). The scales of the two curves are arbitrary. The
uncertainty of these curves and of the position of their outer
edge (defining the maximal chord) was not properly estimated.
Full line - 0 sucrose. Note that in this case the electron
density in the protein and in the hydrocarbon regions are
respectively above and below that of the solvent. Several
features of the function are in excellent agreement with the
model of fig. 2a. a) - The position of the outer edge shows
that at least one dimension of the particle is larger that
100 Å (the maximal chord of the model is 101.6 Å). b) - The
autocorrelation function takes positive values just before
the outer edge ; as a consequence the longuest vectors join
regions whose electron density contrast is of the same sign
(protein-protein in the model). c) - The deep minimum, centred at 31 Å, corresponds to a high concentration of vectors
joining the protein and the hydrocarbon core of the detergent,
in agreement with the model. Dotted line - 49.5% sucrose. In
this solvent the protein (and also the polar group of the
detergent) are obliterated. The position of the outer edge is
not far from the maximal chord of the detergent micelle in
the model of fig. 2a.

the alkyl chains. By mixing the three forms in appropriate proportions, it was possible to show that at one precise D_2O/H_2O ratio the detergent micelles are totally obliterated. The neutron^2 scattering experiments were carried out as a function of the D_2O/H_2O ratio, on both the rhodopsin-detergent complex and the detergent micelles, using two diffe-rent mixtures of the three isotopic forms of the detergent. Only the pa-rameters $I(0,\rho_o)$ and $R^2(\rho_o)$ were determined.

The plots $\{I(0, \rho_o)\}^{1/2}$ $\underline{vs}$ ρ_o and $R^2(\rho_o)$ $\underline{vs}$ $(\rho_o-\bar{\rho}_1)^{-1}$ are linear

and lead to the determination of the parameters discussed above. The values of the binding ratio (determined by chemical analysis), of the molar mass, of the volume inaccessible to deuterium and of the radius of gyration of this volume are reported in the table. It must be noted that the amount of Cemusol bound to rhodopsin is much larger than that of DDAO (although the weight ratio alkyl chains/protein is almost the same for the two detergents). Therefore, if one assumes that the struc-ture of the two rhodopsin-detergent particles is similar, then the dia-meter of the detergent moiety is much larger for Cemusol than for DDAO, as depicted in fig. 2. This conclusion is confirmed by the slope of the plots $R^2(\rho_o)$ $\underline{vs}$ $(\rho_o-\bar{\rho}_1)^{-1}$ (parameter a in eqs. 5 and 9) : the sign of a shows that the outermost regions of the particle is predominantly occu-pied by the detergent in the rhodopsin-Cemusol complex, by the protein in the rhodopsin-DDAO complex.

A striking feature of the complex particles is the small value of the ratio R_v^3/v_1 (0.082, see table), so small in fact as to be incompati-ble with any function $v_1(r)$ whose upper and lower limits are 1 and 0 (the minimum, 0.111, corresponds to a sphere). The most likely explana-tion of this anomaly is that the volume be overestimated. It should be reminded that in neutron scattering studies the determination of the volume, unlike that of the radius of gyration, involves a calibration whose accuracy may sometimes be questionable. Moreover, the unsolvated volume of the rhodopsin-Cemusol particle is 1.65×10^5 $\mathring{A}^3$ (using chemical composition and partial specific volumes as reported in (15) and (17)): this volume is an upper limit since it disregards the presence of exchan-geable protons. In conclusion, and after taking this probable error into account, the ratio R_v^3/v_1 appears to be much smaller than in the X-ray study (see table).

As in the X-ray scattering study (see above), the next question is the structure of the rhodopsin molecule in the complex. The possibi-lity of obliterating the detergent should in principle make this pro-blem easy to solve : it would suffice to use a variety of preparations, each with one particular mixture of the three isotopic forms of the detergent, to measure the radius of gyration of each of those at the D_2O/H_2O ratio at which the detergent is obliterated, and then to extra-polate the radius of gyration to infinite contrast. In fact, this is a cumbersome procedure, and only two points were recorded (see fig. 14 in (17)); the extrapolation involved one additional point obtained with another detergent (DDAO) and the hypothesis that the term b of the

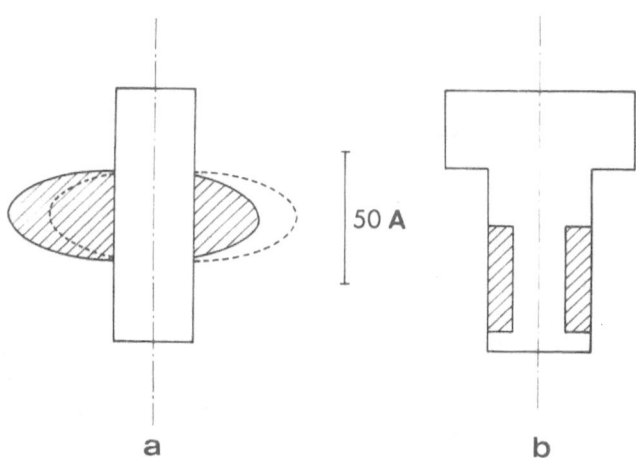

Figure 2. Models of the protein-detergent particles, derived
from the solution scattering studies. These models are meant
to provide a coarse description of the particles. The two
models are postulated to display cylindrical symmetry around
the vertical axes. The protein molecule is represented in
white detergent is hatched.

    a - Rhodopsin-detergent. (model proposed in (15)). The
hydrated molecule of rhodopsin is represented by a cylinder
of diameter 31.4 Å and height 96.6 Å. For DDAO (right half)
the hydrated detergent micelle is supposed to be an ellipsoid
of axes 34x80.2x80.2 Å, with a cylindral hole in the middle.
For Cemulsol LA90 the hydrated detergent micelle is also supposed to be an ellipsoid, of axes 34x113x113 Å (on the assumptions that the short axis is the same for the two detergents
and that the volumes are proportional to the binding ratios).
The agreement of the models with the experiments is discussed
in the text and in the legend of fig. 1. Note that the outermost region of the models is occupied predominantly by the
protein in the case of DDAO, and by the detergent in the case
of Cemusol in keeping with the slopes of the $R^2(\rho_o)$ $\underline{vs}$ $(\rho_o-\bar\rho_1)^{-1}$
straight lines observed in the X-ray and in the neutron
experiments.

    b - Ca^{++}ATPase - DOC - (model proposed in (21)).The model
is highly asymmetric, especially by contrast with the rhodopsin one. The shape and location of the detergent moiety is
speculative. The agreement of this model with the autocorrelation function is shown in fig. 3.

protein (see eqs. 5 and 9) is negligible. In spite of these limitations one can hardly avoid the conclusion that, at infinite contrast, the radius of gyration of rhodopsin is substantially smaller in the neutron than in the X-ray experiments (see table). This difference, consistent with the small value of the ratio R_v^3/v_1 discussed above, strengthens the conclusion that both hydration and exchangeable protons are more highly concentrated in the periphery than in the centre of the molecule (17). It is worth noting that the distribution of hydration and of exchangeable protons is related to the local concentration of different classes of amino acids (18).

Ca^{++} DEPENDENT ATPase

Ca^{++} dependent ATPase from sarcoplasmic reticulum is an intrinsic membrane protein involved in Ca^{++} active transport. Its molar mass is 115,000 g. Electron microscope and X-ray scattering studies on intact membranes suggest that this ATPase is an elongated molecule, with a bulky "head" protruding into the cytoplasm (see a review in (19)).

ATPase can be solubilized in oligomeric or monomeric forms using some ionic or non-ionic detergents (20). The oligomeric form, which seems to be biologically relevant, is enzymatically active ; the mono- meric form can be enzymatically active or not. A common feature of the enzymatically active preparations, both monomeric and oligomeric, is a tendency to aggregate with increasing concentration : this is a major drawback for solution scattering experiments. Using desoxycholate (DOC) two monomeric forms can be obtained (20). One, which contains a few bound lipid molecules is enzymatically active and tends to aggregate with time. The other, obtained by gel chromatography in the presence of DOC, is stable. Although the latter is delipidated and enzymatically inactive, its structure has been shown to be similar to that of the other forms (see references in (21)). For these two reasons - well defined and stable protein-detergent particles, structural analogy with the na- tive form - the delipidated monomeric form was chosen for an extensive X-ray scattering study (22). It is worth mentioning that this was one of the first solution X-ray scattering studies performed using synchro- tron radiation.

An important property of DOC is that the electron density distri- bution inside the detergent micelles is reasonably uniform, and that its value is close to that of a hydrated protein. As a consequence, the internal fluctuation term of eq. 2 may be expected to be small with res- pect to the volume term, at least when the solvent density is far from buoyancy ; therefore the scattering curve recorded in 0 sucrose may be assumed to describe the intensity distribution at infinite contrast (save a scaling factor). The values of the morphological parameters reported in the table correspond to that experimental curve.

The use of improved experimental techniques yielded, in this case, accurate measurements of the scattering curves over an extended range

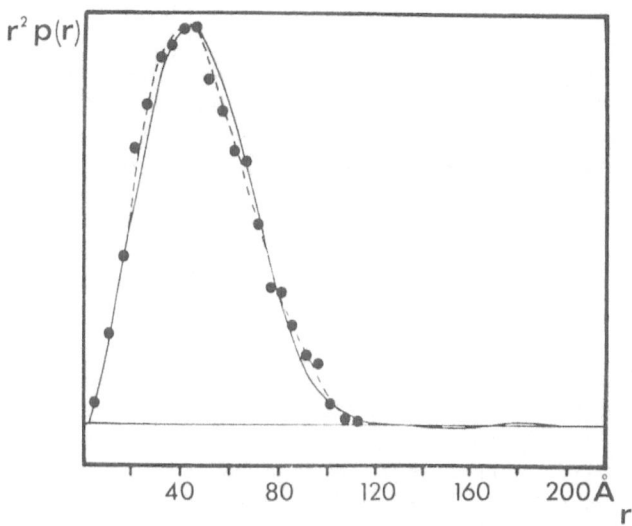

Figure 3. Autocorrelation function of the Ca^{++}ATPase - DOC complex. <u>Full</u> <u>line</u> : experimental curve obtained in 0 sucrose. <u>Dotted line</u> : curve calculated with the model of fig. 2b. The "noise" on the calculated points (black dots) is a consequence of the computational technique (24).

of s : as a consequence the autocorrelation functions are also accurate. The function r^{2}p(r), obtained in 0 sucrose, is plotted in fig. 3. It is easy to show that no simple model - sphere, cylinders, ellipsoids - is compatible with the parameters of the table and with the autocorrelation function. Therefore a more elaborate model had to be chosen ; its form, suggested by previous studies (21), consists of two coaxial cylinders of different diameters and lengths (fig. 2). This model has four degrees of freedom ; imposing the constraints of volume, radius of gyration and maximal chord (see table and fig. 3) that number drops to one. The residual parameter was adjusted by fitting the calculated and the experimental function r^{2}p(r). The final result is shown in figs. 2 and 3. Although it is next to impossible to prove that any model consistent with solution scattering experiments is unique, yet it must be stressed that such a remarkable agreement is most unusual. Therefore one can trust the model of fig. 2 as a reliable low resolution description of the structure of the ATPase-DOC complex (note that the location of the detergent in fig. 2 is speculative).

CONCLUSIONS

Until quite recently solution scattering techniques were sorted more often among the hydrodynamic methods than in the category of the structure-seeking procedures. The latest developments mentioned in the introduction are slowing making an impact and transforming those techniques into valuable tools for low resolution structure analyses. Some concern is also beginning to arise for a definition of the information content of solution scattering studies, especially by comparison with conventional crystallographic analyses (22).

As a consequence of the chemical complexity of the samples, membrane proteins are not ideal objects for solution scattering studies and the two examples above are not the best illustration of the power of those techniques for structure analyses. Yet, the results described here provide the most detailed information available so far on the physical structure of rhodopsin and of Ca^{++}ATPase. From the solution scattering study rhodopsin emerges as an elongated molecule spanning a flat detergent micelle, the Ca^{++}ATPase- DOC complex as an object with a bulky "head" and a thinner "tail". The obvious biological implications is that rhodopsin molecules most likely span the disc membranes in vivo, with one end exposed to the cytoplasm, the other to the extracellular medium and the central part to the hydrocarbon core of the membrane, (this conclusion is in agreement with other studies (23)).Ca^{++}ATPase, instead, is probably a highly asymmetric protein, with a bulky "head" exposed to the intercellular medium, and a longuer and thinner "body" embedded in (and probably spanning) the membrane.

From a technical standpoint, it is worth to point out the different procedures used to go about the problem raised by the presence of the detergent. With the Ca^{++}ATPase system the experimental conditions of the X-ray scattering study were such that the detergent behaved very much like the protein : thereafter the morphological analysis was carried out on the entire protein-detergent particle (the small amount of bound detergent was also a favourable circumstance). With the rhodopsin-DDAO system an analysis of the scattering properties of the detergent micelles led to an estimate of some morphological parameters relevant to the protein moiety of the complex. In the case of the neutron scattering study the detergent was obliterated by using a well poised mixture of three isotopic forms of the detergent.

The importance of measuring the whole of the intensity curves, and of not restricting the analysis to the "Guinier region" must also be stressed. The best example is the Ca^{++}ATPase-DOC complex. In the case of the rhodopsin-DDAO complex the very presence of conspicuous electron density fluctuations in the detergent prevented a detailed analysis of the autocorrelation functions ; yet, the differences between the mean electron density of the protein and that of the detergent could be exploited to retrieve reliable information.

The example of rhodopsin also illustrates the idiosyncrasy of the two low resolution scattering techniques. X-rays are ideally suited for morphological analyses of protein molecules. This is the consequence of two facts : a) - the electron density distribution is fairly uniform inside proteins ; b) - a volume totally impenetrable to the dense component of the solvent is associated to each protein molecule. On the contrary, the morphological information of neutron scattering experiments is deeply entangled with the information relevant to the distribution of diffusible water and of exchangeable protons, which in turn depends upon the spatial distribution of different classes of amino acids (18). Moreover, the fairly large differences of the scattering density of different amino acids (18) may appreciably influence the term independent of ρ_o in eq. 2. The obvious conclusion, amply documented by the example of rhodopsin, is that a most rewarding approach to the structure of macromlecular systems in solution is to use both X-ray and neutron variable contrast experiments : X-rays are likely to provide reliable information regarding size and shape, neutrons to tell something, even at very coarse resolution, about internal structure.

References

(1) Tanford, C. and Reynolds, J.A.R. (1976), Biochim. Biophys. Acta, 457, 133-170.
(2) Garavito, R.M. and Rosenbush, J.P. (1980), 86, 327-329.
    Garavito, R.M., Jenkins, J.A., Neuhaus, J.M., Pugsley, A. and Rosenbush, J.P. (1982), Ann. Microb. (Inst. Pasteur) 133A, 37-41.
(3) Michel, H. and Oesterheldt, D. (1980), Proc. Nat. Acad. Sci. USA, 77, 1283-1285.
(4) Henderson, R. and Unwin, P.N.T. (1975), Nature (London), 257, 28-32.
(5) Fourme, R., Bordas, J. and Koch, M., Nuclear Instruments and Methods, in press.
(6) Stuhrmann, H.B. (1978), Quart. Rev. Biophysics, 11, 71-98.
(7) Ibel, K. (1976), J. Appl. Cryst. 9, 296-309.
(8) Le Maire, M. and Møller, J.V. (1981), Biochimie, 63, 863-866.
(9) Luzzati, V. and Tardieu, A. (1980), Ann. Rev. Biophys. Bioeng. 9, 1-29.
(10) Eisenberg, H. (1981), Quart. Rev. of Biophysics, 14, 141-172.
(11) Tardieu, A., Vachette, P., Gulik, A. and le Maire, M. (1981), Biochemistry, 20, 4399-4406.
(12) Chabre, M. in "Membranes and intercellular communication", R. Belian, : M. Chabre, P. Devaux Editors, North Holland Publ. Comp. (1981), 251-264.
(13) Osborne, H.B., Sardet, C. and Helenius, A. (1974), Eur. J. Biochem. 44, 383-390.
(14) Mc Caslin, D.R. and Tanford, C. (1981), Biochemistry, 20, 5207-5212.
(15) Sardet, C., Tardieu, A. and Luzzati, V. (1976), J. Mol. Biol. 105, 383-407.

(16) Luzzati, V., Tardieu, A. and Aggerbeck, L. (1979), J. Mol. Biol. 131, 435-473.

(17) Osborne, B., Sardet, C., Michel-Villaz, M. and Chabre, M. (1978), J. Mol. Biol., 123, 177-206;

(18) Zaccai, G. and Gilmore, D.J. (1981), J. Mol. Biol. 132, 181-191.

(19) Møller, J.V., Andersen, J.P. and le Maire, M. (1982), Molec. and Cell. Biochem., 42, 83-107.

(20) Le Maire, M., Jørgensen, K.E., Røigaard, H. and Møller, J.V. (1976), Biochemistry, 15, 5805-5812.

(21) Le Maire, M., Møller, J.V. and Tardieu, A. (1981), J. Mol. Biol. 150, 273-296.

(22) Taupin, D. and Luzzati, V. (1982), J. Appl. Cryst., 15, 289-300.

(23) Saibil, H., Chabre, M. and Worcester, D. (1976), Nature (London), 262, 266-270.

(24) Tardieu, A. and Vachette, P. (1982), EMBO Journ., 1, 35-40.

FAST KINETIC SPECTROSCOPIES : LASER FLASH PHOTOLYSIS AND PULSE RADIOLY-
SIS : APPLICATION TO THE STUDY OF THE STRUCTURE, THE DYNAMICS AND
THE INTERACTIONS OF BIOLOGICAL MOLECULES.

R. SANTUS
MUSEUM NATIONAL D'HISTOIRE NATURELLE - Laboratoire de
Physico-Chimie de l'Adaptation Biologique et Laboratoire
de Photobiologie Moléculaire : ERA 951 du C.N.R.S.
43, Rue Cuvier - 75231 PARIS CEDEX 05

ABSTRACT : It is shown, using a few examples, that laser flash spectroscopy and pulse radiolysis are convenient methods for studying the
structure and dynamics of proteins and their interactions with nucleic
acids or polynucleotides. In the case of hemoproteins the short duration of the laser pulse makes it possible to follow conformational
changes in a time scale expanding from the nanosecond to the second.
Selective attack by negatively charged halide radicals formed upon
pulse radiolysis of aqueous solutions of proteins makes it possible to
determine aminoacid residues involved in the active site of the protein
or in interaction with DNA.

INTRODUCTION : The active forms of biological macromolecules always involve more or less perturbed states of the "native" form. These perturbations can be brought about by chemical (ions, molecules or macromolecules) or physical (temperature, pressure, radiation field) effectors.
In this latter case, the electromagnetic wave induces a chemical modification at the level of a particular component of the macromolecule. This
modification leads to a change in the conformation and/or the activity
of the macromolecule. It can also impede or favor interactions with
other molecules or macromolecules. As a result living systems undergo
photoactivation (photosynthesis, vision, pigmentation...), photoinactivation (via alteration of the genetic code or enzyme photoinactivation)
or mutations (via the effects of ionizing or ultraviolet radiations on
DNA).
    Light and ionizing radiations produce very different primary
effects. While the selective absorption of photons in the absorption
band of molecules provokes the formation of excited states, ionizing
radiations act upon the solvent, forming primary radicals which are
able to react with molecules or macromolecules dissolved in the solvent.
However as described below, under appropriate experimental conditions,
these reactions can lead to the same transient radical intermediates.

    The study of these intermediates can be conveniently carried
out in a time scale extending from the nsec. to the second through the
use of two complementary techniques : the pulse radiolysis and the

C. Hélène (ed.), Structure, Dynamics, Interactions and Evolution of Biological Macromolecules, 299–310.
Copyright © 1983 by D. Reidel Publishing Company.

laser flash photolysis techniques. Both of them are based on the same
principle : a short pulse of photons (laser flash photolysis) or
accelerated electrons (pulse radiolysis) excites the sample. In both
cases the resulting changes in optical absorption are followed, as a
function of time, by spectrophotometry.

In this article, we shall present the most recent applications
of these two techniques to the study of the function and dynamics of
biological molecules. We will also give a few examples based on Research
carried out at the Muséum National d'Histoire Naturelle. They concern
studies on nucleic acid-peptide (or protein) interactions or studies on
the active site and the dynamics of hydrolases or hemoproteins.

1. A LASER FLASH PHOTOLYSIS STUDY OF THE STRUCTURE AND THE DYNAMICS OF HEMOPROTEINS.

1.1. Dissociation and recombination of oxygen or carbon monoxyde with hemoglobin or myoglobin.

Historically, the first major experimental studies involving
nano or picosecond laser flash photolysis were carried out on these
two globins by *Lindqvist and coll. (1974)* and *Rentzepis and coll. (1978)*.
They consist to excite the two globins in the visible (530 nm) or
UV (353 nm) absorption bands of their prosthetic group in order to
study the dynamics of molecular events preceding the quaternary conformational changes which trigger the cooperativity of the O_2 binding.
An obvious problem is, of course, to discriminate between tertiary
and quaternary motions of the proteins. This led Rentzepis and coll.
to use myoglobin since in this case one is just concerned with tertiary
conformational changes.

The comparison of the difference spectra obtained 200 ps
after excitation of carboxymyoglobin (MbCO) and oxymyoglobin (MbO_2)
with a 25 ps laser pulse shows that in both cases the final product
is the same (Fig. 1). They correspond to the formation of deoxymyoglobin
(*Eisert et al. 1979*) (*Reynolds and Rentzepis, 1982*). The disappearance
of the liganded species, measured at 420 nm, occurs at comparable rate
(a few picoseconds) for MbCO and MbO_2. Although the dissociation
quantum yield (measured at 420 nm) is independent of the excitation wavelength and is three times greater with MbCO than with MbO_2, (against
30 times under excitation with a conventional flash lamp !), the
deoxymyoglobin formation quantum yield (measured at 450 nm) depends on
the excitation wavelength. It is three times and nine times greater
with MbCO than with MbO_2 upon excitation at 353 and 530 nm respectively.
Under steady state irradiation or conventional flash photolysis
at the same wavelengths, these quantum yield ratios are 3 and 10
respectively. This remarkable molecular adaptation which gives a
greater stability to MbO_2 is due to a very fast recombination rate constant for the binding of O_2 to the protein because O_2 does not escape
from the heme pocket (*Reynolds et al., 1981*).

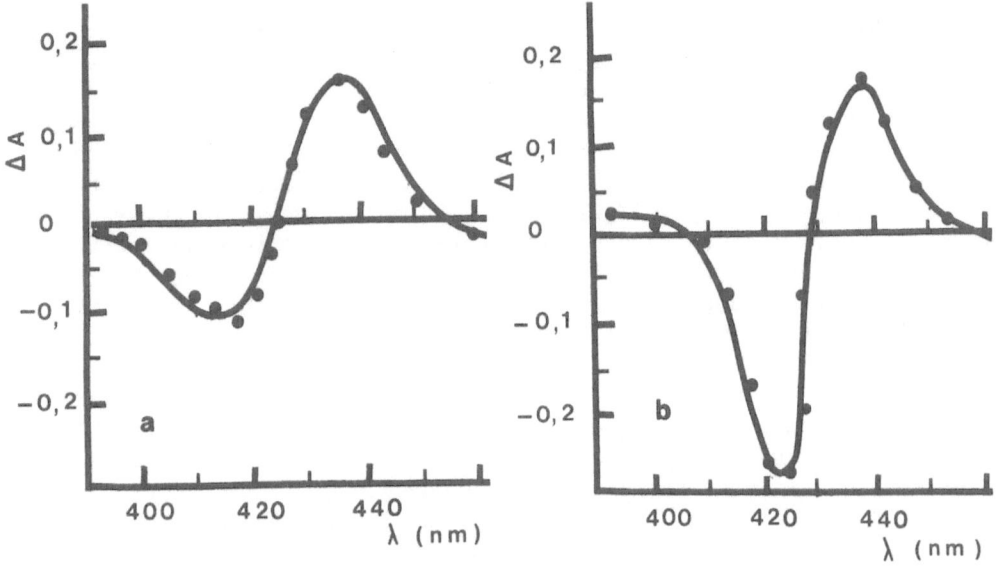

(After Reynolds et al., 1982)
Fig. 1 : (a) Full line, difference spectra taken 200 ps after the
25 ns laser pulse ; dotted line : conventional difference spectrum
MbCO - Mb.(b) same as (a) except MbO₂ replaces MbCO. All spectra have
been taken at room temperature. The details of the experiments are
given in Reynolds et al., 1982.

        As expected from previous considerations the same technique
applied to hemoglobin leads to rather speculative explanations regarding
the complex kinetics (Noe et al., 1978). Their biphasic nature has been
explained as due to a change from the "tense" (T) to the relaxed (R)
hemoglobin state during the recombination following the photodissociation.

1.2. One electron photoreduction of cytochrome P450 by internal electron
     transfer.

        Monooxygenases of the cytochrome P450 - type (Cyt P450) have
been widely studied because they are involved in bacteria or mammalian
cells in the metabolism of various compounds including aromatic or
polyhalogenated hydrocarbons. The most studied Cyt P450 is the bacterial
one because it is easily prepared. It is soluble in aqueous solutions,
where depending upon the spin state of the heme prosthetic group
(high spin (HS) or low spin (LS)) two species can be characterized
(Lipscomb, 1980). It has been shown that a complex formation between
oxygen and the Ferro-Cyt P450 was required for the oxygen activation
and the substrate (camphor) hydroxylation. Moreover, carbon monoxide
binds to the Ferro-Cyt P450 leading to the appearance of a characteristic
absorption band at 450 nm.

In vitro, the Ferri-Cyt P450 reduction can be easily carried
out by photosensitization using a dye and E.D.T.A. or by sodium
dithionite (Eisenstein et al., 1977). However, the oxygen activation
step requires much stronger reductants and therefore cannot be performed
in simple systems. The strongest reducing species which can be formed
"in vitro" is the hydrated electrons (e_{aq}^-). Unfortunately, pulse radiolysis experiments (Debey et al. 1979) have shown that, in constrat
 to cytochrome c, only 10 % of the e_{aq}^- formed in water by the radiolytic pulse, were able to reduce the Ferri-Cyt P450. This unselectivity
precludes, of course, the use of e_{aq}^- in the oxygen activation step.
However it may be thought that, if there exists an electron donor in
the vicinity of the heme, the electron transfer might be possible.

        Cyt P450 contains four tryptophan (Trp) residues. Furthermore,
it is well-known that Trp free or incorporated into proteins can be
photoionized by UV light (Grossweiner, 1976) (Santus et al., 1980).
In principle , it must be possible to produce e_{aq}^- within the protein.
Provided the heme is in the vicinity of the Trp residue(s), an electron
transfer from the Trp to the heme is possible. It turns out that this is
the case with Cyt P450. We showed that the photoionization of, at least,
two Trp does not lead to the Cyt P450 inactivation since after UV
irradiation CO was still able to bind to Cyt_m^{rs} giving rise to the
450 nm absorption band (Pierre et al., 1982). Here is a summary of the
main results (Bazin et al., 1982).
i) The appearance of the Ferro-Cyt P450 camphre-CO complex (Cyt m_{CO}^{rs})
is monophotonic (it is thus possible to photoreduce Cyt P450 by
steady state irradiations). The e_{aq}^- are not involved in the photoreduction although they are produced.
ii) Only the HS species are photoreduccible as observed with conventional methods (such as reduction by dithionite).
iii) Although the photoreduction upon laser flash photolysis occurs in
a few nanoseconds, the growth of the 450 nm absorption takes place on
a 0.1 second time scale via   complex kinetics.

        These complex kinetics can be explained (Bazin et al., 1982)
because of the following equilibria existing before the photolytic flash.

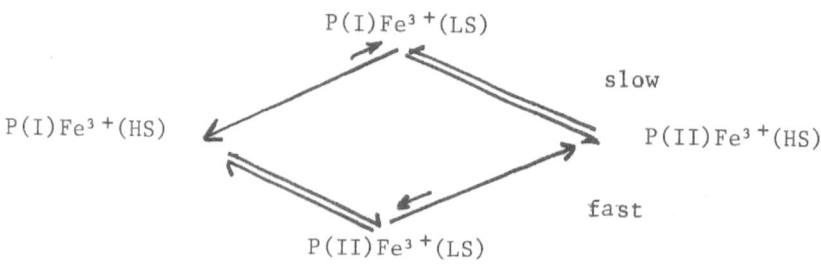

P(I) and P(II) are the conformations of the protein favoring the penta and hexacoordinated heme, respectively. The growth rate of the 450 nm absorption band of the Cyt m_{CO}^{rs} must be different whether the initial state is the penta (P(I) Fe^{3+}(HS)) or the hexa (P(II) Fe^{3+}(HS)) coordinated form.

The fast step (Fig. 2 A - D) corresponds to the reduction of the Ferri-Cyt P450. The growth kinetics at 446 nm is generally biphasic with an "intermediate" risetime ($\simeq$ 290 μs) attributed to the reaction :

$$P(II)Fe^{2+} \text{ (HS)} \rightarrow P(II)Fe^{2+}CO(LS) \qquad \text{(Fig. 2 A, B)}$$

The slower step (risetime 20 ms) obtained at high CO concentration corresponds to the conformational change :

$$P(I)Fe^{2+}CO(LS) \rightarrow P(II)Fe^{2+}CO(LS) \qquad \text{(Fig. 2 D - F)}$$

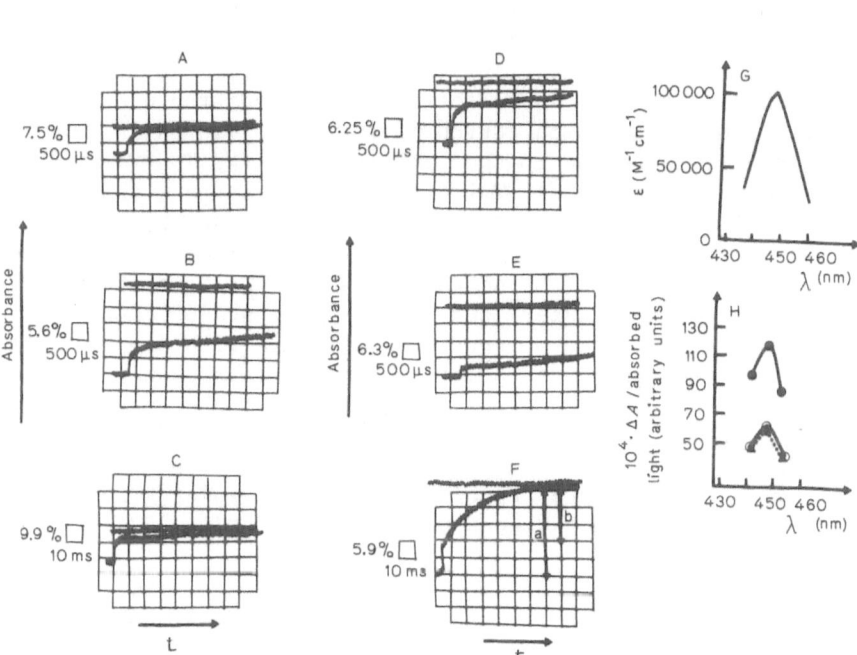

Fig. 2 : Transient absorbance measured at 446 nm (A-F). Conditions in all cases : 19 μM Cyt m^{os} ; 20 mM potassium phosphate buffer, 1.43 mM (100%), temperature 4°C.(A,C) 19 μM camphor, pH 6, 136 mM n-butanol, 27% high-spin form. (D) 19 μM camphor, pH 6, 53% high-spin form. (B,F) 19μM camphor pH 7, 150 mM KCl ; 75% high-spin form. (E) 400 μM camphor, pH 7, 150 mM KCl ; 95% high-spin form. Absorbance readings have been normalized to take into account the laser intensity variations. (G) Difference spectrum ε(Cyt m_{CO}^{rs}) - ε(Cyt m^{os}). (H) Transient difference absorption spectra. Conditions : 21.4 μM Cyt m^{os}, pH 5.4, 20 mM phosphate buffer, temperature 4°C, 1.43 mM CO corresponding to 36,5% high-spin form. (o) a (see F) ; (Δ) b (F) ; (o) (a - b). (After Bazin et al., 1982).

Moreover, the transient absorption spectra taken 1 ms or 100 ms after the laser flash are identical indicating that the species formed during the intermediate state is the Cyt m_{CO}^{rs}.

1.3. Laser flash spectroscopy of complexes of polynucleotides and tryptophan – containing peptides.

The complex formation between a polynucleotide and a peptide such as Lys Trp Lys leads to the Trp fluorescence quenching (*Hélène and Dimicoli, 1972*). The following question may be raised : can these complexed Trp residues still be able to undergo photochemistry, although their first excited singlet state is strongly quenched ?

The laser flash photolysis technique makes it possible to show that the photoionization of Trp residues is not affected by the formation of the poly(rU)-Lys Trp Lys complex. Consequently this biphotonic photoionization occurs from a prefluorescent state (*Le Doan et al., 1981*). Because of the electrostatic repulsion of the polyphosphate backbone, the lifetime of the hydrated electrons which result from the photoionization of the Trp residues is considerably increased. This is due to the inhibition of the $Trp^+ + e_{aq}^-$ recombination (*Bryant et al., 1975*). This result would suggest that Trp residues are not intercalated between the adenine bases because if this was so, the strong electrophilic character of adenine bases (*Anbar et al., 1973*) would lead to an instantaneous scavenging of the photo-ejected electron (phototransfer) and base photoreduction. However the close interaction of the polynucleotidic chain with Trp residues is illustrated by replacing the poly(rU) by a mercurated derivative : the poly(HgU). In this case one observes a heavy atom effect characterized by a strong increase in the formation quantum yield (5 times) and in the decay rate (an order of magnitude) of the Trp triplet while in the case of poly(rU) the disappearance of the triplet transient accompanies the fluorescence quenching (*Le Doan et al., 1981*).

2. A PULSE RADIOLYSIS STUDY OF PROTEIN STRUCTURE AND INTERACTIONS.

2.1. Principles of the method

The radiolysis of deaerated water leads to the formation of three radical species : the hydrated electrons (e_{aq}^-), the hydroxyl radical ($OH^•$) and the hydrogen atom ($H^•$).

These radicals react unselectively with many solutes and these non specific reactions are of little interest. However, in suitable experimental conditions, selective free radicals can be formed (Table I).

In the absence of substrate, these selective radicals decay according to a diffusion controlled second order reaction. On the other

Radical	aqueous systems	Reactions
$X_2^{\bar{\bullet}}$	N_2O Saturation 10^{-2} M X^- $(X^- = SCN^-, Br^-, I^-)$	$e_{aq}^- + N_2O \rightarrow OH^\bullet + OH^-$ $OH^\bullet + X^- \rightarrow X^\bullet + OH^-$ $X^\bullet + X^- \rightleftharpoons X_2^{\bar{\bullet}}$
$CO_2^{\bar{\bullet}}$	N_2O Saturation 10^{-2}M $HCOO^-$	$e_{aq}^- + N_2O \rightarrow OH^\bullet + OH^-$ $OH^\bullet + HCOO^- \rightarrow CO_2^{\bar{\bullet}} + OH^-$
$O_2^{\bar{\bullet}}$	Air or oxygen saturation 10^{-2} M $HCOO^-$	$e_{aq}^- + O_2 \rightarrow O_2^{\bar{\bullet}}$ $HCO_2^- + OH^\bullet \rightarrow CO_2^{\bar{\bullet}} + OH^-$ $CO_2^{\bar{\bullet}} + O_2 \rightarrow O_2^{\bar{\bullet}} + CO_2$

Table 1 : Production of selective radical-anions in aqueous systems

hand in the presence of excess substrate (about an order of magnitude), $X_2^{\bar{\bullet}}$ radicals selectively react, according to a pseudo first order reaction with various substrates such as amino-acids (*Adams et al., 1972*), chlorophylls (*Chauvet et al., 1981*) carotenoids (*Chauvet et al., 1982*). As far as the amino-acids are concerned, the only ones which react with $X_2^{\bar{\bullet}}$ radicals are given in table 2.

Radical-anion	Tryptophan	Tyrosine	Histidine	Phenylalanine	Cysteine
Br_2^-	77	2.0 (pH 7.5)	1.5 (pH 7.6)	< 0.1	18 (pH 6.6)
$(CNS)_2^-$	27	0.5	< 0.1	< 0.1	5 (pH 6.6)
I_2^-	< 0.1	< 0.1	< 0.1	< 0.1	11 (pH 6.8)

Table 2 : Reaction rate constant for the reaction of selected radical-anions with amino-acids at pH 7 unless otherwise stated. The constant are in units of 10^7 M^{-1} s^{-1}.

The $CO_2^{\overline{\cdot}}$ radical-anions react with disulfide bridges found in proteins at rate comparable to the ones given in Table 2. However, in the case of hemoproteins, $CO_2^{\cdot}$ reduces (and $O_2^{\overline{\cdot}}$ as well) the prosthetic groups. Good examples are the cytochrome c (*Land and Swallow, 1971*) and the cytochrome P450 (*Debey et al., 1980*).

2.2. Analysis of the structure of the active site of proteins by $X_2^{\overline{\cdot}}$ radical-anions.

The use of selective free radical-anions allowed the determination of crucial amino-acid residues in several enzymes such as : lysosyme (Trp) ribonuclease (His), trypsin (His), papain (Trp and Cys) (*Adams et al., 1972*). Curiously, one had to wait till 1974 in order to obtain information regarding metalloenzymes such as superoxide dismutase (*Roberts et al., 1974*) (whose reactivity with $O_2^{\overline{\cdot}}$ was previously demonstrated by pulse radiolysis (*Klug et al., 1972*)) and bovine carbonic anhydrase (*Redpath et al., 1975*). In this latter case, we showed that, not only His and Trp but also Tyr residues were involved in the activity of the enzyme. Furthermore we demonstrated that Zn^{2+}, the metal ion required for the enzyme activity, does not modify the enzyme sensitivity towards oxidizing $Br_2^{\cdot}$ radicals.

2.3. Study of the reaction of $Br_2^{\overline{\cdot}}$ radicals with peptides or proteins interacting with DNA.

Literature data on the use of pulse radiolysis for probing interactions between macromolecules or between macromolecules and peptides or molecular assemblies are rather scarce. Let us cite the interaction of penicillin G with bovine serum albumin and lysozyme (*Phillips et al., 1973*), the association of cytochrome c with phospholipid vesicles (*Wainwright et al., 1978*) and the interaction of a cytochrome b_5 fragment with dipalmitoyl phosphatidyl choline vesicles (*Pochon et al., 1981*).

We recently undertook a pulse radiolysis study of the interaction of peptides and proteins with single stranded DNA (*Casas Finet, 1982*). We report here a few data giving the principle and the general results of this study using as model peptide : the lysyl glycyl tryptophyl lysine whose terminal carboxylic group has been blocked with terbutyl alcohol. As a result this peptide brings three positive charges. It follows that it reacts much more rapidly than Trp with $Br_2^{\overline{\cdot}}$ radical-anions (Table 3). On the other hand, the reaction rate constant for the decay of the neutral $Trp^{\cdot}$ radical formed by the reaction :

$$Br_2^{\overline{\cdot}} \; + \; Trp \; \rightarrow \; 2Br^- \; + \; Trp^{\cdot} \; + \; H^+$$

is smaller for the peptide than for the free amino-acid. This is most probably due to the steric hindrance brought about by the large terbutoxyl group which impedes the reaction (Table 3) :

$$Trp^{\bullet} + Trp^{\bullet} \rightarrow \quad \text{Products.}$$

The Trp$^{\bullet}$ radical possesses a transient absorption spectrum peaking at 510 nm (ε = 1,750 M^{-1} cm^{-1}) (*Redpath et al., 1975*) in the visible. This makes it easy to look at its formation or decay since the Br$_2^{\bullet}$ radical strongly absorbs in the UV (λ_{max} = 360 nm).

Conditions	Formation rate constant $(M^{-1} s^{-1})$	decay rate constant
10^{-2} M KBr pH 6.8	2.2 x 10^9	1.3 x 10^8 $M^{-1} s^{-1}$
10^{-2} M KBr + native DNA (10^{-3} M)	4 x 10^8	5.4 x 10^3 (s^{-1})
10^{-2} M KBr + denatured DNA (10^{-3} M)	4 x 10^8	8 x 10^2 (s^{-1})

Table 3 : Reaction rate constants of formation and decay of the trypto- phyl radical formed upon reaction of Br$_2^{\bullet}$ radicals with the Lys-Gly- Trp-Lys-Otbut (8 x 10^{-5} M) in various experimental conditions.

When complexed to native or denatured DNA, the tetrapeptide reacts much more slowly with Br$_2^{\bullet}$ radicals (Table 3). This is obviously due to the electrostatic repulsion exerted by the polyphosphate chain on the Br$_2^{\bullet}$. Once formed within the complex, the Trp$^{\bullet}$ radical disappears much more rapidly (Table 3) than in the absence of complex. The Trp$^{\bullet}$ radical disappearance is a monomolecular process suggesting a chemical reaction involving the Trp$^{\bullet}$ radical and the polynucleotide chain.

These results have been extended to the gene 32 protein from phage T4 which selectively binds to single stranded DNA. However, this protein contains 5 Trp and 7 Tyr. These residues are not at all equiva- lent with respect to the accessibility to the Br$_2^{\bullet}$ radicals. Thus in the absence of complex 2 out of the 5 Trp residues are accessible to the Br$_2^{\bullet}$ attack and only one in the gene 32 protein - single stranded DNA complex (*Casas Finet, 1982*). The presence of various targets (inclu- ding Trp and Tyr) for the Br$_2^{\bullet}$ attack is not very annoying because the Tyr O$^{\bullet}$ and Trp$^{\bullet}$ radicals thus formed do not absorb in the same

spectral range (*Adams et al., 1972*). The interaction of one Trp of the
gene 32 protein with DNA in the complex is confirmed by the increase
in the decay rate of the Trp˙ radical formed during the radiolytic pulse.
This result is in full agreement with the one obtained with model peptides (Fig. 3).

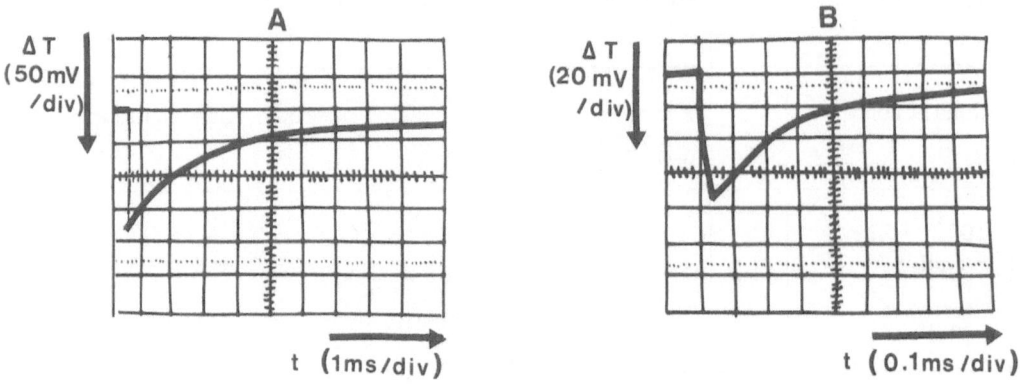

*Fig. 3 : Transient optical absorption (at 510 nm) showing the decay of
Trp˙ (see text) formed after the 700 rad radiolytic pulse (a) in absence
of DNA (b) in presence of DNA. The difference in the ordinate sensitivity scales reflects the change in the Trp˙ yield whether the peptide
Lys-Gly-Trp-Lys Otbut (7.5 x 10⁻⁵ M) is complexed (a) or not (b) to
the DNA (10⁻³ M).*
*This decrease in the radiolytic yield is due to the electrostatic
repulsion of the polyphosphate chain which hinders the approach
of the $Br_2^{\bar{\,}}$ radical (see text).*

3. CONCLUSIONS

        This review illustrates the use of the fast kinetic spectrosco-
pies in the study of the interactions between proteins and nucleic
acids and in the characterization of very fast kinetic steps during the
activation of proteins. Both techniques make it possible to suppress the
gap in time between results obtained using for instance the fluorescence
technique in the nanosecond (or picosecond) time scale and those obtained with other methods such as "T-Jump" or "Stopped-flow" which allow
studies in the millisecond time scale. Although these fast spectroscopic
studies are not always applicable, they provide a priceless tool in many
instances, including the study of the dynamics and interactions of fundamental biological macromolecules.

ACKNOWLEDGEMENTS : I wish to thank Drs M. BAZIN, P. DEBEY and
J.J. TOULME whose help was crucial in many of these studies. I also
thank Drs E.J. LAND and A.J. SWALLOW of the Christie Hospital at
Manchester (U.K.) for using their pulse radiolysis equipment and for
their most valuable collaboration.

REFERENCES :

.Adams, G.E., Redpath, J.L., Bisby, R.H., and Cundall, R.B.: 1972, Isr.
J. Chem. 10, pp.1079-1093.
.Alpert, B., Banerjee, R., and Lindquist, L.: 1974, Proc. Natl. Acad. Sci.
USA, 71, pp.558-562.
.Anbar, M., Bambenek, M., and Ross, A.B.: 1973, NSRDS-NBS 43, US Depart-
ment of Commerce National Bureau of Standards.
.Bazin, M., Pierre, J., Debey, P., and Santus R.: 1982, Eur. J. Biochem.
124, pp. 539-544.
.Bryant, F.D., Santus, R., and Grossweiner L.I.: 1975, J. Phys. Chem.
79, pp.2711-2716.
.Casas-Finet, J.R.: 1982, Thèse 3ème Cycle, Université Paris VI.
.Chauvet, J.P., Viovy, R, Santus, R, and Land, J.: 1981, J. Phys. Chem.
85, pp.3449-3456.
.Chauvet, J.P., Viovy, R., Land, E.J., Santus R, and Truscott, T.G.,
J. Phys. Chem. to be published.
.Debey, P., Land, E.J., Santus, R., and Swallow A.J.: 1979, Biochem.
Biophys. Res. Commun., 86, pp.953-960.
.Eisenstein, L., Debey, P., and Douzou, P.: 1977, Biochem. Biophys. Res.
Commun., 77, pp.1377-1383.
.Eisert, W.G., Degenkolb, E.O., Noe, L.J., and Rentzepis, P.M.: 1979,
Biophys. J., 25, pp.455-464.

.Grossweiner, L.I.: 1976, Current Topics in Radiation Research Quarter-
ly, 11, pp. 141-199.
.Hélène, C., and Dimicoli, J.L.: 1972, FEBS Lett. 26, pp.6-10.
.Klug, D., Rabani, J., and Fridovich, I.: 1972, J. Biol. Chem. 247,
pp. 4839-4842.
.Land, E.J., and Swallow, A.J.: 1971, Arch. Biochem. Biophys., 45,
pp. 365-372.
.Le Doan, T., Toulmé, J.J., Santus, R. and Hélène C.: 1981, Photochem.
Photobiol. 34, pp. 309-313.
.Lipscomb, J.D.: 1980, Biochemistry, 19, pp.3590-3599.
.Noe, L.J., Eisert, W.G., and Rentzepis, P.M.: 1978, Proc. Natl. Acad.
Sci. USA, 75, pp. 573-579.
.Philips, G.O., Power, D.M., Robinson, C., and Davies J.V.: 1973,
Biochem. Biophys. Acta, 295, pp. 8-17.
.Pierre, J., Bazin, M., Debey, P., and Santus,R.: 1982, Eur. J. Biochem.
124, pp. 533-537.
.Pochon, F., Favaudon, V., Ferradini, C., and Pucheault, J.: 1981, Int.
J. Radiat. Biol., 39, pp.207-215.
.Redpath, J.L., Santus, R., Ovadia, J., and Grossweiner, L.I.: 1975,
Int. J. Radiat. Biol. 28, pp.243-253.
.Redpath, J.L., Santus, R., Ovadia, J., and Grossweiner, L.I.: 1975,
Int. J. Radiat. Biol., 27, pp. 201-204.
.Reynolds, A.H., and Rentzepis, P.M.: 1982, Biophys. J., 38, pp.15-18.
.Reynolds, A.H., Rand, S.D., and Rentzepis, P.M.: 1981, Proc. Natl. Acad.
Sci. USA, 78, pp. 2292-2296.
.Robertz, P.B., Fielden, E.M., Rotilio, G., Calabrese, L., Bannister,
J.V., and Bannister, W.H.: 1974, Rad. Res. 60, pp. 441-452.
.Santus, R., Bazin, M., and Aubailly M.: 1980, Rev. Chem. Intermediates,

    3, pp. 231-283.
.Wainwright, P., Power, D.M., Thomas, E.W., and Davies, J.V.: 1978,
    Int. J. Radiat. Biol., 33, pp. 151-159.

FLUORESCENCE DECAY OF DANSYLATED YEAST 3 PHOSPHOGLYCERATE KINASE

Pascal Clerbout, Jean-Paul Privat and Philippe Wahl
Centre de Biophysique Moléculaire, C.N.R.S.,
1A, avenue de la Recherche Scientifique, 45045 Orléans Cedex,
France

ABSTRACT

Fluorescence anisotropy decay of the dansylated yeast 3 phosphogly-cerate kinase has been measured. In addition to the correlation time of the overall molecular rotation, we detected a shorter correlation time of 15 ns which we attributed to an internal molecular motion, namely the relative motion of the two lobes around the hinge region. We suggest that this motion plays a role in the phosphate transfer catalysed by this enzyme.

INTRODUCTION

In the glycolytic pathway, 3 phosphoglycerate kinase (PGK) catalyses the reversible formation of ATP from 1,3 diphosphoglycerate and ADP. This is an important reaction since it is one of the steps in the glycolysis which lead to the formation of ATP.

From sedimentation and translational diffusion measurements (1,2), Smith and Spragg (1) have proposed to consider the molecule of yeast PGK as formed by three spheres corresponding to two lobes linked by a hinge region. Burgers and Pain (3) have found evidences that the thermodynamic stability of the two lobes is different.

This molecular structure of the molecule agrees with the X ray dif-fraction measurements of the yeast PGK structure which is known at 3.5 Å resolution (4,5). The two globular domains are linked by an hinge region wide of 20 Å. The molecular envelope may be approximate by a prolate ellipsoid of 85 Å x 55 Å x 55 Å.

It has been found that Horse PGK showed a strong analogy with yeast PGK. The molecular structure of Horse PGK has been resolved to 2.5 Å (6). In both proteins the nucleotide binding site has been localized on the C lobe. Until now the phosphoglycerate binding site has not been determi-ned experimentally.

C. Hélène (ed.), Structure, Dynamics, Interactions and Evolution of Biological Macromolecules, 311–319.

By considering the molecular structure of the Horse PGK, Banks et al. (6) found that the only possible site of phosphoglycerate was in a positive charged region localized on the N lobe 10 A apart from the nucleotide binding site. Consequently it was assumed that the substrates binding induced a conformational change of the molecule such that the two molecular lobes become closer, allowing the transfer of the phosphate group from a ligande to the other.

These assumptions seem to be corroborate by the small X ray scattering measurements which have shown that the radius of gyration of the yeast PGK decreased of 1 Å in presence of the ligandes (7). In addition, the sedimentation constant of the protein substrates complex is smaller than that of the free protein which according to Roustan et al. means that the substrate binding induces a decrease of the molecular volume (8). Similar changes was observed when the protein binds the sulfate ions (8).

In the present work we have studied the fluorescence anisotropy decay of the dansylated protein in order to detect the conformational chante induced by the ligandes. We were unable to detect such a conformational change, but we found that the protein was very flexible. We suggest that the flexibility of the molecule play an important role in the enzymatic reaction.

MATERIALS AND METHODS

ATP was from Sigma, 3 phosphoglycerate (3PG) and yeast PGK from Boheringer Mannheim. The molecular weight of the protein was checked by gel electrophoresis on polyacrylamide. The enzyme activity was measured by the method of Bucher (9) and found normal.

The PGK was dialyzed against a 0.01 M bicarbonate buffer pH 8.35 in order to release ammonium sulfate contained in the protein suspension.

A dansyl chloride (Koch Light laboratory) solution was added to the protein solution and the mixture left 8 hours in the cold room. The initial dye to protein ratio was 2.5. The unreacted dye was eliminated by dialysis against buffer Tris-HCl 0.05 M pH 7.2 followed by gel filtration on G100 Sephadex. This last operation allowed to eliminate protein aggregates.

The labelling ratio was determined by absorption measurements using the following absorptivity of dansyl : $\epsilon = 4.3 \times 10^6$ cm^{-1} Mol^{-1} at 340 nm and 12×10^6 cm^{-1} Mol^{-1} at 280 nm and for PGK an absorbance of 0.49 cm^{-1} mg^{-1} ml. The dye to protein ratio was equal to 0.4.

The fluorescence measurements were performed at 20°C and pH 7.2, at a protein concentration of 1 mg/ml.

Fluorescence intensity decay and fluorescence anisotropy decay measurements

    These measurements were performed by the single photoelectron method
with an apparatus previously described (10). The exciting flash was filled
with a mixture of nitrogen-helium under pressure (11). Its repetition frequency was 15 kHz. The flash light was filtered by an interference filter
with maximum transmission at 356 nm and band width at 4 nm. The fluorescence was filtered by a wide band filter with a maximum transmission at
485 nm. Excitation and emission light were vertically polarized with
polaroids.

    Data of an experiment were accumulated during 15 hours. The temporal
response function of the apparatus was obtained by measuring a fluorescent standard (a dearated solution of 1,1,4,4 tetraphenyl butadiene in
cyclohexane), the decay time of which was 1.76 ns (12).

    The fluorescence intensity and the fluorescence anisotropy decay
was assumed to be a sum of exponentials the parameter of which were obtained by a least square method which took into account the temporal
response function of the apparatus (13,14). The average fluorescence
life time was defined by

$$<\tau> = \Sigma C_i \ \tau_i$$

where C_i and τ_i were the preexponential and decay times of the intensity
decays.

    The anisotropy at zero time was defined by $R_0 = \Sigma \alpha_j$, where α_j
were the preexponential terms of the anisotropy decay curve $R(t)$.

RESULTS

    The fluorescence spectrum has a maximum at 515 nm with an excita-
tion at 320 nm. This shows that the dansyl moiety is surrounded by a
relatively apolar environment.

    At the experimental accuracy, the fluorescence intensity decays and
the fluorescence anisotropy decays were identical for the free dansylated
PGK, complexed with its substrates or with ammonium sulfate ions. A sum
of three exponential functions were needed to fit satisfactorily the
intensity and the anisotropy decays. The parameters obtained are given
on tables I and II. Fluorescence intensity decays and anisotropy decay
curves of free dansylated yeast PGK are given in Figures 1 and 2.

    The shortest correlation time (that is the short decay time of the
anisotropy decay) was in the subnanosecond range. It must be attributed
to a local motion of the chromophore around its point of attachment.

    In order to see if the two remaining correlation times could be

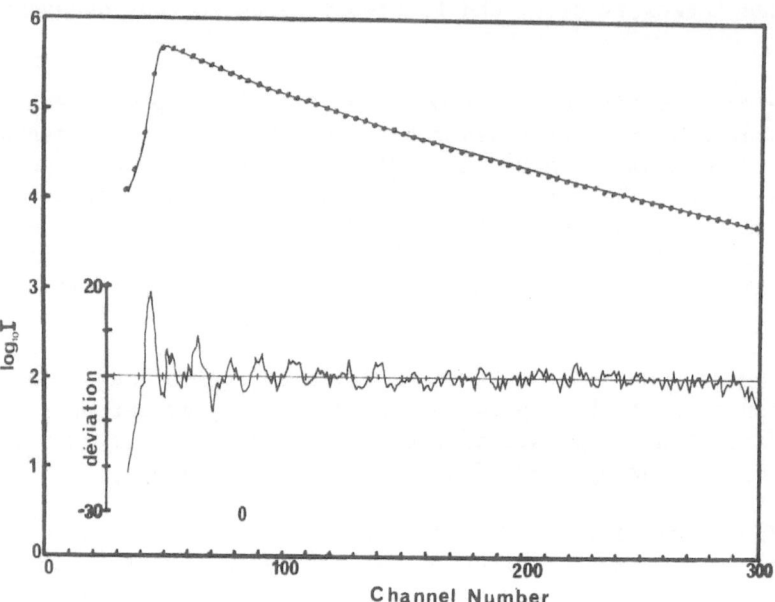

Figure 1 - Fluorescence intensity decay of the dansylated PGK. The points
are experimental values, the solid line represents the computed convolu-
tion function of a sum of three exponentials, the parameter of which were
determined for the least square method. Insert represents the deviation
function which characterizes the fit of the computed with the experimental curve. One channel is 0.321 ns.

Samples	C_1	τ_1(ns)	C_2	τ_2(ns)	C_3	τ_3(ns)	$<\tau>$(ns)
Free PGK	0.30	2.09	0.43	8.26	0.27	20.31	9.59
PGK+ATP+3PG	0.32	2.01	0.41	7.90	0.27	19.61	9.30
PGK+ATP+3PG+MgCl$_2$	0.32	2.13	0.41	8.59	0.27	19.71	9.56
Free PGK	0.30	2.05	0.45	8.16	0.25	20.00	9.35
PGK+10 mM Na$_2$SO$_4$	0.31	2.67	0.46	8.64	0.23	20.23	9.45
PGK+100 mM Na$_2$SO$_4$	0.32	2.85	0.45	8.92	0.23	20.38	9.53

Table 1 - Parameters of the fluorescence intensity decays of the dansylated yeast PGK dissolved in Tris-HCl buffer pH 7.2. Initial concentrations were (ATP) = 13 mM, (3PG) = 3.2 mM, (MgCl$_2$) = 32 mM.

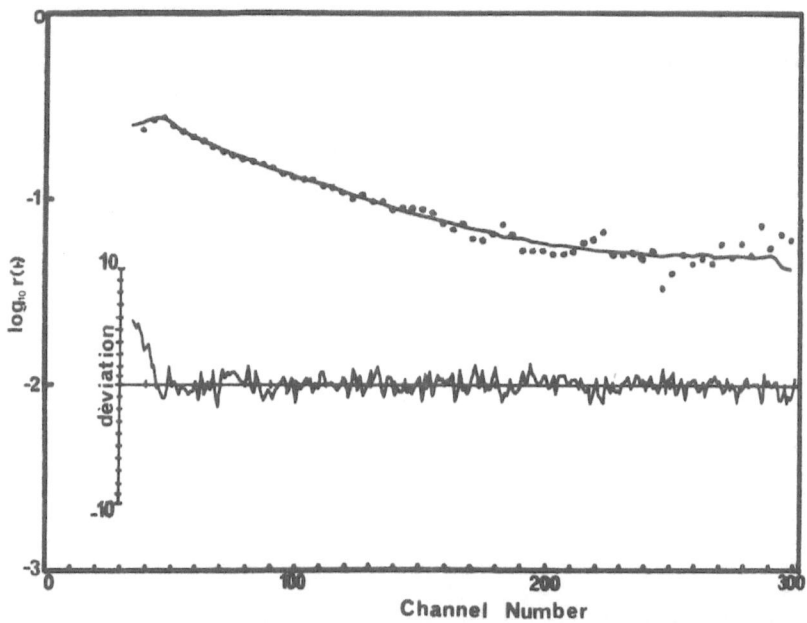

Figure 2 - Fluorescence anisotropy decay of the dansylated PGK. The points are experimental, the solid line is a computed curve which takes into account the apparatus response function and assuming that the anisotropy decay resulting from an infinitely short excitation was a sum of three exponentialsthe parameters of which were determined by an appropriate least square method (14).

Samples	α_1	θ_1(ns)	α_2	θ_2(ns)	α_3	θ_3(ns)	r_0
Free PGK	0.05	1.18	0.18	18.46±3.2	0.08	43.16±6.5	0.31
PGK+ATP+3PG	0.05	0.78±0.35	0.14	15.30±4.0	0.11	40.00±5.7	0.30
PGK+ATP+3PG+MgCl$_2$	0.05	0.68±0.38	0.14	15.50±4.0	0.12	39.90±5.0	0.31
Free PGK	0.06	0.79±0.40	0.15	15.70±3.0	0.10	38.40±5.0	0.31
PGK+10 mM Na$_2$SO$_4$	0.06	0.88±0.6	0.15	15.80±6.0	0.10	38.10±6.0	0.31
PGK+100 mM Na$_2$SO$_4$	0.06	0.73±0.5	0.14	14.97±4.0	0.11	40.10±4.4	0.31

Table 2 - Parameters of the fluorescence anisotropy decay of dansylated yeast PGK. Same experiments as in table 1.

attributed to the rotation of the molecule considered as a rigid body, we computed the theoretical anisotropy decay of a symmetric ellipsoid the dimension of which was given by the molecular envelope obtained by the X ray diffraction studies (4,5).

According to the theory of the Brownian fluorescence depolarization, the anisotropy decay of a symmetric ellipsoid (10, 15) is given by the following formula :

$$r(t) = r_0 \Sigma A_I \exp(-t/\theta_I)$$

$$\theta_I = \theta_D f_I (p)$$

$$\theta_{II} = \theta_D f_{II} (p)$$

$$\theta_{III} = \theta_D f_{III} (p)$$

$$\theta_D = \frac{\eta V}{kT}$$

where V is the ellipsoid volume, η the solvent viscosity, k, the Boltzmann constant, T the absolute temperature, f_I, f_{II}, f_{III} functions of the ellipsoid axial ratio p the analytical expression of which has been given by Perrin (16) and computed by Tao (17) ; A_I, A_{II}, A_{III} depend on the orientation of the absorption and emission transition moments of the chromophore relatively to the molecular axes (10).

In the case of the dansyl chromophore bound to a protein these orientations may be reasonably assumed to be statistical (18) which entails :

$$A_I = A_{II} = 0.4$$

$$A_{III} = 0.2$$

On the other hand, the molecular envelope dimensions leads to $\theta_D = 33.5$ ns.

The anisotropy decays computed with this value of θ_D and assuming various axial ratios p have been drawn on Figure 3. It can be seen on this figure that the theoretical curves can fit the experimental one, only in its final part.

The best fit is obtained for an axial ratio p = 1.5, the value corresponding to the molecular envelope measured by X ray diffraction.

The values of the correlation times are then

$$\theta_I = 32.2 \text{ ns}$$

$$\theta_{II} = 37.9 \text{ ns}$$

$$\theta_{III} = 39.8 \text{ ns}$$

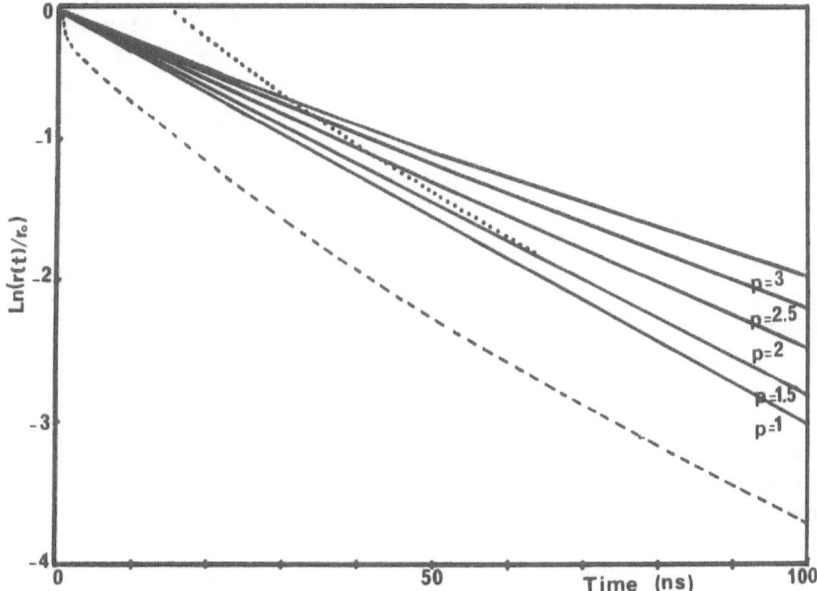

<u>Figure 3</u> - Computed fluorescence anisotropy decay of symmetric ellipsoids
having the same volume as yeast PGK and different axial ratios p (solid
lines). ----- Sum of exponentiels obtained by fitting the experimental
anisotropy decay of dansylated yeast PGK. ····· The same experimental
curve which has been translated in order to fit it to the final part of
anisotropy decay of an ellipsoid of axial ratio p = 1.5.

From this analysis it can be seen that the longest correlation time
θ_3 which appears in table 2 corresponds to an average of θ_I, θ_{II} and
θ_{III}.

The intermediate correlation time $\theta_2 \approx$ 15 ns must be attributed to
an internal flexibility of the molecule. It can be seen that the contribution of θ_2 in the anisotropy decay is important.

DISCUSSION

In the limit of the experimental accuracy, the anisotropy decays of
the dansylated PGK is the same for the free protein, and the complexes
of the protein with its substrates or with sulfate anions.

Roustan et al. (8) found that the binding of the substrates or of the
sulfate ions entailed an increase of the protein sedimentation constant
of 27 % and 24 % respectively. This variation has been attributed to a
decrease of the translational friction of the protein due to conformational change leading to a more compact protein molecule.

According to our analysis θ_3 depends on the rotational friction coefficient of the PGK molecule and consequently should vary more strongly than the sediment constant coefficient when the molecular volume changes. Our measurements show that such a change of θ_3 with substrates or sulfate anions binding cannot be larger than 10 %. These results are then in contradiction with the interpretation of Roustan et al. (8) of their sedimentation measurements. The results of these authors may be due to protein aggregations.

On the other hand, if we assume in a first approximation that the molecule is spherical, the 1Å change of the radius of gyration measured by X ray scattering (7) corresponds to a molecular volume decrease of 10 % which is in the order of our θ_3 value accuracy. Such relatively small conformation change cannot be excluded by our measurements.

Finally we have detected a correlation time θ_2 of 15 ns which cannot be attributed to the overall motion of the PGK molecule. The same correlation time was still present for a preparation having a labelling ratio two times higher. Consequently θ_2 cannot be explained by a fluorescence depolarization due to excitation transfers between chromophores fixed on the same molecule. In addition preliminary experiments in which a fluorescent probe has been attached to the unique cystein residue of the yeast PGK have yield a similar fluorescence anisotropy decay as the dansylated protein.

Therefore we are led to attribute θ_2 to an internal flexibility of the protein molecule. This value of θ_2 corresponds to that of a spherical rigid body having about half the size of the whole protein molecule. This suggests that θ_2 characterizes the relative motion of the two molecular lobes about the hinge region.

If this model is correct, the active site of PGK, which is located precisely between the two lobes, is flexible. Therefore the relative motion of these lobes may bring closer the two substrates, allowing the phosphate transfers to occur.

Recently, the hypothesis has been made that the structure of the active site of enzymes fluctuate in the course of the enzymatic reactions (19). PGK may be an example of this model.

REFERENCES

1. Smith, J., and Spragg, S.P.: 1979, Bioch. Chem. 9, 215-221.
2. Spragg, S.P., Wilcox, J.K., Roche, J.J., and Barnett, W.A.: 1976, Biochem. J. 153, 423-428.
3. Burgess, R.J., and Pain, R.H.: 1977, Biochem. Soc. Trans. 5, 692-694.
4. Bryant, T.N., Watson, H.C., and Wendell, P.L.: 1974, Nature 247, 14-17.
5. Wendell, P.L., Bryant, T.N., and Watson, H.C.: 1972, Nature New Biol. 240, 134-136.

6. Banks, R.D., Blake, C.C., Evans, P.R., Huser, R., and Rice, D.N.,
    Hardy, G.W., Merrett, M., Phillips, A.W.: 1979, Nature 279, 773-777.
7. Pickover, C.A., McKay, D.B., Engelman, D.M., and Steitz, T.A.: 1979
    J. Biol. Chem. 254, 11323-11329.
8. Roustan, C., Fattoum, A., Jeanneau, R., and Pradel, L.A.: 1980,
    Biochemistry 19, 5168-5175.
9. Bucher, T.: 1955, Methods in Enzymology 1, 415-422.
10. Wahl, Ph.: 1975, in New Techniques in Biophysics and Cell Biology,
    R. Pain and B. Smith Eds, Wiley, New-York, vol. 2, p.233.
11. Auchet, J.C., and Wahl, Ph., unpublished work.
12. Brochon, J.C., Wahl, Ph., Jallon, J.M. and Iwatsubo, M.: 1976,
    Biochemistry 15, 3259-3264.
13. Ikkai, T., Wahl, P., and Auchet, J.C.: 1979, Eur. J. Biochem. 93,
    397-408.
14. Wahl, Ph.: 1979, Biophys. Chem. 10, 91-104.
15. Perrin, F.: 1936, J. Phys. Rad. 7, 1.
16. Perrin, F.: 1934, J. Phys. Rad. 5, 497.
17. Tao, T.: 1969, Biopolymer 8, 609-632.
18. Weber, G.: 1952, Biochem. J. 51, n° 2, 145-167.
19. Welch, G.R., Somogyi, B., and Damjansvich, S.D.:1982, Prog. Biophys.
    Mol. Biol. 39, 109-146.

STRUCTURE AND CONFORMATION OF CARBOHYDRATE CHAINS OF N-GLYCOPROTEINS IN
MODEL SYSTEMS

Bernard Gallot, Monique Gervais and André Douy
Centre de Biophysique Moléculaire, C.N.R.S.
1A, avenue de la Recherche Scientifique
45045 Orléans cedex, France

Among proteins glycoproteins are of particular interest because
they play a very important part in the cell life. This importance of
glycoproteins is related to their carbohydrate chains that are antennae
at the cell surface and govern the relations between neighbouring cells
and between cells and the external medium. Namely the carbohydrate chains
of glycoproteins are first receptor sites for hormones, enzymes, proteins
and viruses, then cell-surfaces antigens that are modified in cancerous
cells and they play an important part in intercellular adhesion and recognition and in cell-contact inhibition.

Unfortunately glycoproteins are very complicated molecules : they
have several glycanic chains and all the carbohydrate chains of a glycoprotein are seldom identical. So, in the whole protein it is difficult
to study the specificity of the different saccharidic chains. Furthermore
membranous glycoproteins are very difficult to isolate and purify, they
are often insoluble and it is impossible to obtain them in quantities
allowing detailed physicochemical studies. Therefore to study the structure, the conformation and the specificity of the carbohydrate chains of
glycoproteins it is more advantageous to use models owing only one carbohydrate chain of glycoproteins.

We have chosen, as models of glycoproteins, amphipatic block copolymers in which the hydrophilic block is a carbohydrate chain of a glycoprotein and the hydrophobic block is a peptidic or a lipidic chain, this
hydrophobic block allows us to anchor the carbohydrate chain in a model
membrane and to study its behaviour at an interface.

SYNTHESIS OF MODEL GLYCOPROTEINS

We start from ovomucoid extracted from hen egg white. We perform an
enzymatic degradation of the polypeptidic skeleton by pronase. We fractionate and purify the resulting glycopeptides by column chromatography
and we obtain two well defined glyco-amino acids O_α and O_β (1). The glyco-amino acid O_α contains N-acetylglucosamine, mannose, galactose and

321

C. Hélène (ed.), Structure, Dynamics, Interactions and Evolution of Biological Macromolecules, 321–328.
Copyright © 1983 by D. Reidel Publishing Company.

neuraminic acid. The glyco-amino acid O_β has a molecular weight of 3134 and contains 16 residues : 1 galactose, 5 mannose and 10 N-acetylgluco-samine (Fig. 1) and is terminated by an asparagine residue (2).

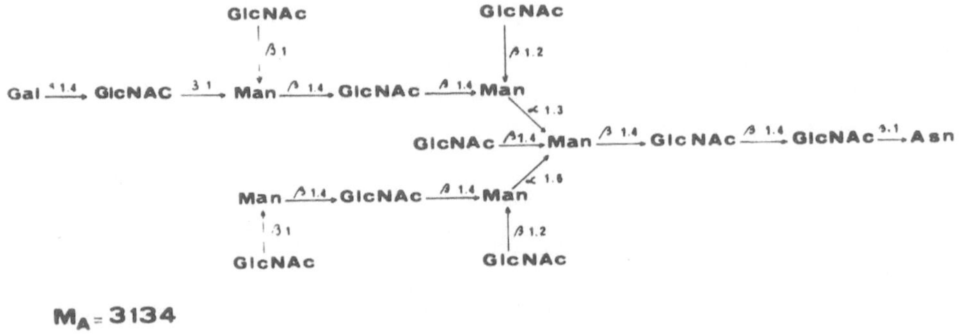

$M_A = 3134$

Figure 1. Primary structure of the glyco-amino acid O_β of hen egg white ovomucoid.

To prepare block copolymers with a hydrophilic saccharide block and a hydrophobic peptide block we use the α primary amine function of the asparagine residue of the glyco-amino acid to initiate the polymeriza-tion of the N-carboxyanhydride of a hydrophobic amino acid (1).

To prepare amphipatic liposaccharides we perform a coupling reaction between the α primary amine function of the asparagine residue of the glyco-amino acid and the carboxylic acid function of a fatty acid (3).

STRUCTURE AND CONFORMATION OF SACCHARIDE-PEPTIDE BLOCK COPOLYMERS

In the amphipatic saccharide-peptide block copolymers O_βEb the hy-drophobic polypeptide block Eb is a poly(benzyl-L-glutamate block) and the hydrophilic saccharidic block O_β is formed of three parts, a rigid core (GlcNAc-Man-(GlcNAc)$_2$ and two antennae, (see Fig. 1) ; molecular models show that the antennae can rotate easily around the Man-Man$_{\alpha 1,3}$ and Man-Man$_{\alpha 1,6}$ linkages and take a T conformation or a Y conformation. Using X ray diffraction and infrared spectroscopy we have demonstrated that the saccharidic chains of O_βEb copolymers adopt a T or a Y confor-mation depending upon the molecular weight of their polypeptide block.

For molecular weight of the polypeptide block higher than 2000 copo-lymers O_βEb exhibit a lamellar structure consisting of plane, parallel and equidistant sheets ; each elementary sheet of thickness d results from the superposition of two layers : one of thickness d_A formed by the saccharidic chains ; the other of thickness d_B formed by the polypeptide chains in an α helix conformation and arranged in an hexagonal array of parameter D.

The lamellar character of the structure is demonstrated by very low angle X ray diffraction : presence of a set of sharp lines with Bragg spacings in the ratio 1, 2, 3 characteristic of a lamellar structure.

The hexagonal organization of the polypeptide helices is demonstrated by low angle X ray diffraction : presence of 3 sharp lines with Bragg spacings in the ratio 1, $\sqrt{3}$, $\sqrt{4}$ characteristic of an hexagonal array.

The α helix conformation of the polypeptide chains is demonstrated by infrared spectroscopy (bands Amide I and Amide II at 1655 and 1545 cm^{-1} respectively) and by X ray diffraction : from the parameter D of the hexagonal lattice and the molecular characteristics of the polypeptide block (molecular weight of the peptide unit and specific volume of the polypeptide block) one can calculate the length h of the projection on the helix axis of the distance between two peptides residues and one finds for h a value in good agreement with an α helix (1).

We have observed two types of lamellar structures as a function of the molecular weight of the polypeptide blocks of copolymers OβEb.

Lamellar Structure LT

For copolymers with a molecular weight of the polypeptide block higher than about 4000, the thickness d_B of the polypeptide layer is slightly higher than the average length L (L = h Pn) of the polypeptidic α helix and the average surface S occupied by a molecule at the interface between the hydrophilic and hydrophobic layer is twice the surface Σ occupied by a polypeptide helix (1). Therefore the polypeptide chains in an α helix conformation are perpendicular to the interface, are assembled in a two dimensional hexagonal array and form a monolayer while the saccharidic chains form a bilayer. The value of 17 Å for the thickness d_A of the saccharide layer shows that the saccharidic antennae are nearly parallel to the interface and that the saccharidic chains adopt a T conformation (Fig. 2a).

Lamellar Structure LY

For block copolymers with a molecular weight of the polypeptide block between about 2000 and 4000, the thickness d_B of the polypeptide layer is slightly higher than twice the average length L of the polypeptide helices and the surface S occupied by a molecule is equal to the surface Σ occupied by a polypeptide helix. Therefore the polypeptide chains in an α helix conformation are perpendicular to the hydrophilic-hydrophobic interface, are arranged in a two-dimensional hexagonal array and form a bilayer while the saccharide blocks also form a bilayer. The value of 35 Å for the thickness d_A of the saccharide layer indicates that the saccharide chains adopt a Y conformation (Fig. 2b).

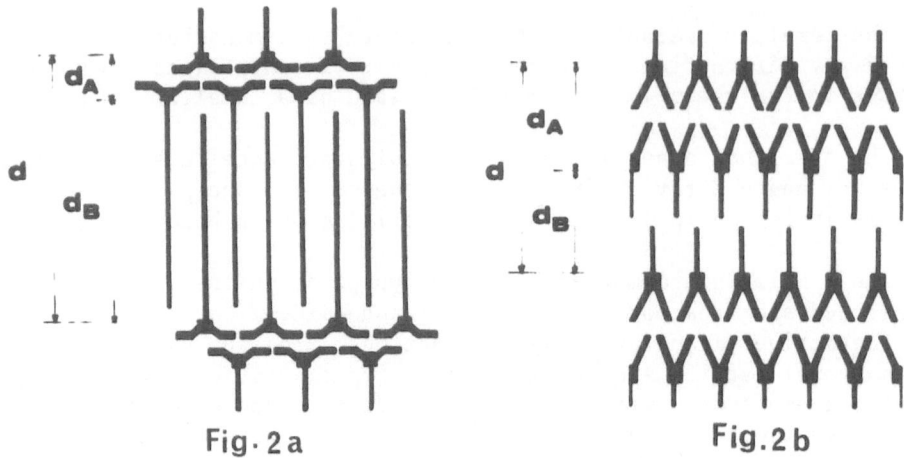

Fig. 2 a   Fig. 2 b

Figure 2. Schematic representations of the lamellar structures of $O_\beta Eb$ copolymers : d = total thickness of a sheet ; d_A = thickness of the saccharide layer ; d_B = thickness of the peptide layer.
2a : Lamellar Structure LT. 2b : Lamellar structure LY. (1)[*]

At the transition between the two lamellar structures the thickness d_A of the saccharide layer suddenly increases from 17 to 35 Å and the glycoprotein saccharide chains exhibit a conformational change from a T to a Y conformation.

STRUCTURE OF AMPHIPATIC LIPOSACCHARIDES

The structure of amphipatic liposaccharides $O_\beta Cn$ formed by coupling the glyco-amino acid O_β from ovomucoid with a fatty acid containing between 8 and 24 carbon atoms has been studied by X ray diffraction, freeze fracture and electron microscopy (3).

We have shown that amphipatic liposaccharides exhibit in the dry state and in water concentrated solution a cylindrical hexagonal structure. In this structure long and parallel cylinders filled by the aliphatic chains of the liposaccharides are assembled on a two-dimensional hexagonal array and are separated by the saccharidic chains swelled by water (3).

CONFORMATION OF GLYCOPROTEIN GLYCANS IN MODEL MEMBRANES

In a first step we have demonstrated the existence of two conformations (a T conformation and a Y conformation) for the saccharidic chains

of glycoproteins. In a second step we wanted, in the one hand to verify
that the two conformations can exist in the presence of a phospholipid
bilayer simulating the cytoplasmic membrane in which membranous glycoproteins are anchored, in the other hand to see how it was possible to
induce a conformational change of the glycoprotein saccharidic chains.
For that purpose we have used model membranes formed by ternary systems
liposaccharide/phospholipid/water. In these systems the liposaccharide
$O_\beta 16$ is formed by the glyco-amino acid O_β from ovomucoid as a hydrophilic
moiety and by the paraffinic chain of palmitic acid as a hydrophobic
moiety and the phospholipid used is dipalmitoyl-phosphatidylcholine
(DPPC) (4).

The study by X ray diffraction, freeze fracture and electron microscopy of ternary systems $O_\beta 16$/DPPC/water has shown that for a constant
water concentration, one observes, as a function of the amount of liposaccharide expressed by the ratio τ between the number of molecules of
liposaccharide and the total number of molecules (number of molecules of
liposaccharide + number of molecules of phospholipid) three periodic
structures : two lamellar structure LT and LY and on hexagonal structure
H separated from one another by a domain of demixtion (Fig. 3).

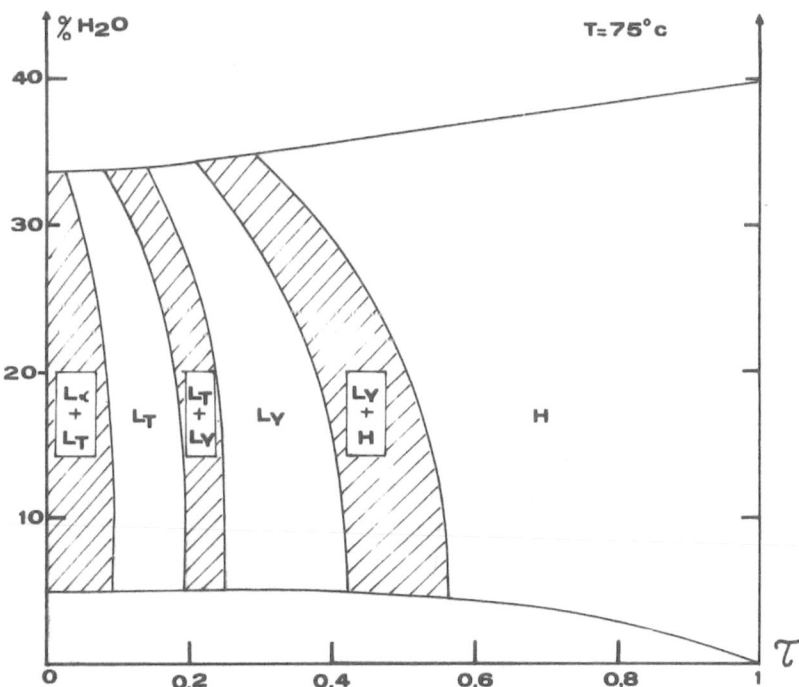

<u>Figure 3</u>.   Phase diagram of the ternary system $O_\beta 16$/DPPC/H_2O. (5)[**]

Lamellar Structures LT and LY

When τ increases from about 0.05 to about 0.45 one observes successively two lamellar structures LT (Fig. 4a) and LY (Fig. 4b) that differ
mainly by the conformation of the saccharidic chains of the liposaccharides.

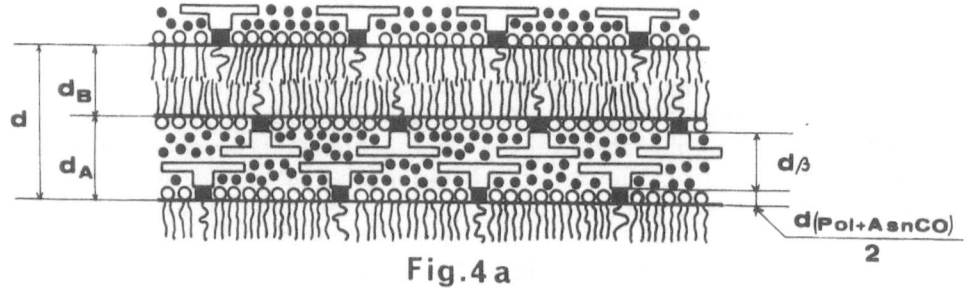

Fig.4 a

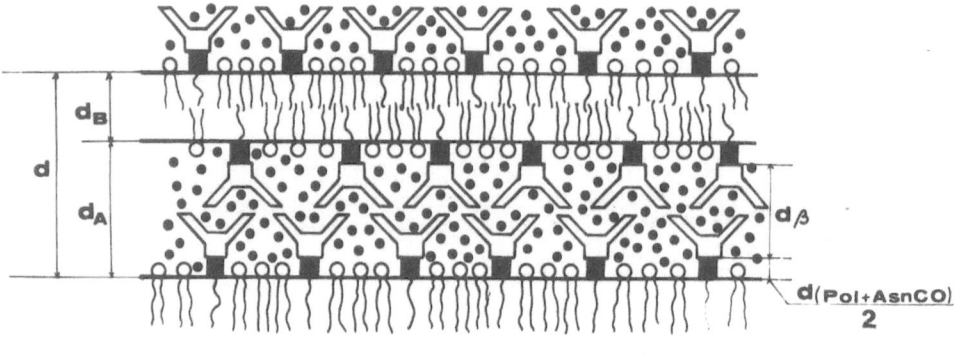

Fig. 4 b

Figure 4. Schematic representations of the lamellar structures of ternary systems $O_\beta 16$/DPPC/H_2O. d_B = thickness of the hydrophobic layer
containing the aliphatic chains of the liposaccharides and the phospholipids. d_A = thickness of the hydrated layer. d_β = thickness of the hydrated lamella containing the saccharidic chains of $O_\beta 16$ (⊏⊐) and the
water (●). $d_{(Pol + AsnCO)} = d_A - d_\beta$ = thickness of the hydrated lamellae
containing the polar head groups of the DPPC (O), the AsnCO groups of
the $O_\beta 16$ (■) and the water (●).
4a : Lamellar structure LT
4b : Lamellar structure LY. (5)[**]

These lamellar structures are formed by the stacking of plane, parallel and equidistant sheets. Each elementary sheet of thickness d results from the superposition of two lamellae :

- a hydrophobic lamella of thickness d_B contains the aliphatic chains (in disordered conformation) of the liposaccharides and of the phospholipids.

- a hydrophilic lamella of thickness d_A contains the polar groups of the phospholipids, the glyco-amino acid and the water ; furthermore, the lamella of thickness d_A exhibit a complex structure, it consists of three layers : two external layers containing the polar groups of the phospholipids and the AsnCO groups forming the linkage between the two types of blocks of the liposaccharides and a central layer containing the saccharidic chains.

Lamellar Structure LT. The lamellar structure LT (Fig. 4a) extends on the domain of τ values from about 0.05 to about 0.19. In the lamellar structure LT the saccharidic chains exhibit a T conformation, i.e., the average direction of the antennae of the glyco-amino acid O_β is parallel to the interface between the hydrophobic and the hydrophilic layer (5).

Lamellar Structure LY. The lamellar structure LY (Fig. 4b) exists for τ values between about 0.23 and 0.45. At the transition LT → LY the thickness d_β of the saccharidic layer increases suddenly of about 50 %, the angle between the average direction of the antennae of the glyco-amino acid O_β and the interface increases suddenly and the saccharide chains exhibit a Y conformation (5).

Hexagonal Structure

For τ values between about 0.5 and 1 one observes (Fig. 5) an hexagonal structure.

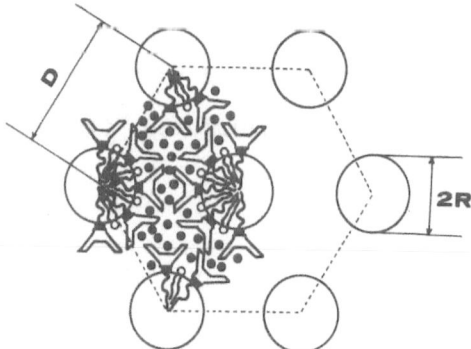

Figure 5. Schematic representation of the hexagonal structure of ternary systems $O_\beta 16/DPPC/H_2O$. For symbols, see Fig. 4. (5)**

The hexagonal structure (Fig. 5) consists of long, parallel cylinders arranged in a two-dimensional hexagonal array ; the cylinders are
filled with the hydrophobic aliphatic chains of the liposaccharides and
of the phospholipids ; the space between the cylinders is occupied by the
polar head groups of the saccharidic chains of the liposaccharides, the
phospholipid and the water.

CONCLUSION

We have demonstrated the existence of two conformations (a T and a
Y conformation) for the carbohydrate chains of glycoproteins. We have
shown the possibility and the conditions of a conformational change
between T and Y conformation first in amphipatic block copolymers with
a saccharidic and a peptidic block, then in model membranes. This conformational change probably modify the specificity of interaction between
glycoproteins and biological effectors such as hormones, enzymes, proteins, lectins... Now we are trying to verify the exactitude of this
hypothesis by studying interactions between our models of glycoproteins
and lectins.

REFERENCES

1. Douy, A., Gallot, B.:1980, Biopolymers 19, 493-507.
2. Debray, H., Decout, D., Strecker, G., Spik, G., Montreuil, J.:1981,
   Eur. J. Biochem. 117, 41-55.
3. Douy, A., Gervais, M., Gallot, B.:1981, Makromol. Chem. 181, 1199-
   1208.
4. Gervais, M., Gallot, B.:1982, Makromol. Chem. Rapid Commun. 3, 77-82.
5. Gervais, M., Gallot, B.:1982, Biochim. Biophys. Acta 688, 586-596.

*Fig. 2 is reprinted by permission from John Wiley and Sons.(see Ref. 1).

**Figs 3,4 and 5 are reprinted by permission from Elsevier/North-
Holland. (see Ref. 5).

WHEAT GERM AGGLUTININ : A REVIEW OF RECENT RESULTS

J.-Ph. Grivet, P. Midoux, P. Gatellier, F. Delmotte, and
M. Monsigny
Centre de Biophysique Moléculaire, C.N.R.S. and Université
d'Orléans, 45045 Orléans cedex, France

1. INTRODUCTION

Many biological molecules contain sugars : glycolipids, lipopoly-saccharides, glycoproteins, polysaccharides, mucopolysaccharides, and proteoglycans. For a long time, the role of the sugar moiety of these compounds has remained unknown. Thirty years ago, it was shown (1) that the antigenic determinants of blood group substances were the oligosaccharide residues of cell surface glycoconjugates. That these substances are highly antigenic is proven by the strong reaction (and sometimes the lethal consequences) which sets in after transfusion of an incompatible blood. Even leaving aside the numerous enzymatic reactions involved in the metabolism of glycoconjugates, there remains a large number of biological phenomena in which sugar-protein interactions are implied. A tremendous amount of work has been devoted to this subject in the last decade. Anti-sugar myeloma antibodies have been characterized (2). Bacterial (3) and plant (4) toxins bind to membrane glycoconjugates before exerting their cytotoxicity. Sugar binding proteins, called lectins (5), have been identified in almost all living things, from viruses to mammals, and bacteria and fungi to plants (6-13). Lectins are proteins, (or glyco-proteins), devoid of enzymatic activity, which selectively bind simple carbohydrates, polysaccharides, and/or the sugar moieties of glycoconju-gates. A lectin comprises several sugar binding sites. Lectins are either soluble or membrane-bound proteins. Although the role of virus and mammalian lectins begins to be understood (12,13), the function of plant lectins is not yet known. Plant lectins are probably involved in the no-dulation of leguminosae,   in bacterial infection processes (14,15), and in viral infection or resistance (16). Whatever role lectins play *in vivo*, they are extensively used *in vitro*, because of their specific binding properties. Current uses of plant lectins include : precipita-tion of glycoconjugates, agglutination of cell populations, studies of membrane glycoconjugates of normal or transformed cells, induction of mitosis and of blastic transformation, selection of variant cell lines, induction of endocytosis of targeted drugs (6-8,17).

The importance of protein-sugar interactions has led us to investi-

C. Hélène (ed.), Structure, Dynamics, Interactions and Evolution of Biological Macromolecules, 329–349.
Copyright © 1983 by D. Reidel Publishing Company.

gate an interesting case, the specific binding of sugars to wheat germ
agglutinin (WGA). This lectin was discovered as a contaminant of wheat
germ lipase (18) and was found to agglutinate preferentially malignant
cells and protease treated cells (19,20). WGA binds N-acetylglucosamine
and its $\beta(1-4)$ oligomers (19-21), and N-acetylneuraminic acid (22-25).
It is used to purify glycoconjugates by affinity chromatography (6,8),
to fractionate cell subpopulation (26-28), and to characterize subpopu-
lations of human blood mononuclear cells using the technique of flow
microfluorimetry (29).

2. THE ENDOGENOUS ROLE OF WGA IN WHEAT

In the dry grain, WGA is associated with the embryo : there is about
1 μg of lectin *per grain*. After 34 days of growth, the amount of lectin
is reduced to 500 ng *per plant* (165 ng in the roots and 335 ng in the
shoot). The actively growing regions of the plant contain the highest
concentration of WGA (31). The agglutinin is located in cells and tissues
that establish direct contact with the soil during germination and growth.
In the embryo, WGA is found in the surface layer of the radicle, the
first adventitious roots, the coleoptile and the scutellum. In adult
plants, WGA is located in the caps and tips of adventitious roots. At
the subcellular level, WGA is located at the periphery of protein bodies
within the cytoplasm and at the cell wall-protoplast interface (32).
These findings, along with the facts that WGA binds the polymer of bacterial cell walls (21) and inhibits the growth of bacterial hiphae (33),
indicate that WGA may function in the defense of the plant against fungal
pathogens. Other plant lectins have been shown to act as receptors for
symbionts (14,34) or pathogens (16). WGA may also be involved in the regulation of protein biosynthesis. Indeed, protein biosynthesis in a cell
free wheat germ extract is reduced when WGA is present (35).

3. STRUCTURE OF WHEAT GERM AGGLUTININ

WGA is a dimeric protein of MW 36000, at neutral pH (36,37). Analogous to what is often found for plant enzymes (38), it is a mixture of
isolectins. Three main constituents have been isolated (39,40), whose
relative abundance depends on the nature of the wheat varieties (30).
Hybrid molecules, made of protomers from different isolectins, are also
found. WGA dissociates into monomers in an acidic medium. Centrifugation
experiments (41) have shown that the pK of this reaction is 4. Succinylated WGA shows very similar properties. It too is a dimer at pH $\simeq$ 7,
with a dissociation half-point of $\simeq$ 4 (41). In acidic media the dissociation equilibrium can easily be shifted towards the dimer by adding the
ligand di-N-acetylchitobiose (42). In view of these results, it is probably best to avoid binding studies at pHs lower than 5, even though the
solubility of the protein is low at neutral pH ($5x10^{-5}$ M). Although the
primary structure of WGA has not been determined, it is known that the
various protomers are very similar (39 and § 6). X-rays diffraction experiments (43-45) have shown that the protomer (164 aminoacids) is orga-

nized in four similar domains (61 aminoacids). There are four disulfide
bridges per domain, a fact which explains the great stability of the
protein. The domains are assembled in a U-shaped structure. It is interesting that the structure of WGA is quite similar to that of several
toxins, which are also organized around a 4 four disulfide core (46).

4. SUGAR SPECIFICITY OF WHEAT GERM AGGLUTININ

WGA binds N-acetylglucosamine and sugar derivatives of related
structure, N-acetylneuraminic acid, N-acetylgalactosamine (Figure 1),

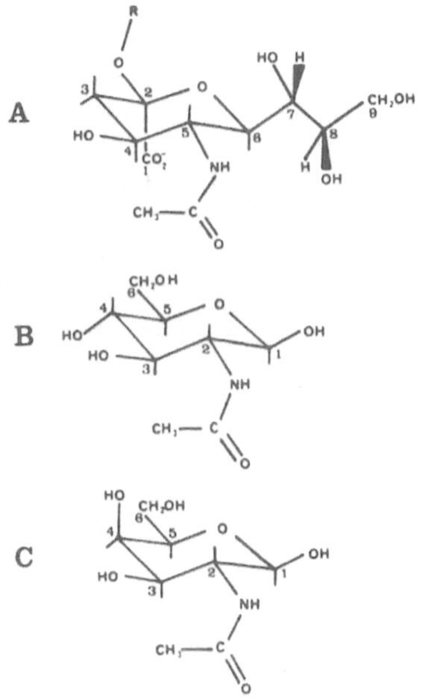

Figure 1 - Structure of sialic acid (NeuNac, A), N-acetylglucosamine
(ClcNAc, B) and N-acetylgalactosamine (C).

oligosaccharides, and glycoconjugates containing these sugars. The sugar
specificity of WGA has been established by the use of various biochemical
methods : inhibition of agglutination (19-21, 24,25,47,48), precipitation
of glycoconjugates (23,49-51), aggregation of vesicle-embedded gangliosides (25,52-54), affinity chromatography on immobilized WGA (24,55-58).

The binding of N-acetylglucosamine and derivatives to WGA involves
the acetamido group on C_2 in GlcNAc and on C_5 in Neu-5-Ac. The elimination of the acetyl moiety of GlcNAc leads to an ammonium group which
impedes the binding. The substitution of the acetyl group of Neu-5-Ac by
a glycoloyl group leads to Neu-5-Gly which does not inhibit the binding
activity of WGA (23,25). Furthermore, glucose or mannose which have an
hydroxyl group on C_2 do not bind to WGA. The hydroxyl on C_3 of GlcNAc
should be free, the 3-O-methyl derivative is not an inhibitor of WGA

(21). The hydroxyl on C_4 of GlcNAc may be substituted as in Man-β-4-
GlcNAc-β-4-GlcNAc → Asn which inhibits the precipitation of neoglycopro-
teins by WGA (49) and inhibits the agglutination of rabbit erythrocytes
(59). When the hydroxyl on C_4 is axial (as in GalNAc) the affinity for
WGA is lower (60). The hydroxyl on C_6 of GlcNAc may also be substituted :
p-nitrophenyl Gal-β-6-GlcNAc glycoside binds to WGA (24). Similarly, the
hydroxyl on C_2 of Neu-5-Ac may be substituted as in the case of Neu-5-
Ac-α-2(3 or 6)-Gal-β-4-Glc (61-63) and O-glycosidic type sialoglycopeptides (24). The carboxylic acid group of Neu-5-Ac seems not to be involved in the binding (23) and the amide derivative of Neu-5-Ac in ganglioside GM_3 has an affinity higher than the native GM_3 (54).

Di-N-acetylchitobiose (GlcNAc-β-4-GlcNAc) and tri-N-acetylchito-
triose (GlcNAc-β-4-GlcNAc-β-4-GlcNAc) have higher affinities for WGA than
GlcNAc and it has been proposed that the binding site of WGA accommodates
three sugar units (21).

Among glycoconjugates, WGA binds the following :

• ganglioside GM_3 : Neu-5-Ac-α-2-3(6)-Gal-β-4-Glc-β-Cer (25). This
may explain the increased agglutinability of protease treated cells (20)
in which gangliosides are accessible to the lectin.

▪ O-glycosidic type sialoglycopeptides

Neu5Acα3(6)Galβ3 ─────┐
                      ↓
    Neu5Acα3(6)GalNAcαOSer(Thr)

when several such glycopeptides are close together, as in fetuin (24).

▪ N-glycosidic type sialoglycopeptides

(Neu5Acα3(6)Galβ4GlcNAcβ4) ─────┐ 0 or 1
                                ↓
    Neu5Acα3(6)Galβ4GlcNAcβ2Manα6 ─────┐   (Fucα6)─────┐ 0 or 1
                   (GlcNAcβ4)₀ ₒᵣ ₁   ↓
                                      Manβ4GlcNAcβ4GlcNAcβAsn
    Neu5Acα3(6)Galβ4GlcNAcβ4Manα3 ─────┘
                                       ↑
    (Neu5Acα3(6)Galβ4GlcNAcβ2) ──┘
                               0 or 1

when several such glycopeptides are present on a glycoprotein (23,57).
Free N-glycosidic type disialoglycopeptides do not bind to WGA (58).
Furthermore, N-glycosidic type glycopeptides without a bisecting GlcNAc
residue (GlcNAc linked to the hydroxyl on C_4 of the β-mannose) were
neither retarded nor bound to WGA. Conversely, various glycopeptides
with a bisecting GlcNAc residue were retarded or bound to WGA, on an
affinity chromatography column (58). Using immobilized WGA to test the
binding of glycopeptide, Man-β-4-GlcNAc-β-4-GlcNAc-β-Asp was not retar-

ded (58) in contrast to the inhibition experiments (49,59).

One should also be aware that the interactions of WGA with gel or membrane bound glycoconjugates can be significantly altered by non specific factors, such as electrostatic interactions with charged sugars (NeuNAc) or involvement of unrelated proteins or oligosaccharidic structures (64-65). Further, the binding of WGA to complex compounds obviously depends on the multivalency and on the detailed geometry of the two reagents. Thus, the retardation of glycosides on an affinity column is controlled by the density of specific residues on the glycoconjugate and by the density of lectin molecules on the gel beads (24). Before turning to solutions studies, it is worth remarking that observed quantities in the assays described above (for instance turbidity in the case of precipitation) are complicated, highly non-linear functions of individual binding constants.

5. SOLUTION EQUILIBRIA INVOLVING WGA

a) General

Binding of small ligands to WGA has been investigated by many methods : equilibrium dialysis, optical spectroscopy using absorption or luminescence, circular dichroism (66), fluorescence polarization (67). NMR investigations (of ^{1}H, ^{2}H and ^{19}F nuclei) have been concerned exclusively with ligand signals since protein spectra are poorly resolved and their assignments impossible. ESR, using labelled substrates could be an interesting approach, allowing one to explore a different time scale. In preliminary experiments, we have shown that 2-acetamido-1,2-dideoxy-β-D-glucopyranosylthioethylamino TEMPO quenched the protein fluorescence, thus proving some interaction (Grivet, J.Ph. and Monsigny, M., unpublished results). However, the ESR spectrum did not change, presumably because the arm between osamine and nitroxyde is too long and the residence time of the spin label in a hindered environment too short.

b) Number of binding sites

Equilibrium dialysis experiments (36,51,68) have shown conclusively that WGA is tetravalent. Further, chemical modification studies prove that one fluorescent tryptophan is associated with each site (69,70). Independent determination of the number of sites is important because spectroscopic methods and specially NMR are not very sensitive to this parameter. This is due to the limited concentration range available and to the necessity of determining simultaneously two pieces of data : equilibrium constant and chemical shift of bound ligand. Whether these four sites are equivalent remains an open question.

c) Size of binding site, requirements for binding

When the interaction of oligomers of *N*-acetylglucosamine with WGA is examined, it is found that affinity constants increase in the series monomer-tetramer (21,71,72). Experiments with different sugars show that the *N*-acetamido group is a prerequisite for binding, that 3-deoxy- or 3-methoxyglycosides do not bind and that β anomers usually have higher affinity than their α counterparts. It has also been shown that sialic acid binds to WGA in solution (73,74).

d) Affinity constants

Up to now, most solution equilibria have been analysed in terms of the simple model of four equivalent and independent sites. This model fits well experimental results concerning monosaccharides, and also, although in a somewhat more restricted concentration range, those concerning 4-methylumbelliferyl-*N*-acetylglucosaminide (MUFGlcNAc). The table below shows values of affinity constants collected from various sources, and obtained with several different experimental methods.

Table I. Monosaccharide affinity constants (M^{-1}) for WGA, at ca. 25°C.

N-acetylglucosamine	240 N (73) ; 250 N WGA I (40) 700 F (41) ; 1320 F 4°C (71)
Sialic acid	560 N (73)
3-deoxy-*N*-acetylglucosamine	small N (73)
N-acetylgalactosamine	small N (73)
3-methoxy-*N*-acetylglucosamine	18 C (75)
α-methyl-*N*-acetylglucosaminide	210-300 N (74)
β-methyl-*N*-acetylglucosaminide	445 C (75) ; 450 N (75)
6-deoxy-*N*-acetylglucosamine	785 C (75)
α-methyl-*N*-trifluoroacetylglucosaminide	420 C (75) ; 700 N (75)
β-methyl-*N*-trifluoroacetylglucosaminide	330 C (75) ; 520 N (75)
N-trifluoroacetylglucosamine	α : 910 N (74) β : 715 N (74)
MUFGlcNAc	2.67×10^4 F (75) ; 2.86×10^4 F (40) 1.98×10^4 D,E,F (68) ; 5×10^4 F (60)

C : competition with MUFGLcNAc ; D : difference absorption spectra of the ligand ; F : fluorescence of WGA or of MUFglycoside ; E : equilibrium dialysis ; N : nuclear magnetic resonance.

There is a temperature dependence of K, which has been estimated, in the case of *N*-acetylglucosamine, as corresponding to $\Delta H° = 34\pm8$ kJ/mole (40).

When one turns to disaccharides or to larger ligands, one finds that

the binding is much more complex than is the case for small haptens
(40,74-76). In such circumstances, spectroscopic methods are at a disadvantage, since there is no simple relationship between the observed signal (some linear combination of chemical shifts or fluorescence intensities) and the fraction of ligand bound, if more than one site (or
class of sites) need to be considered. It is then usual to construct
Scatchard plots under the assumption that all sites exert the same spectroscopic effect. Curved Scatchard plots, such as those of figure 2, can
only be taken as an indication that some assumption(s) of the model is
(are) incorrect.

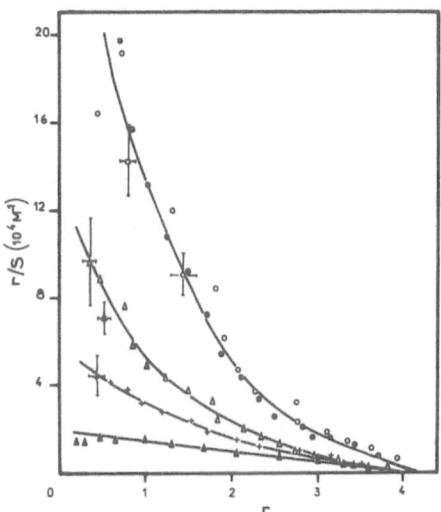

Figure 2 - Scatchard plot of the binding of β-methyl-di-N-acetylchito-
bioside to WGA. The protein concentrations were (in mg/ml) : 0.045 (o),
0.09 (o), 0.18 (Δ), 0.72 (+), and 1.44 (Δ). The protein fluorescence was
excited at 295 nm and observed at 348 nm. From (75).

    In the following, we report "mean" or "apparent" equilibrium cons-
tants derived by forcing the independent, equivalent sites model to fit
the data. One thus gets the data of table II for disaccharides and oligosaccharides.

    It is striking that binding constants derived from optical measure-
ments and from NMR experiments can differ by as much as an order of magnitude. This effect is apparently related to the very different concentration ranges which can be monitored with these two techniques. However,
when one uses increasing protein concentrations in fluorescence spectroscopy, one derives K values which converge towards the NMR result. The
four equivalent, independent sites model becomes decidedly inadequate in
the case of β-methyl-di-N-trifluoroacetylchitobioside. For this hapten,
WGA shows only two sites, with widely varying affinity, dependent on the
protein concentrations. Two very low affinity sites can also be detected

for dilute protein solutions. The only accurate measurement of binding
enthalpies reported so far concerns methylumbelliferyl-chitosides (68)
for which the $\Delta H°$ are (in kJ/M) : - 34.4 (MUFGlcNAc)n,- 41.8 (MUF(Glc-
NAc)$_2$), and - 35.1 (MUF(GlcNAc)$_3$).

Table II. Disaccharide and oligosaccharide affinity constants for WGA
at ca. 25°C (M^{-1}). (See table I for footnotes).

Di-*N*-acetylchitobiose	1.2×10^4 F (71) ; 2.0×10^4 E (36) ; 4.5×10^3 F (72) ; 1.4×10^4 C (40) ; α : 260 ; β : 830 N (40), WGA 1, 55°C
β-methyl-di-*N*-acetyl-chitobioside	1.1×10^3 N (76) ; 670 N (40), WGA 1, 55°C 3.3×10^4 C (40) ; 5×10^3 to 3×10^4 F (75) ; 3.5×10^4 C (75)
N-trifluoroacetyl-*N'*-acetyl-chitobiose (GlcNAc-GlcNTFA)	α : small ; β : 3.8×10^3 F (75) ; 2.8×10^3 N (75)
β-methyl-di-*N*-trifluoroacetyl-chitobioside	1×10^4 to 4×10^4 F (75) ; 1.34×10^4 C (77)
MUF(GlcNAc)$_2$	7.5×10^4 F (68) ; 8×10^4 F (60)
Tri-*N*-acetylchitotriose	2.2×10^4 F (71) ; 2×10^4 F (72) ; 3.3×10^4 C (40) ; 8.3×10^4 C (36)
Reduced tetra-*N*-acetyl-chitotetraose	5.5×10^4 E (51)
MUF(GlcNAc)$_3$	11.7×10^4 F (68) ; 10^4 F (60)
Tetra-*N*-acetylchitotetraose	3.6×10^4 F (71) ; 2.3×10^4 F (72)
Sialyllactose (mixture of isomers)	830 N (74)
Sialyl-α-(2,3)lactose	1220, WGA 1 ; 280, WGA 2
Sialyl-α-(2,6)lactose	190, WGA 1 ; 90, WGA 2 ⎱ N, (61)
Sialyl-α-(2,6)-N-acetyl-lactosamine	105, WGA 1 ; 95, WGA 2

In a very recent and interesting paper, Kronis and Carver (61) report the affinity of three sialooligosaccharides for isolectins 1 and 2.
The two proteins have different affinities, a fact which is related to
the different aminoacid content of the binding site (see section 5).
Also, the α(2,3) isomer of sialyllactose binds more strongly than the
α(2,6) compound. The authors (61) believe that this difference is reflected in a specificity of WGA for cell surface bound NeuNAc-α(2,3).

Final words of caution are in order for the benefit of the reader
who wishes to compare the above K_a values. Many of the earlier measurements were processed using erroneous values of the molecular weight.

The accepted value is now 36000 (i.e. 2 x 18000) (36,43-45). The effect on K_a of these different assumptions is rather difficult to assess. Various schemes have been devised in order to explain the complicated binding results described above. A protein transconformation in the presence of MUF-chitosides has been invoked (40). We have been, up to now, unsuccessful in our attempts to model the binding using two classes of sites. The dependence of the shape of the isotherms on the concentration of WGA suggests some form of ligand mediated protein aggregation. The association of several lectin molecules would entail the disappearance of some sites and possibly a modification of the affinity constant of the remaining ones (75). Preliminary calculations, along the lines presented in (78) seem encouraging.

e) Kinetics of binding

At resonance frequencies of 90 or 220 MHz for protons, ligands such as di-N-acetylchitobiose undergo so called fast to intermediate exchange. The ligand residence time in the bound environment, $\tau_b = 1/k_{off}$, is then easily determined (79,80). For β-methyl-di-N-acetylchitobioside, one finds $k_{off} \simeq 600$ s^{-1} at 25°C (40,76) ; the activation energy is 50.4 ± 8 kJ/M (40). The bimolecular association rate constant is then computed as $k_{on} = 7.5 \times 10^5$ 1 M^{-1} s^{-1}. We have undertaken a study of the binding of N-trifluoroacetylated chitosides by ^{19}F NMR at 254 MHz. We are able to compare total simulated lineshapes to experimental ones, thus avoiding the pitfalls of approximate formulas (81). Preliminary results for methyl-N-acetyl-N'-trifluoroacetylchitobioside (GlcNTFA-GlcNAc-OCH$_3$) at 300 K are $K_a \simeq 3 \times 10^4$ M^{-1}, $k_{off} \simeq 150$ s^{-1}, $k_{on} \simeq 4.5 \times 10^6$ M^{-1} s^{-1}.

The binding of MUFGlcNAc and of MUF(GlcNAc)$_2$ to WGA has been studied in temperature jump experiments (82). These ligands are convenient because they show a large $\Delta H°$ of binding and a strong variation of their fluorescence quantum yield between free and bound states. Two kinetic schemes could fit the observations :

$$\text{sequential :} \quad S + P \rightleftharpoons (SP)_1 \underset{k_{-2}}{\rightleftharpoons} (SP)_{II}$$

$$\text{competitive :} \quad (SP)_a \rightleftharpoons S + P \underset{k_{-b}}{\rightleftharpoons} (SP)_b$$

In order to compare the relaxation and NMR results, we resort to the following rough arguments. In the sequential model, the rate determining step is the conversion of $(SP)_{II}$ to $(SP)_I$ with $k_{-2} = 70$ s^{-1}. The overall time behaviour of the competitive model should be governed by the slowest off rate, from $(SP)_b$ to S+P, that is $k_{-b} = 39$ s^{-1}. These two rate constants are of the same order of magnitude, and are about ten times smaller than the value obtained with NMR. This difference probably reflects the higher affinity (by about the same factor of ten) of MUF-chitosides compared to that of free or methylated saccharides.

f) Ligand dynamics

Closely related to the subject of the last paragraph is the topic of ligand motion in the binding site. Since the site is rather extended, it is quite possible that small ligands could move as a unit between subsites ; T-jump experiments described above give some indication that this could happen. In an elegant experiment, Neurohr *et al.* (83) have measured the deuterium linewidths of *N*-trideuteroacetylglucosamine and of 3-^{2}H-*N*-acetylglucosamine in the presence of WGA. The deuterium relaxation is dominated by the quadrupolar interaction of that nucleus and is directly related to its correlation time. At 35°C, the ligand residence time on WGA is 10^{-4} s and the various correlation times were (in seconds) :

	free	bound
acetamido deuterium	6×10^{-12} s	1.3×10^{-9} s
ring deuterium	1.3×10^{-10} s	3×10^{-8} s

The pyranoside ring and the *N*-acetyl side-chain are immobilized on the protein. Some residual rotation of the methyl group about its symmetry axis remains.

Independently, we have studied the nuclear Overhauser effect ^{19}F [^{1}H] for methyl-di-*N*-trifluoroacetylchitobioside bound to succinyl-WGA (77). At high protein concentrations, the resonance line of the non reducing end trifluoroacetamido group in the bound state can be detected. This signal almost vanishes upon broad band proton irradiation. This effect can be most simply explained by assuming that the glycoside is rigidly bound to the protein and that the correlation time characteristic of proton-fluorine dipolar interactions is longer than 10^{-8} s.

The results described above demonstrate the central role played by *N*-acetamido groups in the specific recognition process.

6. CRYSTALLOGRAPHY OF WGA AND ITS COMPLEXES

a) Two classes of sites

In a long series of difficult experiments, Wright (43-45,62,63,84) has contributed significantly to our knowledge of the structure of WGA. X-ray data show the presence of two "primary" sites, which are located in the contact regions between the protomers. As shown in figure 3, one such site lies between domains C_I and B_{II}. There are also two "secondary" sites, one of which appears in figure 3 between domains D_I and A_{II}. This structure correlates well with the fact that the WGA protomer shows very little affinity for any ligand (69).

The primary sites are readily accessible to *N*-acetylglucosamine, to *N*-acetylneuraminic acid (Neu5Ac, sialic acid), and to sialyllactose. Due to the different orientation of their glycosidic linkages, haptens with terminal sialic acid bind differently compared to GlcNAc oligomers, although the NeuNAc and the first GlcNAc moieties are almost superimposable in the protein. The secondary sites are poorly accessible to $(GlcNAc)_2$ and not at all to Neu5Ac. Since molecular packing prevents the diffusion of trisaccharides into the crystal (except for sialyllactose), the location of a trisaccharide must be modelled, using the experimental coordinates for the dimer. It is found that the reducing end glycoside has almost no interaction with the protein.

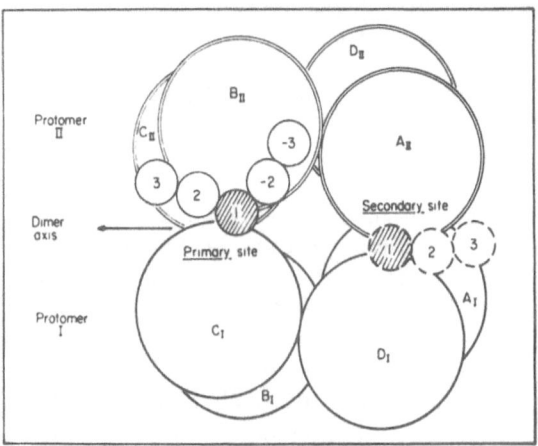

Figure 3 - Structure of WGA, from (63).

b) Main interactions

It is convenient to divide the primary site into subsites 1, 2 and 3. The acetamido methyl group in subsite 1 is in van der Waals contact with an alanine (60_{II}) and a tyrosine (71_{II}). The secondary site is less well defined. It can also be divided in three subsites. In subsite 1, the acetamido methyl group is found very close to a tyrosine (29_{II}) and to a glutamine (148_I). A tryptophan side chain is at about 8 Å from the second and third subsites.

c) Differences between isolectins 1 and 2

The X-ray investigation has also pin-pointed the difference between the two isolectins, WGA 1 and WGA 2. The former lacks histidine, while the latter has two such residues. By Fourier difference methods, the imidazole side-chains were located at positions 57 and 64. His_{64} is part of the binding site and is replaced in WGA 1 by an aromatic aminoacid, Tyr or Phe, thus maintaining a flat hydrophobic surface which is thought to be important for sugar binding.

7. COMPARISON OF SOLUTION AND CRYSTAL DATA

a) General

It is often difficult to decide whether differences between crystal and solution results are real or are technical artefacts. For example, most crystals were reticulated with glutaraldehyde in order to improve their stability in the presence of ligands (our emphasis). The size of the site is not firmly established, since large ligands cannot diffuse into the lattice, while there are conflicting reports on the affinity constants of higher saccharides. Phosphorescence spectra are recorded at 77 K, under conditions where the mode of binding can be altered.

b) Aminoacids of the binding site

The primary sites found in the X-ray work do not comprise tryptophan side-chains, although an arginin residue is nearby, whose electron densi-ty could perhaps be similar to that of tryptophan (62,63). A tryptophan residue is found at some distance (8 Å) from the secondary sites.

In complete disagreement with these results, solution studies show that tryptophans are located close to each binding site. Let us first recall that all the optical parameters of the protein which are affected upon ligand binding are related to tryptophan. These effects could of course be due to a local change of conformation around the Trp residues, due to binding some distance away. When the thiomercuribenzoate derivative of $(GlcNAc)_2$ binds to WGA (41,85), the protein fluorescence is completely quenched, the phosphorescence intensity is greatly enhanced, while the phosphorescence lifetime drops from 5.9 s to 14 ms. This clearcut case of an heavy atom effect on the Trp luminescence will only occur when the mercury atom is in van der Waals contact with the chromophore, or very nearly so. A similar result is obtained with 2-acetamido 1,2-dideoxy-glucopyranosylthiomercurimethane (Monsigny and Hélène, unpublished).

N-bromosuccinimide oxydizes four tryptophans per molecule of WGA ; the fluorescence intensity decreases in linear relation to the amount of Trp oxydized. Two other tryptophan sidechains become accessible and can be oxydized when WGA is denatured in urea (69). Studies of fluores-cence quenching by trichloroethanol (70) and by trifluoroacetamide (Monsigny et al. unpublished) yield quite similar results. These two compounds quench the fluorescence of indole derivatives in solution and of tryptophan in proteins. The bimolecular quenching constants are as follows.

	N-acetyl-tryptophanamide	WGA	fraction of fluorescence quenched
Acrylamide	6.3×10^9 (86)		
Trichloroethanol	6.4×10^9 (70)	1.2×10^9 (70)	1.0
Trifluoroacetamide	2.5×10^9	1.1×10^9	1.0

Both compounds completely extinguish the fluorescence, showing that all the fluorophores are accessible. The effect of trifluoroacetamide is related to the fact that N-trifluoroacetylchitoside cause a decrease of the fluorescence intensity of WGA when bound (N-acetylchitosides, on the contrary, increase the fluorescence yield). Moreover, upon wide-band irradiation of sufficient duration, trichloroethanol undergoes a photochemical reaction with tryptophan, to yield a photoproduct with quite distinct spectral parameters. It has been shown that the protein undergoes no gross structural modification as a result of this reaction. Detailed analysis proves that the number of fluorescent tryptophans and the number of oxydizable chromophores to be both equal to four (70).

The protons which have been examined in the NMR investigations (40,73,74,76) undergo high field shifts, which are presented in the table below.

Table III. Proton chemical shifts for ligands bound to WGA, in ppm, relative to the free ligand, at ca. 25°C, positive down-field.

N-acetylglucosamine	acetamido	-1.35 (73) ; -0.68 (40) WGA 1
Sialic acid	acetamido	-0.4 (73)
α-methyl-N-trifluoroacetyl-glucosamine	methoxy	-0.4 (75)
β-methyl-N-trifluoroacetyl-glucosamine	methoxy	-0.23 (75)
β-methyl-N-acetyl-glucosamine	acetamido methoxy	-0.8 (75) -0.7 (75)
β-methyl-di-N-acetyl-chitobioside	methoxy acetamido, red	-0.72 (76) -0.33 (76) -0.28 (40) WGA 1, 54°C
	acetamido, non red	-0.92 (40), WGA 1, 55°C

These results are quite consistent with the X-ray structure, which places a tyrosine sidechain close to the acetamido group of the privileged subsite and perhaps, also another aromatic ring (Trp ?) near the aglycone. Unfortunately, fluorine chemical shifts cannot, at the present time, be interpreted in any simple manner.

The fluorescence of MUFchitosides is quenched by the protein, as has been discussed previously. It has been customary (87,88) to explain this effect in terms of a local dielectric constant : bound MUF would sense a non-polar, low ε_r environment, as compared to free MUF in aqueous solution. As figure 4 shows, however, the fluorescence intensity hardly correlates with the solvent dielectric constant. The relative range of variation of the intensity is 95 % while the absorption coefficient varies by at most 30 %.

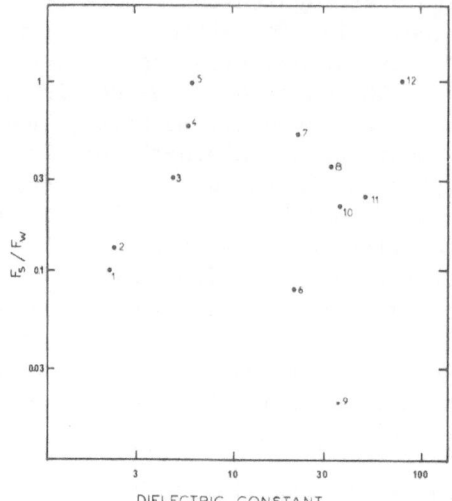

DIELECTRIC CONSTANT

Figure 4 – Fluorescence intensity of MUFGlcNAc (F_s) relative to that found in water (F_w) plotted versus solvent dielectric constant. Code : MUFGlc (5×10^{-6} M) in (1) dioxan, (4) methyl-2-butanol, (5) acetic acid, (7) ethanol, (8) methanol, (10) dimethylformamide, (11) dimethylsulfoxide, (12) water and peracetylated MUFGlcNAc (5×10^{-6} M) in (2) benzene, (3) chloroform, (6) acetone, (9) nitromethane. From (75).

Another explanation could well be the following. We have found that N-acetyltryptophanamide and N-acetyl-tyrosinamide efficiently quench the fluorescence of MUFGlcNAc. When the results are analysed in terms of the formation of a non-emitting complex, the affinity constants are found to be 23 M^{-1} (NAcTrp) and 67 M^{-1} (NAcTyr). We therefore propose that the MUF aglycone interacts (perhaps through formation of a charge-transfer complex) with the postulated Trp side-chain of the WGA binding site. This hypothesis is also supported by the fact that MUFglycosides (and also p-nitrophenylglucosaminide (24)) have a much higher affinity for WGA than their non aromatic counterparts. Thus, both the fluorescence quenching of MUFGlcNAc and its high affinity for the protein could be due to the same interaction with a tryptophyl side chain.

c) Location of the preferred subsite

All the available results imply the existence of a privileged subsite, able to bind specifically an $NHCOCH_3$ group. According to the crystallographic data, this is subsite 1, i.e. it lies at one edge of the total site. Several other models have been proposed, based on spectroscopic work (40,68,76,85), which favor a central location of the preferred subsite ; in the present symbolism, it would be subsite 2. We believe that the arguments presented up to now are inconclusive. Since most trisaccharides cannot reach the sites in the crystal, assertions about them derive from model building, and it is not clear whether alternative mo-

dels were tested and on which grounds they were rejected (62,63,84).
Fluorescence spectroscopy cannot alone solve the problem since we do not
know how and why quantum yields are modified by external perturbations.
^{1}H and ^{13}C NMR signals of the acetamido groups of $(GlcNAc)_n$ are no longer resolved for $n \geqslant 3$. Small ligands may occupy more than one location
in the site (section 4).

At the present time, we feel that the bulk of the evidence favors
a slightly amended form of the model presented previously (76). The site
is comprised of three subsites, the central one, B, having a higher affinity for GlcNAc and Neu5Ac than the others (privileged subsite). A
tyrosine lies close to B, inducing upfield shifts in acetamido methyl
groups. A tryptophan is found in C, which can interact with aromatic
aglycone and be quenched by -S-Hg-X substituents. It can be assumed that
short ligands may sample several locations in the site. A further site
(A') must be provided to accommodate Neu5Ac based trisaccharides.

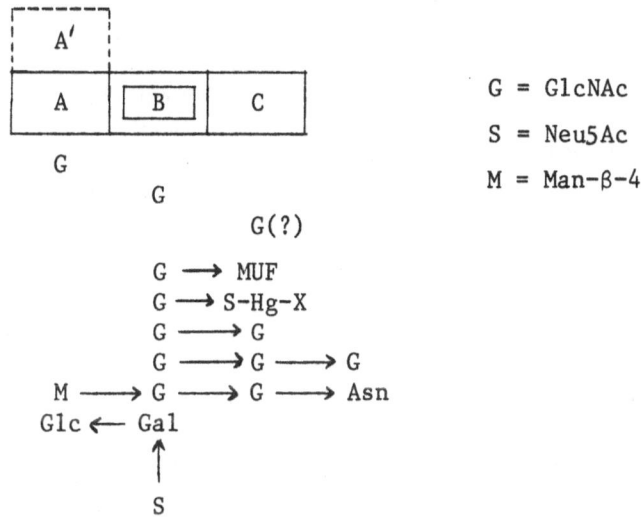

G = GlcNAc

S = Neu5Ac

M = Man-β-4

8. BINDING OF WGA TO CELL SURFACE GLYCOCONJUGATES

WGA binds gangliosides embedded in vesicle membranes ; it induces
aggregation of these vesicles (52-54). The reaction depends on the charge density of the vesicles and on the ionic strength of the surrounding
medium (54). The charge density can be varied by incorporating stearyl-
amine in the membrane. The ligand can be ionized (native GM_3) or neutral
(amide derivative of GM_3). These factors were systematically varied in
a series of aggregation experiments. The results showed that the electrostatic repulsion between charged vesicles was an important factor in
controlling aggregation.

WGA also binds vesicles containing a sialoglycoprotein, glycophorin
(64,65). The interaction is enhanced by the presence of dextran or serum
albumin. Glycophorin bears several O-glycosidic sialoglycopeptides. It is

known to bind WGA (24,89). The binding of WGA to glycophorin held on
liposomes is highly cooperative. This positive cooperativity does not
depend on extensive lateral motions of the receptors in the membrane,
since the same effect is seen at 4°C, a temperature at which the bilayer
is quite rigid. The concentration of WGA used was much higher (60 to 480
µg/ml) than the values required to agglutinate cells or to ellicit various biological responses. Positive cooperativity has also been reported
for the binding of WGA to chinese hamster ovary cells (90,91). In these
experiments, the concentration of lectin ranged from 1 ng/ml to 1 mg/ml,
for 10^6 CHO cells ; the number of bound WGA molecules per cell varied
from 10^3 to $2x10^8$. The positive cooperativity could result from a modification of the membrane (92), or from the interaction of a multivalent
ligand with flexible receptors (64,65), or from a ligand-mediated monomer-
dimer equilibrium of the lectin, as it is known to exist in solution
(§ 3).

The binding of WGA and of succinyl-WGA to normal and to neuramini-
dase treated murine thymocytes has been studied (41). It was shown that
succinyl-WGA does not interact with Neu5Ac, but binds GlcNAc. GlcNAc-
type receptors are about ten times more abundant on the cell surface than
Neu5Ac-type receptors. Other investigations have shown that two types of
ligands exist on the surface : low affinity sites ($K\sim10^5$ M^{-1}) and high
affinity sites ($K\sim10^6$ M^{-1}) (6,8,25,55,93). WGA bound to cell surface
glycoconjugates is rapidly internalized (94) at 37°C. After 2 h WGA is
found located at the perinuclear region of the cell. The internalization
of WGA in HeLa cells induces internalization of a major portion of cell
surface sialoglycoproteins and of membrane protein devoid of affinity to
WGA (94). The endocytosis of WGA bound to cell surface was used to target the toxic moiety of diphteria toxin (95). The toxic moiety (fragment A) linked to WGA, was found to be internalized in 3T3 cells and
to be much more toxic than the isolated fragment A and than free WGA,
but less toxic than the native molecule of diphteria toxin. This relatively low cytotoxicity of the hybrid molecule may be related to the
lack of hydrophobic sequence in fragment A as well as in WGA, making
difficult the translocation of fragment A across the plasma membrane, as
happens for the native toxin (96).

Succinyl-WGA, once bound to the cell surface, is also rapidly internalized. This was demonstrated by the specific internalization of succinyl-WGA substituted by Arg-Leu-daunorubicin in L1210 cells (97). After
internalization, the cytotoxic compounds Leu-daunorubicine and daunorubi-
cine are released (presumbalby through the action of lysosomal cathepsin)
and kill the cells.

9. ACTIVITY OF WGA ON MAMMALIAN CELLS

Wheat germ agglutinin is toxic to cells and has been used to select
resistant lines of chinese hamster ovary cells (90,98,99). The authors
studied the biosynthesis of glycoconjugates and showed that the resistance to WGA is probably linked to a deficiency in specific glycosyl

transferases. Similar work has been reported for murine tumors (100-103). Resistant cells were found to have altered tumorogenic and metastatic capabilities *in vivo*. Contrary to native WGA, the succinylated protein is not toxic for several cell lines (A.C. Roche and M. Monsigny, unpublished results) and is not able to inhibit the PHA or Con A induced human lymphocyte proliferation.

WGA has for a long time been regarded as strictly non-mitogenic or anti-mitogenic for lymphocytes (104,105). However, it was recently shown that WGA could activate various functions of human lymphocytes (106-108). Thus, WGA stimulates synthesis and secretion of immunoglobulins by B lymphocytes (108). The effect depends on the concentration of WGA used, and upon the time of lectin addition after plating peripheral blood. WGA can also activate various functions of macrophages, such as macrophage mediated tumor lysis (109), hydrogen peroxyde release (110), or super-oxyde anion generation (111). WGA also stimulates the tissue thrombo-plastin activity of huma monocytes *in vitro* (112). It binds to the insulin receptors of hepatocytes (113) and adipocytes (114).

REFERENCES

1. Watkins, W.M.: 1972, in The glycoproteins, A. Gottschalk, ed., Elsevier Pub. Co, pp. 830-891.
2. Glaudemans, C.P.J.: 1975, Adv. Carbohyd. Chem. Biochem. 31, pp. 313-346.
3. van Heiningen, S.: 1977, Biol. Rev. 52, pp. 509-519.
4. Olsnes, S. and Pihl, A.: 1981, in The molecular actions of toxins and viruses, Ph. Cohen and van Heinigen, S., eds., Elsevier /North Holland, Amsterdam.
5. Goldstein, I.J., Hughes, R.C., Monsigny, M., Osawa, T. and Sharon, N.: 1980, Nature 285, p. 66.
6. Monsigny, M., Roche, A.C. and Kieda, C.: 1978, in Structure and function of biological membranes. Molecular aspects. Commission of European Communities, Brussels.
7. Goldstein, I.J. and Hayes, C.E.: 1978, Adv. Carbohyd. Chem. Biochem. 35, pp. 127-340.
8. Lis, H. and Sharon, N.: 1981, in The biochemistry of plants. Vol. 6, A. Marcus, ed., Academic Press, pp. 371-447.
9. Gold, R. and Balding, P.: 1975, Receptor specific proteins, plant and animal lectins, Excerpta Medica, Amsterdam.
10. Simpson, D.L., Thorne, D.R. and Loh, H.H.:1978, Life Sciences 22, pp. 727-748.
11. Monsigny, M., Kieda, C. and Roche, A.C.: 1979, Biol. Cell. 36, pp. 289-300.
12. Barondes, S.H.: 1981, Annu. Rev. Biochem. 50, pp. 207-231.
13. Ashwell, G. and Harford, J.: 1982, Annu. Rev. Biochem. 51, pp. 531-554.
14. Bauer, W.: 1981, Annu. Rev. Plant Physiology 32, pp. 407-449.
15. Pistole, T.: 1981, Annu. Rev. Microbiol. 35, pp. 85-112.
16. Sequeira, L.: 1978, Annu. Rev. Phytopathol. 16, pp. 453-481.

17. Lis, H. and Sharon, N.: 1977, in The antigens, M. Sela, ed., IV, pp. 429-529.

18. Aub, J.C., Tieslau, C. and Lankester, A.: 1963, Proc. Nat. Acad. Sci. 50, pp. 613-619.

19. Burger, M.M. and Goldberg, A.R.: 1967, Proc. Natl. Acad. Sci. 57, pp. 359-366.

20. Burger, M.M.: 1969, Proc. Natl. Acad. Sci. 57, pp. 359-366.

21. Allen, A.K., Neuberger, A. and Sharon, N.: 1973, Biochem. J. 131, pp. 155-162.

22. Greenaway, P.J. and LeVine, D.: 1973, Nature New Biology 241, pp. 191-192.

23. Peters, B.P., Ebisu, S., Goldstein, I.J. and Flashner, M.: 1979, Biochemistry 18, pp. 5505-5511.

24. Bhavanandan, V.P. and Katlic, A.W.: 1979, J. Biol. Chem. 254, pp. 4000-4008.

25. Monsigny, M., Roche, A.C., Sené, C., Maget-Dana, R. and Delmotte, F.: 1980, Eur. J. Biochem. 194, pp. 147-153.

26. Hellström, U., Dillner, M.L., Hammarström, S. and Perlmann, P.: 1976, J. Exp. Med. 144, pp. 1381.

27. Boldt, D.H. and Lyons, R.D.: 1979, J. Immunol. 123, pp. 808-816.

28. Lehtinen, T., Perlmann, P., Hellström, U. and Hammarström, S.: 1980, Scand. J. Immunol. 12, pp. 309-320.

29. Boldt, D.H.: 1980, Molecular Immunol. 17, pp. 47-55.

30. Rice, R.H.: 1976, Biochim. Biophys. Acta 444, 175-180.

31. Mishkind, M., Keegstra, K. and Palevitz, B.A.: 1980, Plant Physiol. 66, pp. 950-955.

32. Mishkind, M., Raikhel, N.V., Palevitz, B.A. and Keegstra, K.: 1982, J. Cell. Biol. 92, pp. 753-764.

33. Mirelman, D., Galun, E., Sharon, N. and Lotan, R.: 1975, Nature 256, pp. 414-416.

34. Schmidt, E.L.: 1979, Annu. Rev. Microbiol. 33, pp. 355-376.

35. Abraham, A.K., Kolseth, S. and Pihl, A.: 1982, Eur. J. Biochem. 124, pp. 383-388.

36. Nagata, Y. and Burger, M.M.: 1974, J. Biol. Chem. 249, pp. 3116-3122.

37. Rice, R.H. and Etzler, M.E.: 1974, Biochem. Biophys. Res. Commun. 59, pp. 414-419.

38. Gottlieb, L.D.: 1982, Science 216, pp. 373-380.

39. Rice, R.H. and Etzler, M.E.: 1975, Biochemistry 14, pp. 4093-4099.

40. Lacelle, N.: 1979, Ph. D. Thesis, University of Toronto, Canada.

41. Monsigny, M., Sené, C., Obrenovitch, A., Roche, A.C., Delmotte, F. and Boschetti, E.: 1979, Eur. J. Biochem. 98, 39-45.

42. Roche, A.C. and Monsigny, M.: 1979 in Glycoconjugates, R. Schauer, P. Boer, E. Buddekke, M.F. Kramer, J.F.G. Vliegenthart and H. Wiegandt, eds, Thieme, Stuttgart, pp. 130-131.

43. Wright, C.S.: 1974, J. Mol. Biol. 87, pp. 835-841.

44. Wright, C.S., Nagata, Y., Burger, M.M. and Langridge, R.: 1974, J. Mol. Biol. 87, pp. 843-846.

45. Wright, C.S.: 1977, J. Mol. Biol. 111, pp. 439-457.

46. Drenth, J., Low, B.W., Richardson, J.S. and Wright, C.S.: 1980, J. Biol. Chem. 255, 2652-2655.

47. Lotan, R., Sharon, N. and Mirelman, D.: 1975, Eur. J. Biochem. 55,

pp. 257-262.

48. Debray, H., Decout, D., Strecker, G., Spik, G. and Montreuil, J.: 1981, Eur. J. Biochem. 117, pp. 41-55.

49. Goldstein, I.J., Hammarström, S. and Sundblad, G.: 1975, Biochim. Biophys. Acta 405, pp. 53-61.

50. Cederberg, B.M. and Gray, G.R.: 1979, Anal. Biochem. 99, pp. 221-230.

51. Privat, J.P., Delmotte, F. and Monsigny, M.: 1974, FEBS Letters 46, pp. 224-228.

52. Redwood, W.R. and Polefka, T.G.: 1976, Biochim. Biophys. Acta 455, pp. 631-643.

53. Maget-Dana, R., Roche, A.C. and Monsigny, M.; 1977, FEBS Letters 79, pp. 305-309.

54. Maget-Dana, R., Veh, R.W., Sander, M., Roche, A.C., Schauer, R. and Monsigny, M.: 1981, Eur. J. Biochem. 114, pp. 11-16.

55. Adair, W.L. and Kornfeld, S.: 1974, J. Biol. Chem. 249, pp. 4696-4704.

56. Bhavanandan, V.P., Umemoto, J., Banks, J.R. and Davidson, E.A.: 1977, Biochemistry 16, pp. 4426-4437.

57. Lotan, R. and Nicolson, G.L.: 1979, Biochim. Biophys. Acta 559, pp. 329-379.

58. Yamamoto, K., Tsuji, T., Matsumoto, I. and Osawa, T.: 1981, Biochemistry 20, pp. 5494-5499.

59. Bouchard, M.: 1979, Thèse de 3ème Cycle, Université d'Orléans, France.

60. Privat, J.P., Delmotte, F. and Monsigny, M.: 1974, FEBS Letters 46, pp. 229-232.

61. Kronis, K.A. and Carver, J.P.: 1982, Biochemistry 21, pp. 3050-3057.

62. Wright, C.S.: 1980, J. Mol. Biol. 132, pp. 53-60.

63. Wright, C.S.: 1980, J. Mol. Biol. 141, pp. 267-291.

64. Ketis, N.V., Girdlestone, J. and Grant, C.W.M.: 1980, Proc. Natl. Acad. Sci. USA 77, pp. 3788-3790.

65. Ketis, N.V., and Grant, G.W.M.: 1982, Biochim. Biophys. Acta 685, pp. 347-354.

66. Thomas, M.W., Walborg, Jr. E.F. and Jirgensons, B.: 1977, Arch. Biochem. Biophys. 178, pp. 625-630.

67. Stein, P.J. and Heehn, K.G.: 1980, Biochem. Biophys. Res. Commun. 95, pp. 547-552.

68. van Landschoot, A., Loontiens, F.G., Clegg, R.M., Sharon, N. and de Bruyne, C.K.: 1977, Eur. J. Biochem. 79, pp. 275-283.

69. Privat, J.P., Lotan, R., Bouchard, P., Sharon, N. and Monsigny, M.: 1976, Eur. J. Biochem. 68, 563-572.

70. Privat, J.P. and Charlier, M.: 1978, Eur. J. Biochem. 84, 79-85.

71. Lotan, R. and Sharon, N.: 1973, Biochem. Biophys. Res. Commun. 55, pp. 1340-1346.

72. Privat, J.P., Delmotte, F., Mialonier, G., Bouchard, P. and Monsigny, M.: 1974, Eur. J. Biochem. 47, 5-14.

73. Jordan, F., Bassett, E. and Redwood, W.R.: 1977, Biochem. Biophys. Res. Commun. 75, pp. 1015-1021.

74. Jordan, F., Bahr, H., Patrick, J. and Woo, P.W.K.: 1981, Arch. Biophys. Biochem. 207, pp. 81-86.

75. Midoux, P.: 1980, Thèse de 3ème cycle, Université d'Orléans, France.

76. Grivet, J.P., Delmotte, F. and Monsigny, M.: 1978, FEBS Letters, 88,

pp. 176-179.

77. Midoux, P., Grivet, J.P. and Monsigny, M.: 1980, FEBS Letters 120,
     pp. 29-32.
78. Nichol, L.W., Jackson, W.J.H. and Winzor, D.J.: 1967, Biochemistry
     6, pp. 2249-2256.
79. Sutherland, I.O.: 1971, Annu. Rep. NMR Spectroscopy 4, pp. 71-235.
80. McLaughlin, A.C. and Leigh, J.S.: 1973, J. Mag. Res. 9, pp. 296-
     304.
81. Feeney, J., Batchelor, J.G., Albrand, J.P. and Roberts, G.C.K.:
     1979, J. Mag. Res. 33, 519-529.
82. Clegg, R.M., Loontiens, F.G., van Landschoot, A., Sharon, N., de
     Bruyne, C.K. and Jovin, T.M.: 1980, Arch. Int. Physiol. Biochem.
     88, pp. B69-B70.
83. Neurohr, K.J., Lacelle, N., Mantsch, H.H. and Smith, I.C.P.: 1980,
     Biophys. J. 32, pp. 931-938.
84. Wright, C.S.: 1981, J. Mol. Biol. 145, pp. 453-461.
85. Monsigny, M., Delmotte, F. and Hélène, C.: 1978, Proc. Natl. Acad.
     Sci. USA 75, pp. 1324-1328.
86. Eftink, M.R. and Ghiron, C.A.: 1976, J. Phys. Chem. 80, pp. 486-493.
87. Dean, B.R. and Homer, R.B.: 1973, Biochim. Biophys. Acta 322, pp.
     141-144.
88. Harina, B.M., Bothner-By, A.A. and Gill, III T.J.: 1977, Biochemis-
     try 16, pp. 4504-4512.
89. Boldt, D.H., Speckart, S.F., Richards, R.L., Alving, C.R.: 1977,
     Biochem. Biophys. Res. Commun. 74, pp. 208-214.
90. Stanley, P., Suoo, T. and Carver, J.P.: 1980, J. Cell. Biol. 85,
     pp. 60-69.
91. Stanley, P. and Carver, J.P.: 1977, Proc. Natl. Acad. Sci. USA 74,
     pp. 5056-5059.
92. Bornens, M., Karsenti, E. and Avrameas, E.: 1976, Eur. J. Biochem.
     65, pp. 61-69.
93. Obrenovitch, A., Sené, C., Roche, A.C., Monsigny, M., Visher, P.
     and Hughes, R.C.: 1981, Biochimie 63, pp. 169-175.
94. Kramer, R.H. and Canellakis, E.S.: 1979, Biochim. Biophys. Acta
     551, pp. 328-348.
95. Gilliland, D.G., Mannhalter, H. and Collier, R.J.: 1981 in Receptor
     mediated binding and internalization of toxins and hormones. Middle-
     brook, J.L. and Kohn, L.D., eds. Academic Press, New York, pp. 311-
     327.
96. Pappenheimer, A.M. Jr. and Moynihan, M.R.: 1981 in (ref. 95) pp. 31-
     51.
97. Monsigny, M., Kieda, C., Roche, A.C. and Delmotte, F.: 1980, FEBS
     Letters 119, pp. 181-186.
98. Briles, F.B., Li, E. and Kornfeld, S.: 1977, J. Biol. Chem. 252, pp.
     1107-1116.
99. Yogeeswaran, G., Murray, R.K. and Wright, J.A.: 1974, Biochem.
     Biophys. Res. Commun. 56, pp. 1010-1016.
100. Jumblatt, J.E., Tien-Wen Tao, Schlup, B., Finne, J. and Burger, M.M.:
     1980, Biochem. Biophys. Res. Commun. 95, pp. 111-117.
101. Yogeeswaran, G. and Tien-Wien Tao: 1980, Biochem. Biophys. Res.
     Commun. 95, pp. 1452-1460.

102. Finne, J., Tien-Wien Tao and Burger, M.M.: 1980, Cancer Res. 40, pp. 2580-2587.
103. Dennis, J.W. and Kerbel, R.S.: 1981, Cancer Res. 41, pp. 98-104.
104. Karsenti, E.,Bornens, M. and Avrameas, S.: 1975, Eur. J. Immunol. 5, pp. 74-79.
105. Greene, W.C. and Waldman, T.A.: 1980, J. Immunol. 124, pp. 2979-2987.
106. Gordon, L.K., Hamill, B. and Parker, C.W.: 1980, J. Immunol. 125, pp. 814-819.
107. Udey, M.C., Chaplin, D.D., Wedner, H.J. and Parker, C.W.: 1980, J. Immunol. 125, pp. 1544-1550.
108. Greene, W.C., Goldman, C.K., Marshall, S.T., Fleisher, T.A. and Waldmann, T.A.: 1981, J. Immunol. 127, pp. 799-804.
109. Kurisu, M., Yamazaki, M. and Mizuno, D.: 1980, Cancer Res. 40, pp. 3798-3803.
110. Tomioka, H. and Saito, H.: 1980, Infect. Immun. 29, pp. 469-476.
111. Kayashima, K., Ondue, K., Nakagawara, A. and Minakami, S.: 1980, Microbiol. Immunol. 24, pp. 449-461.
112. Lyberg, T. and Prydz, H.: 1980, Thrombosis Haemostasis 42, pp. 1574-1579.
113. Cuatrecasas, P.: 1974, Annu. Rev. Biochem. 43, pp. 169-214.
114. Livingston, J.N. and Purvis, B.J.: 1981, Biochim. Biophys. Acta 678, pp. 194-201.

III - EVOLUTION OF BIOLOGICAL MACROMOLECULES

THE ORIGIN AND EVOLUTION OF LIFE AT THE MOLECULAR LEVEL[*]

Manfred EIGEN and Ruthild WINKLER-OSWATITSCH
Max-Planck-Institut für biophysikalische Chemie
Am Fassberg
3400 Göttingen, F.R.G.

BIOLOGICAL COMPLEXITY

The most conspicuous attribute of biological organisation is its complexity. We see this especially clearly when we come down to the level of molecular detail. The physical problem of the origin of life can be reduced to the question : "Is there a mechanism by means of which complexity can be generated in a regular, reproducible way ?". One way of regarding the origin of biological complexity as a process obeying natural law was put forward, 120 years ago, by Charles Darwin. His theses can be formulated in modern terms as follows :

- Complex systems arise by an evolutionary process.
- Evolution is based upon natural selection.
- Natural selection is a physical consequence of self-reproduction.

The third thesis is in fact neo-Darwinian. It follows from the quantitative treatments of population genetics developed -particularly by Haldane, Fischer and Wright- in the first half of this century. The molecular-biological revolution of the Fifties led to the euphoric expectation that the laws of genetics would prove reducible to the magic formula

DNA $\longrightarrow$ RNA $\longrightarrow$ Protein $\longrightarrow$ everything else

This dogma of molecular biology postulates that every detail of a complex structure is governed by information which flows irreversibly from the genotypic legislative to the phenotypic executive at the somatic level of an organism. In the Eighties, after the discovery of reverse transcriptases, restriction endonucleases, exons and introns -in short, of Nature's tools for the processing of genotypic information- we phrase our statements a little more cautiously :

- All organisms have to reproduce their genetic information.
- Only nucleic acids can be reproduced true to sequence.
- Reproduction provides a basis not only for the conservation of

353

C. Hélène (ed.), Structure, Dynamics, Interactions and Evolution of Biological Macromolecules, 353–370.

information but also for a selective processing of information
and for optimisation.

Thirty years of molecular-biological research have shown us how to
ask the "right" questions. First of all, we must be able to observe accurately the basic process, viz., the reproduction of genetic information. From our results we construct an abstract, general biological principle with experimentally verifiable logical consequences. Finally we
search known biological structures for a "fossil" record, in order to
find out whether the historical origin of life could in principle have
followed the basic schema which we have formulated. Answers must be
sought to the questions :

- Can a causal connection be shown between molecular self-reproduction, selection and evolution ?
- Is self-organization, based upon self-reproduction and selection,
  an inevitable process, whose pre-requisites and consequences can
  be found in natural systems ?
- Are there "fossil" relics of molecular evolution ?

EXPERIMENTS IN MOLECULAR EVOLUTION

We shall first examine in detail the mechanism of reproduction of a
virus whose genetic information lies encoded in a single-stranded RNA
molecule. This virus is able to perform basically four functions, each
of which is associated with a different protein unit : a capsid, which
encloses the RNA and protects it from degradation by hydrolysis ; a penetration enzyme, to inject the genetic material into the host cell ; a
factor to break down the host cell ; finally, a mechanism which re-
programmes the complex machinery of the host cell in accordance with
instructions from the virus. In the case which we shall discuss, the
last function is carried out by a protein molecule that associates with
three ribosomal proteins to give an enzyme which recognizes the viral
genome and replicates it very rapidly. The entire apparatus of metabolism
and gene translation is turned over to the exclusive purpose of producing
new virus particles. This factor thus embodies the essence of viral infection.

The reproduction mechanism of Q_β, an RNA virus capable of infecting
bacteria, is shown schematically in figure 1.

During the replication of RNA, the enzyme, which consists of four
subunits, moves from the 3' to the 5' end of the template. The newly-
formed RNA replica has an internally folded structure and this prevents
the formation of a double strand. Sol Spiegelmann (1), who was the first
to isolate the reproduction enzyme of this virus and, using it, to synthesize infectious viral RNA, has shown that annealing destroys the reproductive capacity of the virus. The reproduction does not take place at a
steady speed ; the enzyme pauses at so-called "pause sites". Presumably
it has to wait for the next portion of the matrix to melt apart before

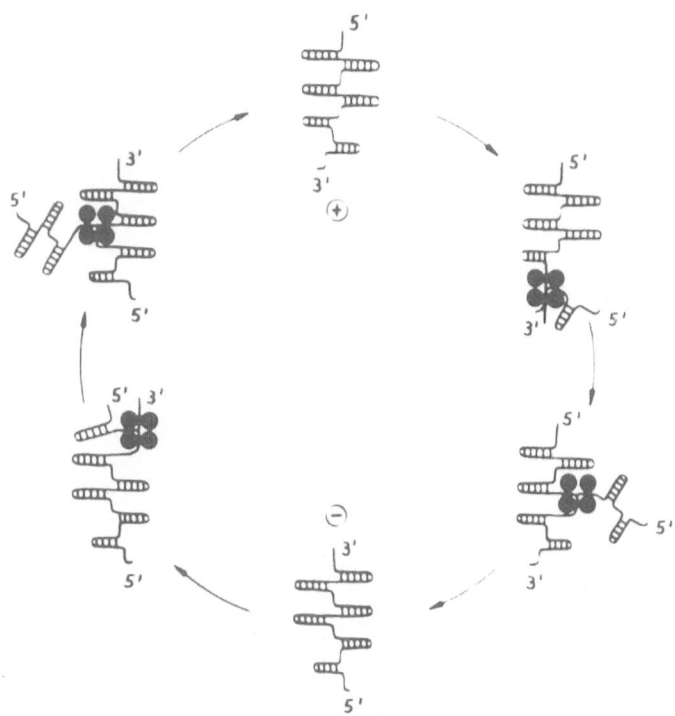

Figure 1 - The single-stranded RNA of the bacteriophage Qβ is reproduced
with the assistance of an enzyme, called Qβ replicase, that consists of
four subunits (black dots). The enzyme recognizes the matrix specifically
and during synthesis, it moves from the 3' to the 5' end of the template
strand. The replica formed (-) is complementary to the template (+). The
3' and 5' ends are symmetrically related in such a way that both plus-
and minus-strands have similar 3' ends ; both are recognized by the replicase, and the minus-strand thus acts as a template for the formation
of a plus-strand. Internal folding of the two strands prevents the formation of a plus-minus double helix.

it can proceed to copy it. The discovery and isolation of a non-infec-
tious RNA component, 220 nucleotides long, are also due to Spiegelmann
(2) ; this RNA molecule, known as "minivariant", contains the 3'- (and
the 5'-) end needed for recognition by Qβ replicase and is therefore
replicated by the enzyme - in fact much faster than the real Qβ-RNA.
"Midivariant" is thus a "scrounger", which cannot infect a cell alone,
since it cannot produce the specific reproduction enzyme.

    In our laboratory the mechanism of RNA replication by Qβ replicase
has been studied quantitatively (3-6). The experiments were begun by
Manfred Sumper and Rüdiger Luce and continued by Christof Biebricher and
Rüdiger Luce. Our knowledge of the mechanism of the replication reaction

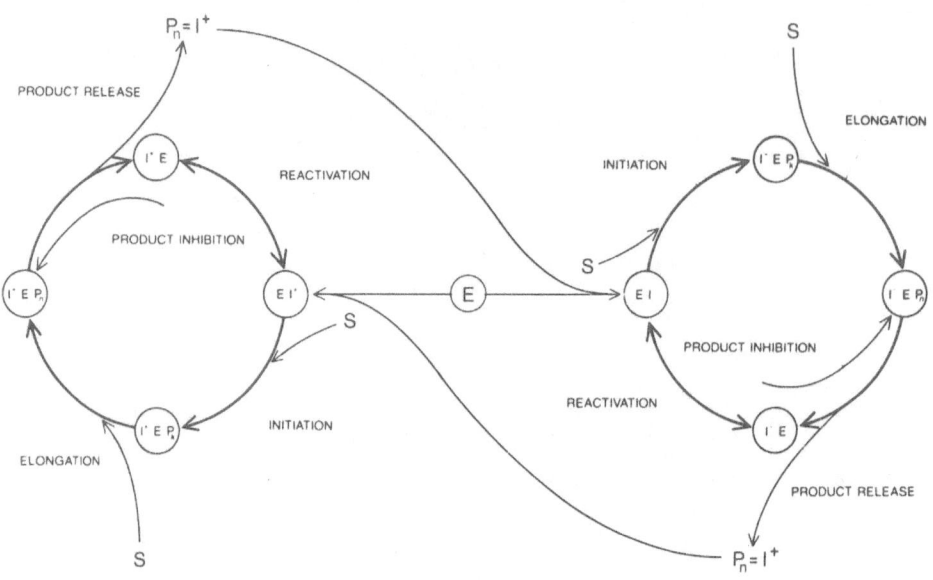

Figure 2 - A characteristic of the mechanism of RNA replication is the coupled pair of cycles of synthesis, for the plus- and minus-strand respectively. A catalytically active complex consists of the enzyme (replicase) and an RNA template. Four phases of each cycle can be distinguished : (i) the commencement of replication by the binding of at least two substrate (nucleoside triphosphate) molecules ; (ii) elongation of the replica strand by successive incorporation of nucleotides ; (iii) dissociation of the complete replica away from the replicase ; (iv) dissociation of the enzyme from the 5' end of the template and its reassociation with the 3' end of a new template. The matrix is represented by I (information), the enzyme by E and the reaction product by P. The ultimate reaction product P_n is then used as a template (I). The substrate, S, is the triphosphate of one of the four nucleosides A, U, G, and C.

results from a combination of :

(i) experimental investigation of the rate of replication as a function of the concentrations of substrate, enzyme and RNA-templates,

(ii) the mathematical analysis of a model for the replication (see figure 2) and,

(iii) the computer simulation of this model using realistic values for parameters, obtained from experimental data.

A typical experiment is sketched out in figure 3. A ^{32}P-labelled nucleotide is used to measure the amount of newly-synthesized RNA as a function of time. The concentrations of the substrates (the four nucleo-

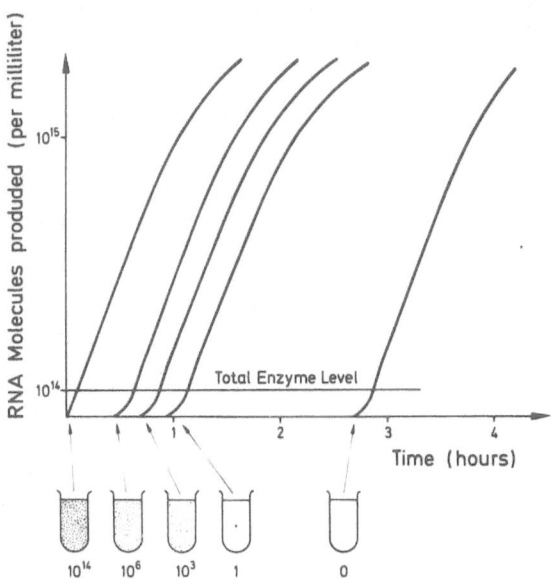

Figure 3 - If a synthesis mixture (buffer, salts, Qβ replicase, nucleo-side triphosphates and RNA templates) containing equal numbers of enzyme molecules and template strands is incubated, then the rate of production of RNA molecules is linear in time, since the number of catalytically effective complexes is equal to the (constant) number of enzyme molecu-les.(The flattening of the reaction curves at high RNA concentration is due to inhibition of the enzyme by binding of excess RNA to the active site). If the template concentration is lowered in steps of a particular factor, the growth curves are displaced along the time axis by equal intervals. This logarithmic dependence of the induction period indicates an exponential growth law which holds as long as the enzyme is in excess over the template RNA, during which time the number of catalytically active complexes perpetually increases (autocatalysis). Even if no tem-plate is present at the beginning, an RNA arises "de novo" after a long induction period. This acts as a template and multiplies rapidly. Such synthesis of RNA "de novo" is a particular property of Qβ replicase.

side triphosphates of A, U, G, and C) and that of the enzyme are cons-tant. The initial template concentration is varied by serial dilution with a constant dilution factor. The increase in RNA concentration in the course of time can be divided into three phases :

(1) An induction period, which increases logarithmically with in-creasing dilution. Repeated dilution by a constant factor pro-duces a series of constant displacements along the time axis.

(2) A linear increase in RNA concentration, starting at the moment
    when the RNA concentration becomes (approximately) equal to the
    enzyme concentration.
(3) A plateau, which is not reached until the concentration of the
    RNA greatly exceeds that of the enzyme.

This behaviour can be explained by the following kinetic considerations. The reaction leading to new RNA templates is catalysed by a complex of enzyme and template. The affinity between these partners is so
high that at the concentrations used every RNA molecules binds to an
enzyme. The number of catalytically active complexes then rises exponen-
tially until the RNA concentration becomes equal to the concentration of
enzyme. At this point the enzyme is saturated with RNA. From now on the
number of catalytically active RNA-enzyme complexes remains constant and
the synthesis enters the linear phase – that is, the rate of appearance
of new RNA molecules becomes constant. The new RNA molecules, now in
excess over the enzyme, bind not only as templates but also –less strongly– at the site of synthesis. This leads to inhibition of synthesis by
the excess RNA molecules present in the linear phase, and finally the
synthesis comes to a standstill.

By altering the reaction conditions (e.g. substrate concentration,
enzyme concentration, initial ratio of plus- to minus-strand) it proved
possible to measure the kinetic parameters and to verify the reaction
scheme of figure 2. The principal conclusions of this investigation
were :

- The replication is catalysed by a complex consisting of one enzyme and one RNA molecule.
- The growth rate is proportional to the lower of the two total
  concentrations (enzyme or RNA). This means for low and high RNA
  concentrations exponential and linear growth respectively.
- If two different mutants compete, then even in the linear region
  the mutant with the selective advantage continues to grow expo-
  nentially, until this mutant saturates the enzyme.
- A selective advantage in the linear phase is given solely by differences in the kinetics of enzyme-template binding. In the exponential phase, in contrast, the competition is based upon overall
  rates of production.
- The overall growth rate of plus- and minus-strands is given in
  the exponential phase by the geometric mean of their rate parameters and in the linear phase by the harmonic mean. The ensemble of
  plus- and minus-strands grows up in the ratio of the square roots
  of their respective rate parameters of production.
- The replication rate depends on the length of the RNA chain to be
  synthesized and on the concentrations of the substrates (A, U, G,
  C). The latter dependence is weaker than linear as successive
  substrate molecules are frequently pre-bound to the enzyme.

We now arrive at the actual question which we wished to answer :
what are the consequences of the laws of replication ? Can new properties

be attained by the evolution of a replicating system ? Are the characteristics of the replicative system sufficient, or do we need to look for
additional necessary properties ?

Spiegelmann (7) has provided an important stimulus for research in
this direction. By serial transfer of RNA templates from one nutrient
glass to another he obtained, at the end of such a series, selected variants of the Q_β genome which were no longer infectious but which showed
a higher replication rate (measured per nucleotide). They made use of
their higher replication rate to escape the selection pressure of dilution. The significant finding was however that these new variants not
only had a higher rate of chain elongation but also possessed 500 instead of the original 4500 nucleotides, which enabled them to complete a
round of replication in a fraction of the original time. Evolution of
this sort, resulting in a loss of information, might well rather be termed "degeneration". However, the experiments showed that this replication system is highly capable of adaptation - an indispensible prerequisite for evolutionary behaviour.

These experiments became particularly relevant when, in 1974, a
surprising observation was made by Manfred Sumper. In series which were
diluted so far that in each sample the probability of finding a template
molecule was very small, there arose none the less, reproducibly and homo-
geneously, a molecule with size and structure similar to the "minivariant"
isolated by Spiegelman. This phenomenon is indicated in figure 3. In
contrast to the template-instructed replication, this synthesis was associated with a disproportionately long induction period which depended
critically upon the reaction conditions. Sumper was convinced that he
had found a variant which the Q_β enzyme had "invented" and synthesized
"de novo". (Other experts in the field believed almost unanimously that
he was seeing an impurity carried by the enzyme). Sumper was able by
further experiments to refute the "impurity" postulate, and since then
the existence of synthesis "de novo" has been amply proven - above all
in experiments conducted by Christof Biebricher and Rüdiger Luce. Kinetic measurements have shown that the induction of template-instructed
and "de novo" synthesis are subject to entirely different rate laws.
Thus, for example, the template-instructed synthesis proceeds on a single
enzyme molecule and the substrates are introduced successively into the
growing chain. Synthesis "de novo", on the other hand, requires a complex   of several enzyme molecules, and the rate-determining step is a
nucleation by three or even four substrate molecules. However the principal evidence lies in the demonstration that, in the early phase of synthesis, variants appear which lengthen under selection pressure and
which lead under different experimental conditions to different end-
products. The last observation is also due to Manfred Sumper. He obtained
different "minivariants", which grew normally under conditions -e.g.
in the presence of reaction inhibitors- in which the wild-type was no
longer capable of existence.

The decisive experiment of Biebricher and Luce is shown in figure
4. A synthesis medium containing highly purified enzyme and substrates

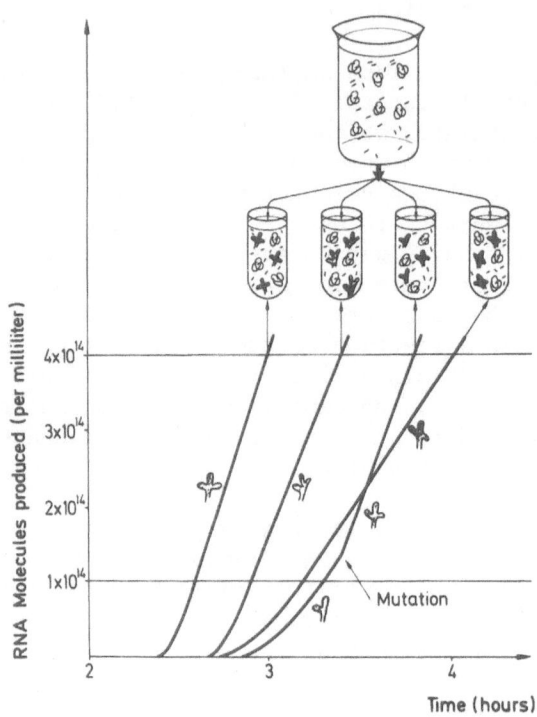

Figure 4 - A solution of nucleotide triphosphate is incubated in the
presence of Q_β replicase for just long enough to assure the manifold
replication of any templates which may contaminate the enzyme. The incubation is interrupted before even one template has time to arise "de
novo". The solution is then divided up into portions and the incubation
is continued, this time long enough to allow products to arise "de novo"
and to multiply. The RNA formed in each portion is analysed by the fingerprint method ; various different reaction products are found. Sometimes the growth curve displays the appearance of a new mutant. While
the incubation time of template-instructed synthesis is determined unambiguously (because of the superposition of many individual processes)
the synthesis "de novo" shows a scatter of induction times. This indicates that the initiation step is a unique molecular process which is then
rapidly "amplified".

is incubated and maintained at a suitable temperature, for a time adequate to allow the multiplication of any templates present but too short
to enable products to arise "de novo". Then the solution is divided into
portions. Each portion is incubated long enough to allow synthesis "de
novo" and the products are compared by the fingerprint method. If the
"impurity" hypothesis is correct then multiplication of the impurity in
the first phase should lead to the same product from each portion of the
incubated medium. If the "de novo" hypothesis is correct then the pro-

ducts should be different, since at the beginning different enzyme molecules were working on different products. Selection -that is, preferential reproduction of one rudimentary strand- could not yet take place,
since in the first, short incubation none of the products "de novo" was
complete.

The experiment gave many different products. Only if these were
mixed and again incubated did a single, homogeneous, reproducible variant
grow up. The earliest products that could be detected directly were about
70 nucleotides long. In the course of evolution there appeared longer
chains, e.g., at high salt concentration the "minivariants", with about
220 nucleotides.

The important result of these experiments is, however, not just an
explanation of this unusual feature of the Q_β system. More important, we
now have a flexible replication system at our disposal, with which a
series of further interesting studies can be set up. Above all, it can
be shown that selection and evolution are inevitable consequences of
self-replication, and can as such be investigated quantitatively. For
example, the question of rapid optimisation under extreme experimental
conditions has been answered in detail. The results of the evolution
experiments described can be summarized in four principal statements :

- The synthesis "de novo" of RNA by Q_β replicase proceeds by a
  mecanism fundamentally different from that of template-instructed
  RNA synthesis. The active reaction complex contains at least two
  enzyme molecules and requires nucleation by a seed of three or
  four substrate molecules.
- The rate-limiting step is nucleation, while elongation and reproduction follow rapidly. The singular nature of the molecular process which initiates the reaction is reflected in the scatter in
  induction times for synthesis "de novo".
- Synthesis "de novo" produces a broad spectrum of mutants of
  varying length, containing sequences capable of adaptation to a
  great variety of environmental conditions.
- Initiation of self-reproduction is clearly sufficient to set the
  process of evolutive optimisation in motion.

This finding allows the development of an evolution reactor, in
which optimally reproducing RNA sequences may be produced in a relatively
short time. This in turn makes possible the development of a principle
for the evolution of RNA structures with optimised translation products.
Experiments in this direction are in progress.

Self-replication and mutagenicity in an open system far from equilibrium are thus sufficient to produce behaviour-patterns including selection and evolution. Even in relatively simple replication systems properties optimal with respect to the wild-type can be produced in vitro in a
few generations. Such effects must be the consequence of a physical principle. Can such a principle be formulated quantitatively ?

SELECTION AND EVOLUTION GOVERNED BY NATURAL LAW

In earlier publications we have shown that the principle of selection can be deduced from the premisses of a self-replication system as an extremum principle. It states that inherent linear autocatalysis causes the relative population numbers to take on values which correspond to the highest reproductive efficiency of the system as a whole. The distribution of relative concentrations in the stationary population is, after a short induction period, independent of changes of the system as a whole. The population consists of a uniquely defined wild-type (or several equivalent, i.e., "degenerate" variants) and a spectrum of mutants. The wild-type is most frequently represented in the distribution, but in a well-adapted population it comprises only a small fraction of the total. The quotient of the population number (x_i) of the individual mutant i and that of the wild-type (x_m) is given by a function of the rate parameters for mutation $(W_{im}$, for $m \rightarrow i)$ and reproduction $(W_{mm}$ and W_{ii} for $m \rightarrow m$ and $i \rightarrow i$ respectively) :

$$\frac{x_i}{x_m} = \frac{W_{im}}{W_{mm} - W_{ii}}$$

The following variables are also of importance :

$\bar{q}$ = average accuracy of copying of a nucleotide
$1-\bar{q}$ = average error rate per nucleotide
ν_i = number of nucleotides in the sequence i

$Q_i \approx \bar{q}^{\nu_i}$ = fraction of sequences of type i correctly copied
σ_m = superiority of the wild-type over its spectrum of mutants (corresponding in general to the ratio of the replication rate of the wild-type and the average replication rate of the mutants).

The consequences of this extremum principle, valid for Darwinian systems, are :

- Selection of a distribution of mutants dominated by the wild-type. This is only stable as long as the conditions $\sigma_m > 1$ and $Q_m > \sigma_m^{-1}$ are fulfilled.
- Evolution by the selection of newly-appearing mutants which by virtue of a selective advantage disobey the condition $\sigma_m > 1$ and thus destabilize the dominance of the wild-type.
- Restriction of the information content due to the conditions $Q_m > \sigma_m^{-1}$. The upper limit is given by γ_{max} $\nu_{max} = \ln \sigma_m / 1 - \bar{q}_m$. It corresponds roughly to the reciprocal of the average error rate $(1-\bar{q}_m)$, as long as σ_m is sufficiently larger than unity. If the information content ν_m of the wild-type approaches the upper limit ν_{max}, Q_m becomes approximately equal to σ_m^{-1}. The proportion of wild-type in the total population is then very small :

$$\frac{x_m}{\sum\limits_{k} x_k} = \frac{Q_m - \sigma_m^{-1}}{1 - \sigma_m^{-1}}$$

Both selection – the stabilization of a particular distribution – and evolution – the establishment of one new population after another – result from an "inward compulsion". They are the inevitable consequence of self-reproduction behaviour.

The fact that the evolutionary optimisation process indeed reaches the "mountain peaks" and does not get stuck in the "foothills" lies in the topology of multidimensional mutation space. Consider for example a binary sequence with ν members. We can give each position in the sequence a co-ordinate axis with two points and thus obtain a ν-dimensional phase space in which each of the 2^ν points represents a mutant. The process of evolution can then be regarded as a route in this space, characterized by a continually rising selection value. The topology of such a multi-dimensional space is not easily imaginable ; the "mountains" are extremely bizarre, for although these are 2^ν points, the greatest separation (in terms of mutation steps) is only ν. There are saddle-points of various orders, at which movement in k directions leads uphill and in $(\nu - k)$ directions downhill. This provides a sufficient basis for relatively small mutational jumps to enable the system always to find an uphill route. There is a best value of ν, for which the number of routes is large enough and the probability of multiple mutations (depending on population size) is great enough for any optimal "peak" to be reached.

Let us summarize : selection, evolution and optimisation are processes which follow regular physical laws and which can be formulated quantitatively. This of course does not mean that the actual, historical process of evolution can be deduced from theory ; the starting-point, the complex boundary conditions and the multitude of superimposed perturbing influences are all more or less unknown. Theory tells us simply what follows when certain premisses are set up and certain boundary conditions are imposed. It explains the reproducible, regular phenomenon of Nature in an "If – then" description. This generalization applies to the theory outlined here, which has helped us to interpret and explain the experimental results described above. It has been confirmed by quantitative measurement under the exactly-defined initial and boundary conditions of the laboratory. For processes occurring in Nature, however, it reveals only trends, minimal requirements, limitations and, perhaps, some consequences. It must also be shown by experimental investigation whether conclusions from the theory have any relevance for naturally-occurring processes. Two such conclusions are especially worthy of mention :

- The quantity of information which can be selected in a molecular population depends upon the average error rate and the average selective advantage of the wild-type. Crossing the critical error threshold leads to such an accumulation of errors that the infor-

mation in the wild-type sequence is irretrievably lost.
- The capacity of the wild-type for adaptation is greatest close
  to the error threshold. The quantity of information compatible
  with a stable distribution is then in optimal relation to the
  variety in its spectrum of mutants. Such a system responds very
  flexibly to changes in its environment. The wild-type is the
  predominant individual sequence, but it makes up only a small
  fraction of the complete mutant spectrum.

MOLECULAR RECORDS OF EVOLUTION IN NATURE

The predictions of the theory of evolution can be tested on natural
systems. Charles Weissmann and his co-workers (10,11) have obtained the
following results for Q_β viruses.

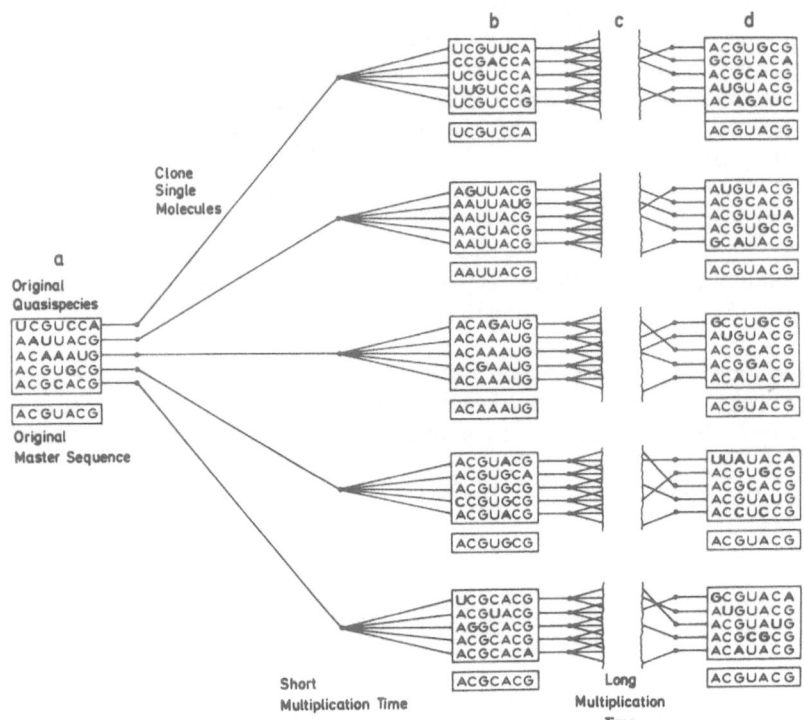

Figure 5 - In the experiment, carried out by Charles Weissmann and his
co-workers, single Q_β-RNA molecules (or viruses) from a wild-type distribution (a) were cloned in E.coli bacteria. After rapid multiplication the
clones of individual RNA molecules (b) were analysed and compared by the
fingerprint method. Differences were noticed in one or two positions in
the sequence. After a further, long period of reproduction (c) the wild-
type distribution (d) was found in every clone, i.e., the average sequences had become identical again.

The wild-type has a defined sequence, with, however, does not mean that the majority of viruses share exactly the same sequence. It means merely that the superposition of all sequences gives an unambigous "majority" or master sequence, namely, that of the wild-type.

The cloning of single viruses or single viral RNA molecules followed by their rapid multiplication leads to populations with various sequences. The sequences generally deviate in one or two positions from the wild-type, which is itself found in hardly any clone (see figure 5). The fact that the wild-type makes up only a (nearly negligibly) small fraction of the mutant spectrum implies that the information content of the wild-type has very closely approached the threshold value ν_m.

Deliberately-produced extra-cistronic single mutations (these are non-lethal mutations in portions of the sequence that are not translated) revert to the wild-type. They correct their errors with a probability around 3×10^{-4}. Quantitative analysis of such data allows the determination of the error rate, the selective advantage of the wild-type and thus the establishment of the critical error threshold for the transmission of the information in the genome. This value agrees within experimental error with the quantity of information present (4500 nucleotides). The fact that the cloning of individual mutants is possible at all is due to the dominance in the distribution, around the wild-type, of mutant whose growth rates are very similar to that of the wild-type itself. These are preferentially "fished out" in the serial dilution steps needed for the cloning of single molecules. Since they multiply nearly as rapidly as the wild-type, they begin by producing a spectrum of mutants with an average sequence identical to their own. At some point, the wild-type will reappear in this spectrum. However it can only assert itself slowly, the speed with which it does so corresponding to the (small) difference between the growth rates of the wild-type and the cloned mutant. Finally each clone is dominated by the wild-type (see figure 5).

Yet again the conclusion can be drawn that all single-stranded RNA viruses are subject to similar restrictions with regard to information content. In Nature there are no (single-stranded) RNA viruses whose replicative unit contains more than the order of 10^4 nucleotides. All larger viruses possess double-stranded nucleic acids or are composed of several replicative units. These in turn are subject to analogous relationships between the error threshold and the maximum reproducible information content. DNA polymerase in general work more accurately than RNA polymerase. This is due to their additional facility for recognition and correction of errors.

What do these results signify for our understanding of early evolution ?

The first replicative units must have possessed considerably less information than the RNA viruses, which work with an optimised RNA-copying machinery. In the absence of efficiently adapted enzymes the accuracy of reproduction depends solely upon the stability of the base pairs. Under these conditions of the GC pair has a selective advantage over the AU pair of a factor of about 10. Model experiments show that for GC-rich polynucleotides the error rate per nucleotide can hardly be reduced below a value of 10^{-2}. The first "genes" must accordingly have been polynucleotides with a chain length around 100 bases or less.

Molecular evolution demands inherent self-reproductivity. RNA seems to fulfil this function best of all known macromolecules. On account of its complex structure RNA must first have appeared in Nature long after proteins or protein-like structures. A protein can by chance fulfil a particular function, but this fulfilment is determined by purely structural and not at all by functional criteria. Adaptation to a particular function, however, demands an inherent mechanism of self-reproduction. The only logically justifiable way of exploiting the immense functional capacity of the proteins in evolution lies in an intermarriage between these two classes of macromolecule, that is, in the translation into protein of the information stored in the self-reproductive RNA structures.

This at once raises the question : "Could RNA ever have arisen without the help of enzymes -without replicases ?". Experiments by Leslie Orgel and his co-workers (12) suggest that this was possible. It was found that zinc ions -found today as co-factors in all replicases- are excellent catalysts for the 3'-5' union of nucleotides, thus they allow the template-instructed synthesis of polymers. This was first demonstrated with poly(C) as template. If activated G and A nucleotides are offered, in equal concentration, then G is preferentially incorporated into the product by a factor, depending on reaction conditions, between 30 and 200.

This suggests strongly that in a suitable medium GC-rich strands with a chain length around 100 nucleotides will arise spontaneously, reproduce themselves and undergo adaptation by evolution.

Can we today find a record of these first "genes" ?

The information content of such "genes" suffices only for relatively small proteins, certainly not optimally adapted. This means, however, that these are by now long outdated as information carriers and have been displaced by better ones. The displacement proceeded hand in hand with the development of the machinery of translation. In the translation apparatus RNA structures carry not only information but also their own function and in addition they represent targets for processing functions. It is therefore more likely that they have survived up to the present as functional entities, for example as transfer RNA (tRNA) or as ribosomal RNA (rRNA) than as message carriers. Since the functional RNA molecules did not have to store genetic information, they were hardly subject to selection pressure once they had arrived at an adapted structure. The functional nucleic acids were recruited to begin with from the same reservoir as their information-carrying sister molecules, the mRNAs. We may thus expect for example tRNA, as an eyewitness of the early evolution ot the translation apparatus, to retain some memory of the structure of the earliest "genes". This molecule, with a chain length of about 76 nucleotides, fulfils exactly the criteria which theory requires, which present-day structures confirm and which are relevant for prebiotic conditions.

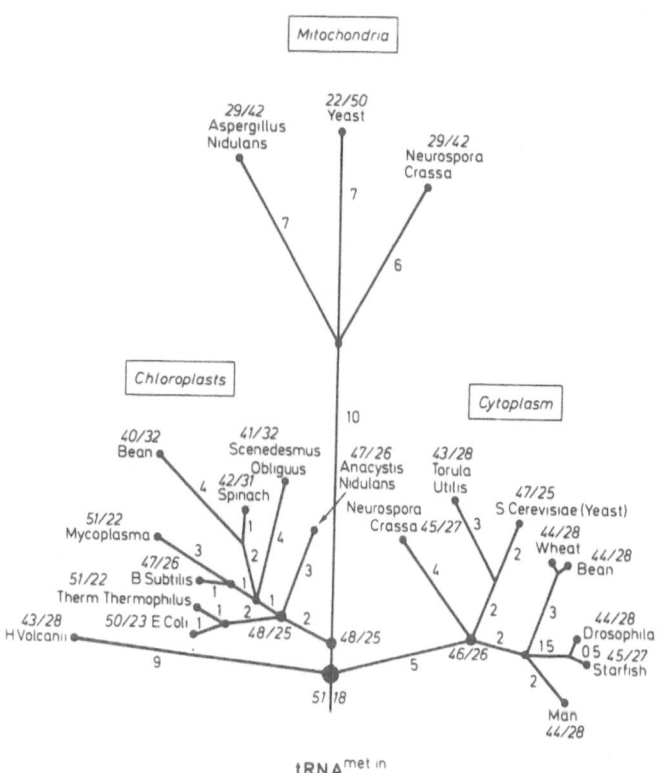

Figure 6 - The family tree of a tRNA, here the one involved in the initiation of translation, shows a few changes in nucleotide sequence (indicated by numbers of the branches), even after thousands of million of years. All such known mammalian sequences are practically identical. The quotients give the ratio of (guanidine plus cytosine) to (adenine plus uracil). This ratio is greatest near the earliest branching-points and smallest at the ends of the long branches ; it is almost reversed in the mitochondria (the "power stations" of the cell), which have a ratio around 1/2, vis-a-vis the early branching points, with a ratio around 2/1. Four groups emerge clearly : the archaebacteria (only one representative ; H. Volcanii), the eubacteria and blue algae (hardly distinguishable from the chloroplasts), the eucaryotes and the mitochondria. The long distance to the mitochondria reflects a high rate of replacement of G and C by A and U. The purine-pyrimidine succession shows the mitochondria to be close relatives of the eubacteria, while their distance from the archaebacteria and the eucaryotes is relatively large.

Many tRNA sequences are known today, both for a given codon at a variety of phylogenic levels and for a given species and many different anticodons. Each such category is interesting (13,14) for comparative analysis. Phylogenic analysis shows whether tRNA has retained information from prebiotic times, or this information has been lost in the course of evolution. The comparison of different tRNA molecules in a single species may then lead to a reasonably complete reconstruction of the early forms and allow statements about the early evolution of the translation apparatus.

The phylogenic family tree of tRNA$^{Met}_{init}$ in figure 6 shows that tRNA belongs to the most highly-conserved structures that we know. Of the examples shown the tRNAs of the fruit-fly (Drosophila) and the starfish differ only in a single nucleotide pair, and these differ from human tRNA at only four (weighted) positions. Organisms which parted company thousands of millions of years ago, such as the eubacteria, the chloroplast and the archaebacteria, appear on the tRNA scale as close relations, with closely-connected sequences.

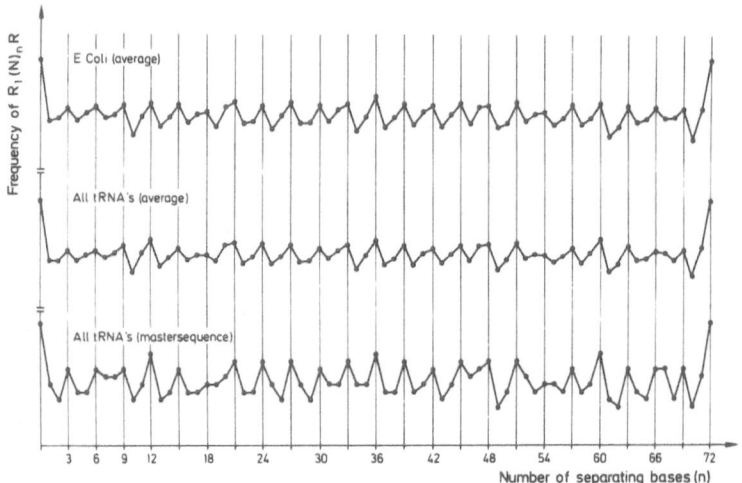

Figure 7 - Correlation analysis of the repetition of purine in tRNA. A tRNA sequence is divided into triplets, beginning at the 5' end and in phase with the anticodon. The frequency with which a purine (R) in the first position of the triplet occurs n positions later is counted and plotted against n. The period of three which emerges indicates clearly a triplet structure of the form RNY. The curves show values for the averaged sequences of E. coli and of all tRNAs investigated to date ; these are compared with that of the master sequence arising from the superposition of all tRNAs. The fact that the correlation is clearer in the master sequence suggests that this may represent a "memory" of the earliest phase of evolution.

This brings a reconstruction of the primaeval structure into the realm of the possible. If the sequences of different tRNAs -for example for E.coli bacteria, for yeast cells or for archaebacteria- are compared then they all show a high GC content, which increases further when the sequences are superimposed. However, it can be seen from the mitochondrial sequences, which are all rich in A and U, that G and C could be displaced in the course of evolution, presumably on account of a rich supply of the metabolite A in mitochondria ; G and C are thus not demanded by considerations such as stability. Further, the reconstructed precursor sequences suggest a periodic triplet structure (see figure 7) which in turn suggests a primaeval code pattern GNC (N = any of the four nucleotides A, U, G, and C). This tRNA was not only the primitive adaptor but also a primitive carrier of genetic information -a function now lost in the course of evolution. A comparative analysis of tRNA sequences leads to conclusions which can be summarized as follows :

- tRNA is an "ancient" adaptor which has changed relatively little in the course of phylogenesis.
- Different tRNA molecules within a species appear as mutants of a master sequence.
- The original master sequence is largely capable of reconstruction.
- It is characterized by a high GC content and a code pattern RNY.
- All findings to date are compatible with a primaeval code GNC for the amino acids most common in Nature : glycine, alanine, aspartic acid and valine.

SYNOPSIS

We indeed find a congruence between theory, model experiment and "historical" record. Irrespective of this, we should still pay attention to the following caveat's :

The physical theory of evolution -like every other physical theory- describes no more than an "if-then" behaviour pattern. If the theory is correct then it predicts the consequences resulting from particular initial conditions.

The value of the theory is to be assessed exclusively by its capacity for experimental test. Model experiments give quantitative "standards" by means of which the probability of each stage in the emergence of life may be estimated.

Neither theory nor experiment allows a conclusion about the actual historical process of evolution. This requires a specific historical record.

The congruence between theory, model experiment and historical record enables us to regard the principle of "life" as one of Nature regularities.

On the basis of these principles the evolution process may be simulated and reproduced in the laboratory.

LITERATURE

1. Spiegelman, S. et al., Proc Natl. Acad. Sci. USA :
    (1963) 50, 905 ; (1965) 54, 579, 919 ; (1968) 60, 866 ; (1969) 63,
    805.
2. Mills, D.R., Kramer, F.R., Spiegelman, S.: 1973, Science 180, 916.
    Kramer, F.R., Mills, D.R.: 1978, Proc. Natl. Acad. Sci. USA 75, 5334.
3. Sumper, M., Luce, R.: 1975, Proc. Natl. Acad. Sci. USA, 72, 162.
4. Biebricher, Ch. K., Eigen, M., Luce, R.: 1981, J. Mol. Biol. 148,
    369.
5. Biebricher, Ch. K., Eigen, M., Luce, R.: 1981, J. Mol. Biol. 148,
    391.
6. Biebricher, Ch. K., Eigen, M., Gardiner, W.C. Jr., to be published.
7. Mills, D.R., Peterson, R.I., Spiegelman, S.: 1967, Proc. Natl. Acad.
    Sci. USA 581, 217.
8. Eigen, M.: 1971, Naturwiss. 58, 465.
9. Eigen, M., Schuster, P.:1977, Naturwiss. 64, 541.
                             1978, Naturwiss. 65, 7, 341.
10. Domingo, E., Flavell, R.A., Weissmann, Ch.: 1976, Gene 1, 3.
    Batschelet, E., Domingo, E., Weissmann, Ch.: 1976, Gene 1, 27.
11. Weissmann, Ch., Feix, G., Slor, H.: 1968, Cold Spring Harbor Symp.
    Quant. Biol. 33, 83.
12. Lohrmann, R., Bridson, P.K., Orgel, L.E.: 1980, Science 208, 1464 ;
    1981, J. Mol. Evol. 17, 303.
    Bridson, P.K., Orgel, L.E.: 1980, J. Mol. Biol. 114, 567.
13. Eigen, M., Winkler-Oswatitsch, R.: 1981, Naturwiss. 68, 217.
14. Eigen, M., Winkler-Oswatitsch, R.: 1981, Naturwiss. 68, 282.

*This manuscript is based on a lecture first given by M. Eigen at the
meeting of the "Schweizerische Naturforschende Gesellschaft" at Davos
on 25th September, 1981, and was published in the proceedings of the
meeting.

      We are most grateful to Dr. Paul Wooley for translation of the ma-
nuscript into English.

CONDENSATION–POLYMERIZATION AND MORPHOGENESIS IN AQUEOUS MEDIUM AS
A MODEL FOR THE CHEMICAL EVOLUTION

Fujio EGAMI
Mitsubishi-Kasei Institute of Life Sciences,
Machida-shi, Tokyo 194, JAPAN

(1) Formation of peptide bonds is considered to be possible even in an aqueous medium in the primeval sea.

(2) Characteristics of marigranules are presented and compared with those of other protocell-like structures.

The first organism when life first began must have had the following functions, although to an extent far less efficient than the extant primitive organisms.

1. Directed transformation of materials and energy or metabolism in an open system.

2. Growth and production of more or less similar phylogenies or descendants.

Through natural selection, functions were established in the course of chemical evolution and evolved to more and more teleonomic[1] (purposeful) for the maintenance of individuals and phylogenies in the course of late chemical and early biological evolution. Thus, organisms with self-reproducing activity based on the transfer of genetic information finally appeared.

From a chemical point of view, it can be said that chemical and early biological evolution was no more than the evolution of multicomponent-multicatalytic systems or chemical systems at the molecular level.

The expression, "Chemical system at the molecular level", used by me for the past several years,[2] has not found general acceptance as yet. Thus, I should like to explain the meaning of this phrase through some familiar examples.

A yeast extract containing an alcoholic fermentation system and glutathion synthesizing system can perform the following reactions:

Alcoholic fermentation,

C. Hélène (ed.), Structure, Dynamics, Interactions and Evolution of Biological Macromolecules, 371–382.
Copyright © 1983 by D. Reidel Publishing Company.

$$C_6H_{12}O_6 + 2ADP + 2Pi \rightarrow 2C_2H_5OH + 2CO_2 + 2ATP + 2H_2O \qquad (1)$$

Glutathion synthesis,

$$Glu + Cys + Gly + 2ATP \rightarrow Glu-Cys-Gly + 2ADP + 2Pi \qquad (2)$$

The sum of these reactions may be expressed as,

$$C_6H_{12}O_6 + Glu + Cys + Gly \rightarrow 2C_2H_5OH + 2CO_2 + Glutathion + 2H_2O \qquad (3)$$

Here, thirteen enzymes, several coenzymes and cofactors in the alcoholic fermentation system and at least two enzymes in the glutathion synthesizing system constitute a chemical system at the molecular level collaborate with each other to realize the final objective of the synthesis of glutathion.

Thus, depending on the degradation of glucose, becomes the thermodynamically improbable formation of peptide bonds possible. This is a simple instance of a chemical system at the molecular level. (Synthesis in chemical laboratories or industries might be regarded as systems carried out in reaction vessels or reactors.)

I. CONDENSATION-POLYMERIZATION IN AQUEOUS MEDIUM.

It has been often said that the peptide bond formation involves highly endergonic reaction,

$$\underset{\overset{|}{R_1}}{H_2NCHCOOH} + \underset{\overset{|}{R_2}}{H_2NCHCOOH} = \underset{\overset{|}{R_1}\overset{|}{R_2}}{H_2NCHCONHCHCOOH} + H_2O \qquad (4)$$

$$\Delta G^{o'} = \text{about 3-4 Kcal/mole}$$

and the abiotic formation of peptide bonds in the primeval sea must have been impossible.[3] However, as mentioned above, chemical systems at the molecular level must have been established and evolved in the course of chemical evolution. More advanced chemical systems at the molecular level such as protein synthesis and advanced energy transformations are carried out depending on organized particles or membranes. Thus, it may be expected that peptide bonds would have been formed in the primeval sea in the course of chemical evolution although less efficiently than in extant organisms. As is well known, in extant organisms, the energy barrier was overcome by ATP and other so-called energy-rich phosphate bonds. At the early stage of chemical evolution prior to the accumulation of such compounds, the other routes leading to the formation of peptide bonds, although less effective, must have been utilized.

Indeed, simple amino acid amides formed as intermediates in

Strecker's amino acid synthesis and other amides must have played a role in the early stage of chemical evolution.

It is generally accepted that formaldehyde and formose, a mixture of sugars formed by the polymerization of formaldehyde, accumulated in the primeval sea. As shown by the enzymatic synthesis of glutathion which depends on the chemical energy of glucose, the possibility of the formation of peptide bonds depending upon the oxidation or degradation of formaldehyde and related derivatives in the primeval sea cannot be excluded.

I should like to talk briefly about our experiments on the formation of peptide bonds in an aqueous solution.

For several years, we have been engaged in the experimental approach toward chemical evolution in primeval sea.[4,5,6,7] The assumptions underlying the execution of this project are the following:
1. The origin of life, the chemical evolution just preceding it, and early biological evolution occurred in the primeval sea under an anoxygenic atmosphere.
2. Chemical evolution in the primeval sea was accelerated by the presence of transition metals. At least six transition metals (Mo, Fe, Zn, Mn, Co, Cu) essential to most prokaryotes must be taken into consideration.

We thus set up a modified sea medium using these transition metals: HPO_4^{2-}, SO_4^{2-}, Mg^{2+}, Ca^{2+}, each at 0.01 M, Na^+ 0.015 M, K^+ 0.05 M, Cl^- 0.07 M, NO_3^- 0.0005 M, Zn^{2+}, MoO_4^{2-}, Fe^{3+}, Cu^{2+}, Co^{2+}, Mn^{2+}, each at 0.0001 M, and the pH was adjusted to 5.5 following the addition of starting materials.

The media, provided with appropriate starting materials, were kept at 105°C for several days or even for several weeks under anoxygenic atmosphere. The particular temperature that would accelerate the reaction and exclude any microbial contamination was chosen.

Initially, we chose a one-carbon compound, formaldehyde, and a one-nitrogen compound, hydroxylamine, as starting materilas (generally CH_2O 0.03 M and NH_2OH 0.05 M and for glycylglycine formation CH_2O 0.5 M and NH_2OH 0.5 M). A series of amino acids were produced over a period of several days. These were tentatively identified by the retention time in automatic amino acid analysis (Gly, Ala, Ser, Thr, Ile, Asp, Glu, Val, Leu, Arg, Lys, Pro, His etc). Among them, the main protein amino acids, also confirmed by other methods, were Gly, Ala, Ser, Asp and Glu. It should be pointed out that not only free amino acids but also amino acid polymers (degree of polymerization 2∿7) which give rise to amino acids by acid hydrolysis were produced. As shown in Fig. 1, the formation of glycylglycine followed that of glycine and attained a maximum value after about 25 days and, as expected from thermodynamics, this compound finally disappeared within about 100 days (T. Ochiai and F. Egami, unpublished). In contrast to this, in the natural chemical

evolution of an open system, it might be possible to maintain stable
concentrations of peptides. Nevertheless, the formation of glycylglycine, from formaldehyde and hydroxylamine is remarkable and I believe
that it was initially made possible by a primitive chemical system at
the molecular level. As expected, glycylglycine was not produced from
glycine itself under the same conditions.

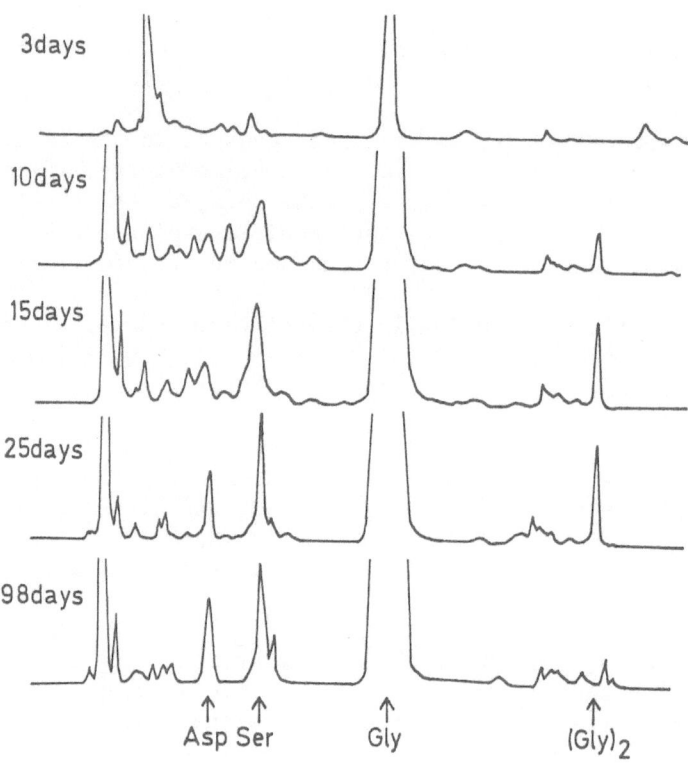

Fig. 1. Amino acid analysis of the acid hydrolysate
of the reaction mixture at various reaction times.

Although the mechanism for the peptide bond formation in an
aqueous medium remains to be elucidated, two probable mechanisms can
be considered: 1) Certain agents with dehydrating activity were con-
comitantly produced, by which peptide bonds were formed. As probable
condensing agents, cyanamide[8] and most of all, aminoacetonitrile[9,10]
could be considered. 2) Peptide bonds can be formed by the transamidation of other amide bonds, capable of being formed as intermediates in
Strecker's synthesis or being coupled with energy supplying reactions
such as oxidation of aldehydes, oxidative decarboxylation of α-oxoacids
etc. The second mechanism seems more apt to apply than the first.

Thus, we have actually observed amide formation in a dilute neutral or slightly acidic aqueous medium.

$$2 \text{ glyoxylate} + NH_3 \rightarrow N\text{-oxalylglycine} + H_2O \qquad (5)[11,12]$$

$$2 \text{ pyruvate} + NH_3 \rightarrow N\text{-acetylalanine} + CO_2 + H_2O \qquad (6)[11,12]$$

$$2 \text{ phenylpyruvate} + 2NH_3 \rightarrow$$
$$N\text{-phenylacetylphenylalanineamide} + CO_2 + 2H_2O \qquad (7)[13,14]$$

Once amide bonds are formed, no energy barrier exists for the formation of peptides by transamidation in a prebiological aqueous environment

$$GlyNH_2 \rightarrow (Gly)_2, (Gly)_3 \qquad (8)[15]$$

$$GlyNH_2 + (Gly)_3 \rightarrow (Gly)_{4 \sim 6} \qquad (9)[15]$$

$$\text{phenylacetylphenylalanineamide} + Gly \rightarrow$$
$$N\text{-phenylacetylphenylalanylglycine} \qquad (10)[15]$$

Formation of such oligoglycine is promoted by ammonia or other basic substances.[15]

It may be concluded that the formation of peptide bonds in an aqueous medium cannot be excluded, contrary to generally accepted concepts.

II. MORPHOGENESIS IN AN AQUEOUS MEDIUM

Phase-separation from the medium is generally prerequisite for the initiation of protocell formation. The evolution of phase-separated particles (probionts according to A.I. Oparin), the late stage of chemical evolution, was designated as "prebiological evolution" by Oparin and Gladilin.[16] According to them, evolution to the formation of probionts was achieved in an open system at the molecular level in accordance to chemical laws. In contrast to this, in prebiological evolution under the new law of natural selection, novel biological properties (the capacity to counteract an increase in entropy, purposefulness of structure and heredity) come about and finally a primitive organism was generated. Although it has not been generally accepted to distinguish "prebiological evolution" from chemical evolution, I agree with the above authors and believe that typical Darwinian evolution based on the natural selection played a predominant role in the evolution of protocells producing more or less similar phylogenies (descendants). However any protocell-like models so far presented, including coacervate droplets of Oparin, cannot be regarded as evolving particles subjected to natural selection prior to the origin of life. The reason for this opinion is that the capacity for synthesizing constituent macromolecules from materials in the environments and growth and production of similar phylogenies are prerequisite to natural selection. The evolution of a protocell structure or morphogenesis in chemical evolution must be distinguished from those of inanimate and living systems. Morphogenesis in inanimate systems is characterized by an increase in entropy as in the case of the separation of crystals or liposomes from the medium. Formation of coacervate droplets and proteinoid microsphers belong essentially to this type of morphogenesis. In contrast, characteristic morphogenesis in living systems, the formation of particles such as chromatin, ribosomes, cell membranes etc. is directly or indirectly based on genetic information. Thus, morphogenesis to evolve to a living system in chemical evolution or in prebiological evolution must be morphogenesis accompanied with the growth and formation of phylogenies and at a later stage, with the introduction of genetic information.

Thus, in consideration of the above, we have directed our efforts toward bringing about the formation of protocell-like structures in an aqueous medium.

Several protocell-like models have so far been postulated and among them, Folsome's organic microstructure,[17] Kenyon's melanoidin microsphere,[18] and Heinz's lumisphere[19] seem to warrant particular interest. However, little is known in regard to the chemical nature of these models. Coacervate droplets of A.I. Oparin[20] and proteinoid microspheres of Fox[21] have been extensively studied for more than 40 years and 20 years, respectively, and generally highly appreciated, although the significance of these particles as protocell models is denied by Folsome[17] and Day[22].

We presented "marigranules" as a novel protocell model in 1978.[23,24,25] Although far less studied than coacervate droplets and proteinoid microspheres, I believe that marigranules have several advantages probably essential for further evolution. Thus, I should like to briefly summarize our studies on marigranules and related particles prepared in the modified sea medium mentioned above, but with different starting materials.

Marigranules

Starting materials: An amino acid mixture consisting of glycine and acidic, basic, and aromatic amino acids (Gly 0.05 M, L-Glu, L-Asp, L-Lys, L-Arg, L-His, L-Trp, L-Phe, L-Tyr, each 5 mM) was dissolved in the modified sea medium.

Formation of particles: The mixture was heated at 105°C. Well organized particles, shown in Fig. 2, were separated out from the medium gradually over a period extending from several days to several weeks.

Properties: Melting point > 300°, if any. Elemental analysis; C: 58.32, H: 3.76, N: 14.23, ash: 7.62%. IR(KBr); 3,560-2,000, 1,710, 1,680-1,610, 1,550, 1,510, 1,450, 1,380, 1,260, 1,160, 1,130, 850, 750 cm^{-1}. The existence of peptide bonds is suggested by the bands at 1,680-1,610 and 1,550 cm^{-1}. These granules consist of two parts: a surface layer soluble in ethanol and an interior part soluble in 1N KOH. The solubilized marigranules were hydrolyzed by elastase and more than 30% of the nitrogen content was liberated in the form of an amino group. This suggests that the interior part of the marigranules consist of elastin-like macromolecules. The maximum molecular size of solubilized marigranules was found to exceed 80,000 daltons. Fig. 2 shows ubiquitous marigranules observed by scanning electron microscopy. Many of the marigranules show budding-like junctions suggesting the possibility of reproduction.

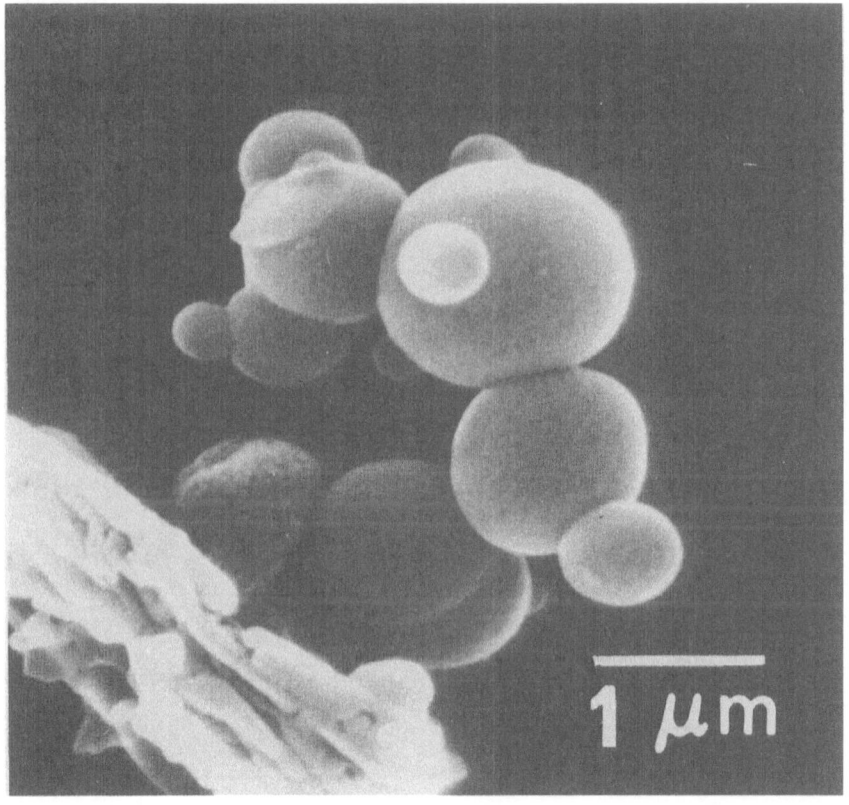

Fig. 2. Scanning electron micrograph of marigranules
(by Miss Y. Ogawa)

Particles from a mixture without tryptophan.

For the formation of marigranules, tryptophan is essential.[26] Without tryptophan no well-defined particle can be formed. However,

for the formation of marigranules, certain reducing sugars can replace
tryptophan, although the form of particles produced are not as beautiful
and more irregular. A typical example from the amino acid mixture
without tryptophan but with D-erythrose (5 mM) is shown in Fig. 3.

    Properties: Elemental analysis; C: 27.73, H: 2.06, N: 5.68, ash:
47.71%. Amino acids used as starting materials were detected in the
acid hydrolysate of the particles. These particles were 0.3–1 μm in
diameter, smaller than normal marigranules (0.3–2.5 μm in diameter),
quite stable, and their surfaces were not damaged by 1N KOH (unpublished
data).

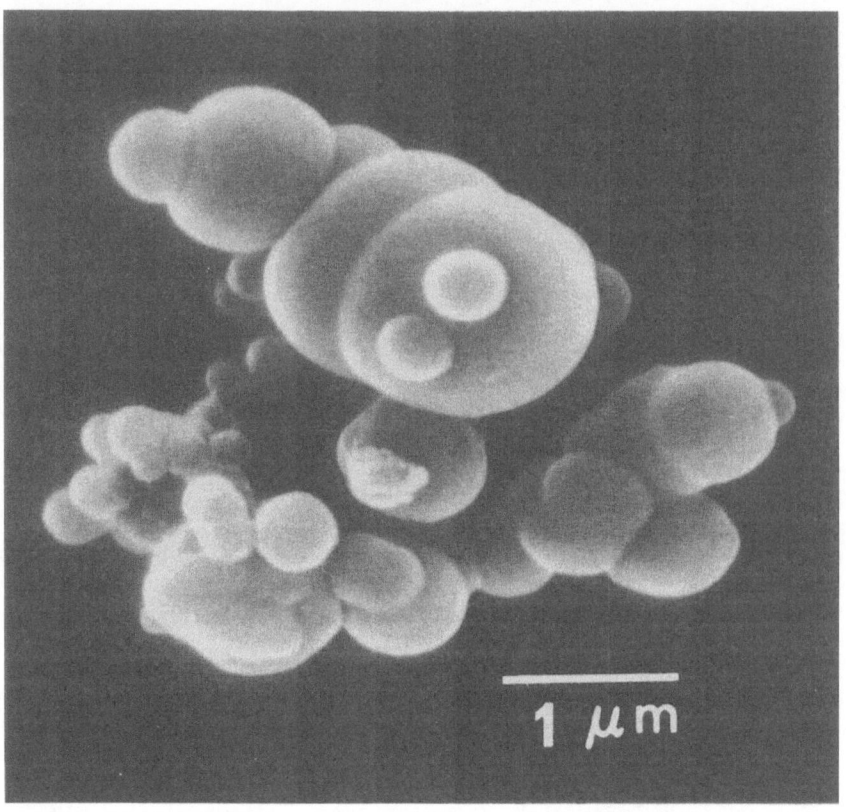

Fig. 3. Scanning electron micrograph of particles
formed from an amino acid mixture containing D-erythrose
instead of tryptophan (by Miss Y. Ogawa).

Particles from D-etythrose

    D-erythrose alone in the modified sea medium was found to give

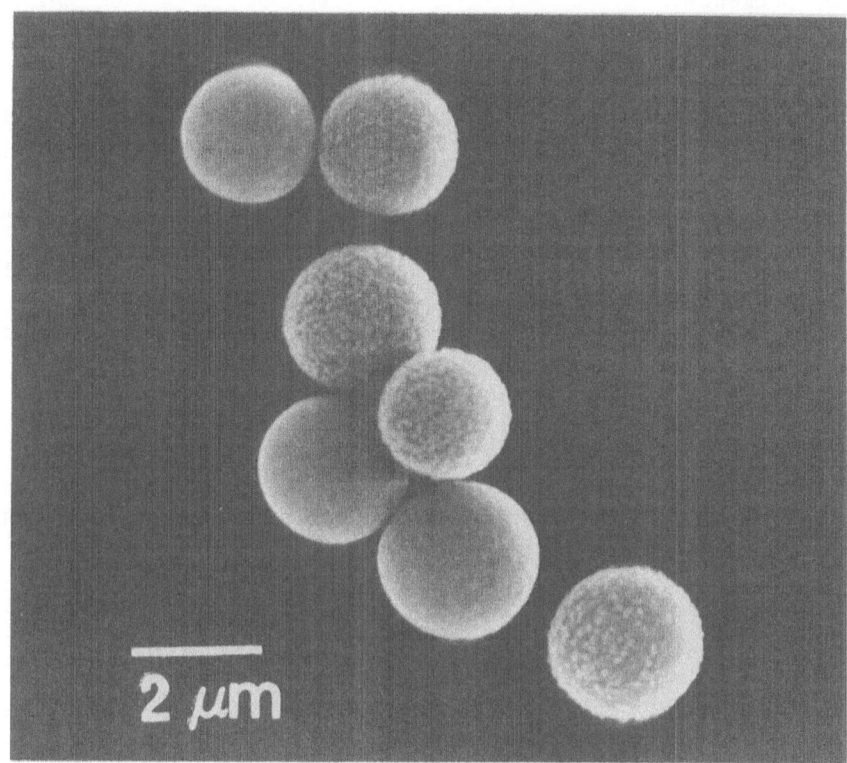

Fig. 4. Scanning electron micrograph of the particles
from D-erythrose (by Miss Y. Ogawa).

beautifil spherical particles (Fig. 4). In contrast to the normal
marigranules, the particles from D-erythrose had no budding-like
structure.

     Properties: Elemental analysis; C: 59.54, H: 4.72, N: < 0.3, ash:
2%. Note that the particle formed from D-erythrose contained almost
no nitrogen. The particle size was 0.8-3 μm in diameter, and slightly
larger than the normal. These particles were stable, but soluble in
1N KOH (unpublished data).

III. CONCLUDING REMARKS

     The formation of peptide or similar amide bonds in an aqueous
medium without dehydrating agents has been regarded as thermodynamically
difficult. However these bonds probably formed in the early stages of
chemical evolution. The degradation of sugars known as formoses or
oxidoreductive decomposition of oxoacids may have been the sourse of

energy for such reactions. Some experimental examples for this have been presented.

In the course of chemical evolution, these processes were probably replaced by more efficient processes through the participation of clays[27] and other solid surfaces, and finally the processes in which the so-called energy rich phosphate bonds participate must have been established.

Various protocell models have been postulated. Among them, coacervate droplets of A.I. Oparin and proteinoid microspheres of S.W. Fox have been extensively studied over long periods of time. We presented "marigranules" as a novel protocell model in 1978. Although far less studied than coacervate droplets and proteinoid microspheres, this model has advantages probably essential to the latter stages of chemical evolution.

Table 1. Comparison of marigranules, coacervate droplets, and acidic proteinoid microspheres

		marigranules[a]	coacervate droplets[b]	proteinoid microspheres[c]
Stability				
dilution	20°C	no effect	no effect	no effect
	100°C			soluble
1N HCl	20°C	no effect	soluble	damage (+)
	100°C			soluble
1N KOH	20°C	soluble	soluble	soluble
	100°C			
pH 4	20°C	no effect	no effect	damage (+)
	100°C		damage	soluble
pH 6	20°C	no effect	soluble	damage (++)
	100°C			soluble
pH 8	20°C	no effect	no effect	damage (+++)
	100°C		damage	soluble
		marigranules are only damaged in 1N KOH		in all cases microspheres are soluble at 100°C

Formation: a) an amino acid mixture was heated in a modified sea water.
b) ready made polymers, (histone and gum arabic) were kept in an appropriate aqueous medium.
c) acidic proteinoids prepared under unhydrous conditions were dissolved in water at higher temperature and cooled.

The properties of marigranules are compared with those of typical coacervate droplets and proteinoid microspheres in Table 1. The essential strong points of marigranules may be summarized as follows.

1) In contrast to coacervate droplets and proteinoid microspheres, marigranules are stable over a wide range of naturally conceivable pH, ionic strength, temperature, and dilution.

2) Marigranules contain lipid-like membranes.

3) Marigranules are not the result of a simple association product of macromolecules ready-made or produced elsewhere. In this respect, marigranules differ from coacervate droplets and proteinoid microspheres. Marigranules are produced in an aqueous medium (modified sea medium) *in situ* from small molecules (amino acids).

4) The formation of marigranules is accompanied by the polymerization of amino acids added and small polymers produced in the medium; that is, marigranules synthesize their own component polymers and accomplish the formation of an organized structure. Thus marigranules perform the active maintenance of the organized structure.

Consequently I believe that marigranules may be regarded as a thermodynamically open phase-separated system (in Oparin's sence), which may evolve further to a level closer to a living system through natural selection. Marigranules synthesizing the more "teleonomic (purposeful)" polymers are more capable of surviving and evolving through natural selection. In contrast to this, the coacervate droplets of Oparin are a static product and not an open system. After the incorporation of ready-made appropriate enzymes such as phosphorylase or polynucleotide phosphorylase into the coacervate, these droplets may react as an open system. But the basic structure of the coacervate is always static.

The morphogenesis of marigranules is quite different from crystallization or liposome formation (formation of equilibrium structure), and from the self-assembly of ribosomes or viruses (morphogenesis based on genetic information). It is a non-equilibrium self-organization possibly leading to the formation of an ordered structure through natural selection.

Finally I hope that formation and evolution of protocell-like structures including marigranules will receive further discussion on the basis of the physical and mathematical theories of Prigogine[28], Haken[29], Thom[30], and others.

REFERENCES

1. Monod, J.: 1970, *Le hasard et la nécessité* (Éditions du Seuil, Paris).

2. Egami, F.: 1982, *From Cyclotrons to Cytochromes*, edited by Kaplan, N.O., and Robinson, A., Academic Press, New York, pp.135-144.

3. VanHolde, K.E.: 1980, *The Origin of Life and Evolution*, ed. by Halrorson, H.O., and VanHolde, K.E., Alan R. Liss, Inc. New York, pp.31-46.

4. Hatanaka, H., and Egami, F.: 1977, Bull. Chem. Soc. Jpn. 50, pp.1147-1156.

5. Ochiai, T., Hatanaka, H., Ventilla, M., Yanagawa, H., Ogawa, Y., and Egami, F.: 1978, *Origin of Life: Proc. 5th Inter. Conf. Origin of Life*, ed. by Noda, H., Center for Acad. Publ. Jpn., pp.135-139.

6. Kamaluddin, Yanagawa, H., and Egami, F.: 1979, J. Biochem. 85, pp.1503-1507.

7. Yanagawa, H., Kobayashi, Y., and Egami, F.: 1980, J. Biochem. 87, pp.359-362.

8. Miller, S.L.: 1955, J. Am. Chem. Soc. 77, pp.2351-2360.

9. Chadha, M.S., Replogle, L., Flores, J., and Ponnamperuma, C.: 1971, BioOrganic Chemistry 1, pp.269-274.

10. Kamaluddin, Yanagawa, H., and Egami, F.: 1981, Ind. J. Biochem. Biophys. 18, pp.215-218.

11. Egami, F., Makino, Y., Sato, K., Nishizawa, M., and Yanagawa, H.: 1981, Viva Origino 9, pp.66-67.

12. Yanagawa, H., Makino, Y., Nishizawa, M., and Sato, K., and Egami, F.: 1982, J. Biochem. 91, pp.2087-2090.

13. Egami, F., Makino, Y., Sato, K., and Nishizawa, M.: 1981, Proc. Jpn. Acad. 57 Ser. B. No.9, pp.329-332.

14. Egami, F., Makino, Y., Nishizawa, M., and Sato, K.: 1982, Nippon Nōgeikagaku Kaishi (J. Agr. Chem. Soc. Jpn.) 56, No.7, pp.537-543.

15. Nishizawa, M., Makino, Y., and Egami, F.: To be published.

16. Oparin, A.I., and Gladilin, K.L.: 1980, BioSystems 12, pp.133-145.

17. Folsome, C.F.: 1979, *The Origin of Life*, W.H. Freeman and Co. San Francisco, pp.82.

18. Kenyon, D.H., and Nissenhaum, A.: 1976, J. Mol. Evol. 7, pp.245-251.

19. Heinz, B., Ried, W. and Pflug, H.D.: 1980, Naturwissenschaften 67, pp.178-181.

20. Oparin, A.I.: 1968, *Genesis and Evolutionary Development of Life*, Academic Press, New York.

21. Fox, S.W., and Dose, K.: 1977, *Molecular Evolution and Origin of Life*, Marcell Decker, New York and Basel.

22. Day, W.: 1979, *Genesis on planet earth*, The House of Talos Publ., Michigan, pp.319.

23. Yanagawa, H., and Egami, F.: *Origin of Life: Proc. 5th Interm. Conf. Origin of Life*, Ed. Noda, H., Center for Acad. Publ. Tokyo.

24. Yanagawa, H., Kobayashi, Y., and Egami, F.: 1980, J. Biochem. 87, pp.855-869.

25. Yanagawa, H., and Egami, F.: 1980, BioSystems 12, pp.147-154.

26. Yanagawa, H., and Egami, F.: 1981, *Origin of Life: Proc. 6th Interm. Conf. Origin of Life*, Ed. Wolman, Y., D. Reidel, Holland, pp.309-312.

27. Bernal, J.D.: 1949, Proc. Phys. Soc. 62, pp.527-

28. Prigogine, I.: 1967, *Introduction to thermodynamics of irreversible processes*, New York, John Wiley & Sons.

29. Haken, D.: 1978, *Synergetics*, Springer Verlag, Berlin, Heidelberg, New York.

30. Thom, R.: 1977, *Stabilité structurelle et morphogénèse, 2e éd.*, Inter Editions, Paris.

REFLECTIONS ON MOLECULAR ASYMMETRY AND APPEARANCE OF LIFE

Gérard SPACH and André BRACK
Centre de Biophysique Moléculaire, C.N.R.S.,
1A, avenue de la Recherche Scientifique,
45045 Orléans cedex, France

SUMMARY

Molecular asymmetry and its spreading are first examined in present
Life, then in prebiotic systems. Further, the need is emphasized for a
comparative investigation of chemical and configurational evolutions
with regard to the major components of the biological asymmetric molecules as a whole. Some proceeding ideas are outlined such as configuratio-
nally simplified models for a DNA ancestor. Also the importance of configurational segregation is stressed within its different aspects including
covalent compounds or intermolecular complexes. The formation of small
asymmetric volumes bounded by a semipermeable membrane composed of chiral
molecules is suggested as a stereoselecting and concentrating device.
Life could thus have evolved in small asymmetric milieu, without calling
for a "racemic" Life, difficult to understand and to accept without prior
configurational segregation and amplification.

INTRODUCTION

Whereas the atomic theory can be tracked back to the greek and latin
philisophers, the notion of molecular asymmetry, although so tightly
bound to life phenomena in present Science, does not seem to have been
conceived before the last century discoveries. Still, nature has afforded
many macroscopic models such as the helical twists of a snail shell or of
the pine cone scales. But most of the living beings share a morphology
which symmetry is either radiated or bilateral. Moreover, no human recei-
ving organ is sensitive to the orientation of circularly polarized light
and much intuition and observation are needed to analyse the quartz crystal hemihedry.

Asymmetry is however common in human activity. The writing direction
has hesitated in times past between right and left, asymmetric letters
being eventually replaced by their mirror image. Right and left have
always symbolized for men auspiciousness or dexterity and ominousness
or clumsiness, as well as certain symbolic or ornemental motifs, spirated

C. Hélène (ed.), Structure, Dynamics, Interactions and Evolution of Biological Macromolecules, 383–394.
Copyright © 1983 by D. Reidel Publishing Company.

and thus chiral in a plane, such as the fylfot (indian swastika), the
greek key pattern, the chinese ying and yang.., perhaps of a common origin. It may be interesting to seek for the meaning of the manichaean
evolution that supplanted these more ancient symbols by symmetrical ones
such as the cross, the star, the crescent.. ! (Figure 1).

Figure 1 - From left to right. Upper row : swastika, greek key pattern,
yang. Lower row : cross, star, crescent.

As for the mirror which turns a chiral form into its enantiomer, it
has fascinated novelists and backed up countless marvellous stories
bringing into play another life sheltered in the virtual image. It is to
the mirror that Pasteur refers when defining molecular asymmetry (1).
Before Pasteur, asymmetry in living systems was more a morphological or
anatomical contingency than a molecular obligation.

MOLECULAR ASYMMETRY IN PRESENT TERRESTRIAL LIFE

The present terrestrial life is dominated by two biopolymers, nucleic acids which are the support of genetic information and proteins
which catalyze biochemical reactions, and by a lipidic micellar system
which forms the cellular protecting, selectively permeable, membranes.
Some of their constituents, i.e. sugars, amino acids and lipid polar
heads enjoy at least one asymmetric carbon atom (Figure 2).

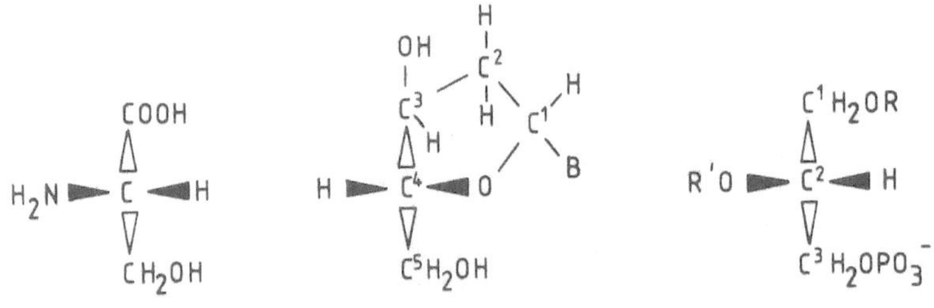

Figure 2 - From left to right : L-serine ; desoxy-β-D-ribofuranose ;
3-sn-(or L-) phosphatidic acid (R = R' = H : sn-glycerol-3-phosphoric
acid).

In addition to these main constituents living cells also contain
chiral terpenoid derivatives some of which are organic substances active
in trace amount, the vitamins.

The biopolymers themselves form well known asymmetric helical structures and superstructures, like the α and β conformations of polypeptides,
the A, B or Z-forms of nucleic acids and the helical conformations of
polysaccharides.

The nineteen chiral amino acids present in proteins belong without
exception to the L-configuration class and the two sugars found in nucleic acids were related to the D-series. This assignement occurred by a
mere convention that consisted to connect the configuration of the sugar
asymmetric carbon atom the most distant from the carbonyl group to that
of the amino acid asymmetric carbon atom through lactic acid CH_3-CHOH-
COOH by a series of chemical reactions. If another sequence of reactions,
possibly more complex, would be selected, the same configurational correlation could have been chosen for natural amino acids and sugars. This
conventional character is quite glaring for the phospholipid polar heads
as L-glycerol 3-phosphoric acid is equivalent to D-glycerol 1-phosphoric
acid.

Sugar molecules have several asymmetric carbon atoms of which two,
atoms 3 and 4 in case of desoxy-D-ribose, belong to the phosphodiester
chain of nucleic acid (Figure 3).

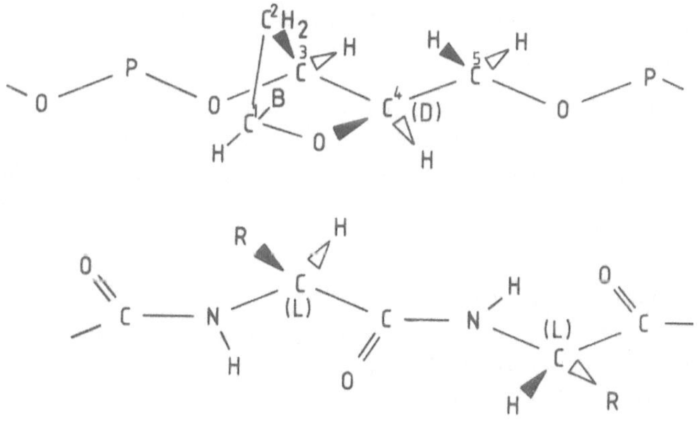

Figure 3 - Stereochemistry of nucleic acid (up) and protein chains
(down).

Less attention is given in text books to the stereochemistry of dihydrogenated isoprenyl (phytyl) derivatives, the configuration of which
is of the same type as the "tactic" nature of synthetic polymers. These
motives occur not only in vitamins (E, K_1,..), but also in phospholipids
of halobacterium cutirubrum [Fig. 2, R = R' = $(CH_2CH_2CHCH_2)_4H$], or in
chlorophyll (phytol), etc... CH_3

Opitcal Purity of the Main Constituents

The question of the optical purity of the main constituents of life molecules and the origin of "non natural" enantiomers have been examined (for review, see reference 2).

Amino acids belonging to the D-series are commonly found in bacteria, more precisely, in some cell wall components, murein and teichoic acids. The murein glycan is cross-linked by peptide fragments containing D-alanine and D-glutamic acid, whereas teichoic acids, polymers of Gram positive bacteria, are formed by long chains of glycerol or ribitol

$$\begin{array}{ccccc} OH & OH & OH & OH & OH \\ | & | & | & | & | \\ CH_2-C & - & C & - & C & - & CH_2 \\ | & | & | \\ H & H & H \end{array} \quad = \quad ribitol$$

molecules linked in 1-3 or 1-5 positions respectively by phosphodiester bridges. Eventually, the free glycerol hydroxyl groups are bound alternately to D-alanine and D-glucose. Interestingly, these polymers also bear asymmetric carbon atoms.

Other D-amino acids are found in antibiotics (polymixins, gramicidins, ...) which operate at different sites of an opponent organism. D-amino acids are not coded by nucleic acids and the first step of the murein biosynthesis is an enzymatic racemization of L-alanine. In the same way, the biosynthesis of antibiotics make use of a kind of protein matrix including a racemase, the evolutionary pathway of which is an intriguing problem.

Sugars of the L-series are also found in nature. Among the most important are L-fructose (6-desoxy L-galactose) and L-rhamnose (6-desoxy L-mannose) both present in polyosides of bacterial shells. The biosynthesis of these two sugars proceeds through reduction of a nucleoside diphosphate of D-mannose or D-glucose respectively.

Thus, the enantiomers of the unnatural series make more resistant to enzyme degradation the cell protecting devices, at the borderline between inanimate and living matters, and the defensive tools of the cell. The question is still open to know whether they are fossils or improvements.

The presence of D-amino acids is however not restricted to the preceding examples. By carefully scrutinizing the optical purity of proteins it has been found that a spontaneous racemization of residues can occur, although with a very low yield estimated at 0.14 % of D-residues per year in the most favorable case of aspartic acid. Fortunately most of the proteins undergo a turn over within hundred days or so and probably the D-amino acid oxydase plays an important role in maintaining the optical purity and removing the poisonous D-amino acids. However some proteins in bones, teeth enamel, crystalline lens, are not subjected to

turn over and D-amino acids may accumulate and may play a role in ageing
or disease processes. As for the asymmetric structures, the α helices or
the β sheets are destabilised or dirupted respectively by the introduction of D-residues in an all-L chain (3). Thus the vital importance of
amino acids optical purity in present life is easily grasped. It may be
even more crucial for sugars of nucleic acids.

Epimerization of ribose and desoxyribose does not seem to have been
looked for. Moreover the importance of optical purity on the existence
or stability of nucleic acids secondary structures is poorly understood.
It is apparent from model examination that sterically sensitive asymmetric spots are the C-3 and C-4 atoms of the phosphodiester chain, and the
C-1 atom bearing the base moiety.

The C-2 position is also important. Indeed the A, B and Z forms
are accessible to double stranded DNA whereas only the A form has been
observed for duplex RNA (4). The epimerization of C-2 leads to D-arabino-
furanose (Figure 4) whose adenine triphosphate derivative has antiviral

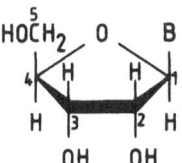 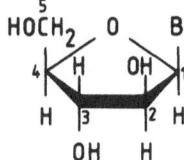

Figure 4 - β-D-ribofuranose and β-D-arabinofuranose.

and antibacterial activities (5). This substrate combines to the growing
DNA chain and stops the replication process. Repair enzymes can remove
this defect in higher organisms but not in microorganisms. However it is
not known whether the break-down of the chain growth is due to steric
hindrance at the level of the polymer conformation or of the enzyme system. Interestingly, oligomers of arabinofuranosyladenine derivatives have
been found to form left handed double stranded complexes with oligomers
of arabinofuranosyluracil derivatives, but not with poly(U) or poly(A)
respectively (6).

No report seems to have been published on the optical purity of
lipids.

MOLECULAR ASYMMETRY BEFORE LIFE

As a rule, any chemical reaction ending with the formation of chiral
compounds in large quantities, run in a symmetrical environment, yields
a racemic mixture, i.e. a mixture of equal quantities of right and left
handed enantiomers. Are there any reasons or conditions that this did
not occur on the primeval earth acting as a huge reactor when the first
fundamental constituents of biomolecules were formed? The problem thus

set has two sights, i) the origin of an excess of a prevalent enantiomer,
ii) its further amplification until appearance of life, which will be
discussed first.

Enantiomer Enrichment

Amplification of asymmetry can be examined in practice independently
of its origin. Some limiting factors must be considered when discussing
that problem : the minimum amount of a chiral germ -or the lack of enantiomer balance in the initial mixture composition- that will push the
system out of equilibrium ; the life time of chiral compounds against
spontaneous racemization reactions.

The amplification mechanisms proceed generally from broken symmetry
processes. These are illustrated at the macroscopic scale by the buckling
of a symmetrically compressed beam and by the asymmetric coiling of a
viscous liquid thread over a flat surface (Figure 5). At the molecular
level, open -or far from equilibrium- systems may be considered, many of
which based on covalent or noncovalent stereoassociations have been described experimentally or theoretically (for review, see reference 7).

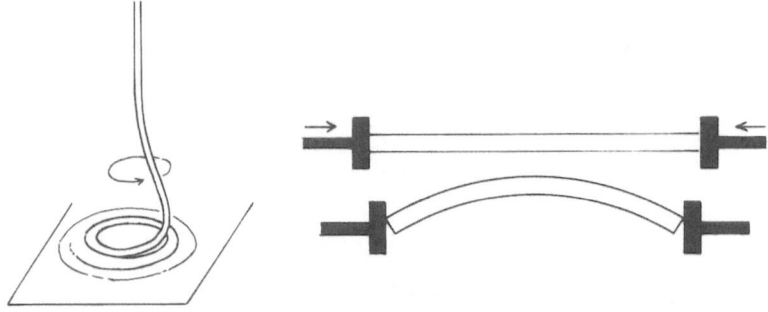

Figure 5 - Buckling phenomena of a viscous flow over a flat surface
and of a compressed thin beam.

One of the earliest process is the seeding of a supersaturated solution of a racemic mixture by one of the antipodes or by any chiral
germ. If in addition the two enantiomers are in equilibrium in solution
only, then the system will produce by concentration a unique crystallized
antipode. An interesting effect is the so-called "inversion rule" stating
that chiral crystals may induce preferential crystallization of a parent
product of opposite absolute configuration from a racemic mixture solution (8), for example, D-glutamic acid in presence of L-aspartic acid.

Open systems have been described that include the autocatalytic
formation of soluble enantiomers. If the racemic mixture is insoluble,
the reciprocal neutralisation of equal quantities of each enantiomer
allow the self amplification of a molecular fluctuation which acquires
a time irreversible character.

The stereoselective formation of covalent compounds has been approached through polymerization of chiral monomers such as α-amino acids
or their derivatives. For example, a right handed α helical seed, even
if not fully optically pure, can induce the stereoselective polymerization of the monomer whose configuration is the same as that of the seed
dominant residue (9). This effect is lasting only over a few residues
distance, until a helical sense reversal occurs. Under certain conditions
the formation along the polypeptide chain of domains enriched alternately
in one than the other enantiomer has been described (9).

A slight dissymmetry existing in a racemic mixture of α-amino acid
monomers can be amplified in the eventually α-helical polymer by partial
polymerization (10). Likewise, when the β sheet structure is present in
the polypeptide chain, then segregation of residues of the dominant
configuration occurs in the sheets (3). Repeated secondary processes of
enrichment, including for instance a segregation step as just described,
then isolation of the enriched domains followed once more by a polymerization, can lead to a high level of enrichment but with a low yield (3).
Other secondary processes can be realized or conceived as soon as segregation manifests itself : stereoselective adsorption, formation of stereospecific catalytic sites... This last mean is probably by far the
most potent for chirality propagation.

Origin of Asymmetry

Theoretical models on the origin of asymmetry on earth can be divided into two classes, those which call merely for a matter of chance and
other which call for an asymmetrical environment originating from the
Universe or from the Earth.

The tenants of the first scheme point out that the principle of
equimolarity of a racemic mixture is only a relative notion. If the number of molecules is small, random fluctuations may favor one enantiomer
over the other. For instance, it has been calculated that for a population of ten millions molecules, which is about the amount of chiral constituents in the smallest living cell, there is an odd probability to find
a 0.02 % or more excess of one enantiomer. But it is not yet known whether
such a small amount is sufficient for further amplification by mechanisms
as described precedingly.

Symmetry conditions that can influence the course.of a reaction in
a chiral sense are known (helical symmetry). Necessary conditions are
that the acting forces interact efficiently with the system which must
not include equilibrium reactions (11).

Parity non conservation has raised much hopes and disappointments
too. This fundamental asymmetry of the Universe, the kind Pasteur dreamed
for (1), has been examined from various aspects, e.g. circularly polarized radiations emitted by the slowing down of longitudinally polarized
electrons (Bremsstrahlung), inducing degradation reactions or stereoselective crystallization of racemic mixtures ; energy difference between

optical isomers producing a difference in their reaction rates. It must
be recognized that to-day no experiment has convincingly supported these
theoretical considerations for the origin of a prevalent enantiomer on
earth. Either the results were shown to be artefacts or to be so weak
that they are doubtfull (12).

Other chiral force fields that could have been acting on the earth
surface were looked for. Asymmetric synthesis and degradation have been
realized with success when using circularly polarized light. An original
approach using earth gravity and macroscopic vortex has been tested (13).
Unfortunately there are not so many reasons to find at the earth surface
a prevalence of a given chirality for that kind of physical forces. So,
even when efficient, they were probably acting in limited domains.

Still the question remains to know at which step of the evolution,
either chemical or biological, these phenomena of enantiomer segregation
and enrichment have developed and how large should the optical purity
be for life to emerge.

EVOLUTION

The problem of asymmetry evolution is generally treated independently of chemical evolution. For some authors it is not thought as a fundamental problem as any chance event could explain the emergence of asymmetry at any level of evolution. However, in view of the importance of
optical purity in present life, it is hard to believe that, at the beginning, a completely "racemic" life, based on biomolecules of both configurations in the same protocell, or on meso-biopolymers, has raised.
We think rather that, even if life at its beginning may have bear some
racemic character, a joint examination of molecular and configurational
evolutions, going to more and more complexity, should bring new insights
and suggest new experiments. To define a more self-consistent system,
this approach should be developed simultaneously on the whole of biomolecules. Indeed, the choice made by nature of L-amino acids together with
D-sugars, chiral lipids and other compounds of defined chirality is certainly not fortuitous. Interactions at different levels of structural
complexity between the main chiral constituents or components, leading
to covalent derivatives or intermolecular complexes must have help to
this selection, together with kinetic processes such as diffusion, adsorption, micellization, not yet widely studied from a prebiotic point
of view.

Another feature which should be kept in mind is that present life
occurs within limited volumes. This may be important as either chance
events or asymmetric force fields that are or were acting at the earth
surface cannot have created anything else than microquantities, at the
earth scale, of an enantiomer excess, probably limited in space and time.

Chirality and molecular evolution deserves a unifying theory. We
will here only examine a few examples illustrating these views.

Chiral Molecular Evolution

Natural α-amino acids are readily obtained under prebiotic conditions together with other amino acids. The recently described formation
of amino acid N-carboxyanhydrides in aqueous solutions in presence of
carbon dioxyde or bicarbonate ions, and their subsequent polymerization
into peptides may be a clue for the selection of natural amino acids in
the mixture. At pH near neutrality :

$$NH_2\text{-CHR-COOR'} \quad \xrightarrow[\text{HCO}_3^-]{\text{CO}_2 \text{ or}} \quad
\begin{array}{c} NH-CO \\ | \qquad \diagdown O \\ CHR-CO \diagup \end{array} + R'OH$$

probably through carbamate ions : $^-$OOC-NH-CHR-COOR'. Disubstituted α-
amino acids and β- or ω-amino acids are discared from the reaction as
they do not form easily carbamate ions or N-carboxyanhydrides respectively. Furthermore, this reaction has the great advantage over thermal polymerization of amino acids to preserve any optical activity, and for
this reason is more plausible (14).

Reciprocally, the prebiotic formation of nucleotides and polynucleotides is still a challenging problem, at both chemical and configurational points of view, complicated by the presence of many asymmetric centers in the sugar moiety. A good candidate for an ancestor of nucleic
acids may have a reduced number of asymmetric atoms per monomeric constituent. The most simple model with only one chiral center could for example derive from the already mentioned teichoic acids built up with glycerol. The polymer chain should possess isotactic stereoregularity. A link
could thus be also provided between phospholipids and polynucleotides.

The suggestion, already many times formulated, that lipid bilayer
membranes could have been present before the accumulation of proteins
and polynucleotides sounds reasonable. However limited attention has been
given to the optical purity of lipids and to the possible stereoselective
role of membranes. Vesicles with 0.4 µm diameter, the order of magnitude
of a blue-green algae cell, contain each about one million lipid molecules ; if these are chiral, there is an odd chance to find an excess of
0.07 % or more of one enantiomer per vesicle. But if interactions between
homochiral molecules are favored over interactions between heterochiral
ones, then the formation, in equal number, of vesicles enriched in one
enantiomer or the other will be expected. While surveying the literature
we found that the idea of spontaneous two dimensional resolution of chiral lipids as a mean to provide an asymmetric and stereoselective envelope to small volumes was first expressed in 1981 (15). The experimental
findings revealed however that dipalmitoylphosphatidyl choline molecules
behave as non chiral. The asymmetric carbon atom of the phosphoglycerol
pole appears to be buried in the rest of the molecule. But evidence for
a partial resolution in a chiral monolayer has been provided with a ste-
aramide bearing an optically active amide moiety (16).

The lipid family is large and it may be worthwile to examine other molecules. Some have a polar head which bears more than one asymmetric center such as phosphatidyl serine or phosphatidyl inositol, other have aminated hydrophilic poles such as sphingolipids, and so on.

This promising approach may be also substantiated further by the existence of somehow exotic but perhaps indicatory lipids which have asymmetric carbon atoms in the fatty fraction. Let us mention chaulmoogric acid, lactobacillic acid with a cyclopropanyl group :

$$CH_3(CH_2)_5 - \underset{\underset{\displaystyle CH_2}{\diagdown \diagup}}{\overset{\displaystyle \overset{H}{C'} \ \overset{H}{C'}}{C' - C'}} - (CH_2)_9 \ COOH,$$

tuberculostearic acid :

$$CH_3(CH_2)_7 \underset{\underset{\displaystyle CH_3}{|}}{-CH-}(CH_2)_8 \ COOH,$$

etc...

Another interesting case, above-mentioned, is the fatty alcohol radical of halobacterium cutirubrum lipids which derives from dihydro-oligoisoprene. Indeed, the isotactic nature of dihydrogenated isoprene oligomers represents perhaps the most ancient trend in chirality, as this kind of stereoregularity can in principle be obtained abiotically by stereospecific inorganic catalysts.

Interactions and Stereoselection

Stereoselection can be sought for in solutions or in the solid state, among chiral bioconstituents of similar or different kind. It may result new molecules by reaction, or intramolecular complexes. The relevance of such studies to the chiral and molecular evolution will be illustrated by a few examples.

The spontaneous resolution of enantiomers by crystallization, i.e. the formation of a conglomerate instead of a racemate, is a process difficult to predict on molecular considerations. Detailed studies have been made on some amino acids derivatives. It was found for instance that among N-acetyl methionine derivatives, the dimethylamide resolves spontaneously upon crystallization whereas the diethylamide derivative crystallizes in a racemic form. This behaviour appears mainly governed by optimal packing forces in the crystal, rather than by prefered associations preexisting in solution (17). Homochiral and heterochiral associations have been found for leucine dipeptides in carbon tetrachloride solutions (18).

Within the class of hybrid molecules we have already noticed phosphatidyl serine and phosphatidyl inositol. Many other mixed compounds are of interest. One may for instance wonder whether aminoacyladenylate anhydrides are formed more readily between an L-amino acid and a desoxy-

D-ribose adenosine triphosphate or between a D-amino acid and a desoxy-
D-ribose ATP. Likewise, examination of atomic models reveals that in a
3'-ester of an L-amino acid and a "D"-nucleotide, the asymmetric α-carbon
atom of the amino acid residue is brought in close vicinity of the asymmetric C-3 and C-4 atoms of the sugar moiety by ammonium-phosphate electrostatic interactions. This may represent a favorable situation for the
coselection of L-amino acids and D-sugars.

Such hybrid molecules have also lead to interesting speculations on
amino acid directed nucleic acid synthesis. Polymerization of a nucleotide substituted with an amino acid in a 2' position has been postulated as
a possible pathway (19).

The interaction of D- and L-alanine with an optically active model
membrane system has been examined (20). The methyl nuclear magnetic resonance showed clearly that the D- and L-amino acids were ordered differently in a decyl-2-sulfate phase.

It will be fruitful to undertake similar experiments on more different compounds and to include dynamic processes.

CONCLUSION

When comparing the complexity of chiral molecules and the importance of their optical purity in the most simple organism presently living,
to the state of chiral matter as it can be thought to have existed in the
primordial soup, it is rather difficult to imagine the transition from
inanimate to animate states through a purely "racemic" life, although
peculiar peptides like gramicidin A may be taken as indicatory for it.
The idea of a racemic life can be dismissed by the postulate that life
occurred in sufficiently large asymmetric volumes. Possibly these volumes were bounded by an asymmetric stereoselective envelope or membrane
so that the inside could be enriched in primeval enantiomers generating
stereospecific catalysts and stereoregular biopolymers.

A striking feature of present life is the existence of fine stereo-
interactions between biomolecules of different kinds. Interactions of
that sort must have play a role in the simultaneous choice made by nature
of biomolecules belonging to defined stereochemical series.

Thus, a joint examination of chemical and chiral evolutionary processes, first at the molecular level, of the different biomolecules as a
whole appears a necessity for selecting the most plausible pathways which
have evolved in a relatively short period of time in view of the limited
stability of the chiral components.

ACKNOWLEDGEMENTS

We are indebted to many of our colleagues for helpful discussions.

APPENDIX - QUESTIONABLE RESULTS

    Needless to say, it is difficult to be fully confident in some
findings reported in the literature concerning the stereoselective synthesis, degradation or adsorption of chiral compounds when very small
differences are measured between two enantiomers. This is particularly
true for the effect of weak interactions arising from parity non conservation in the selective degradation of α-amino acids, or for the role of
clays in their preferential adsorption (especially in view of the fact
that clays have no known asymmetric centers).

    For more inquires, the reader is requested to refer to the latest
critical reviews on these subjects (12,21).

REFERENCES

1. Pasteur, L.:1884, Rev. Scient. 7, 2-6.
2. Ulbricht, T.L.V.:1981, Origins of Life 11, 55-70.
3. Brack, A., and Spach, G.:1981, Origins of Life 11, 135-142.
4. Arnott, S., and Chandrasekaran, R.:1981, Biomolecular Stereodynamics,
    Sarma, R.H. edit., Adenine Press, N.Y., 99-122.
5. Privat de Garilhe, J., de Rudder, J.:1964, C.R. Acad. Sci. 259, 2275-
    2279.
6. Ikehara, M., and Tazuka, T.:1973, J. Am. Chem. Soc. 95, 4054-4056.
7. Spach, G.:1981, Symbioses 13, 72-86.
8. Addadi, L., Van Mil, J., Gati, E., and Lahav, M.:1981, Origins of
    Life 11, 107-118.
9. Spach, G.:1974, Chimia 28, 500-503.
10. Bonner, W.A., Blair, N.E., and Dirbas, F.M.:1981, Origins of Life
    11, 119-134.
11. De Gennes, P.G.:1970, C.R. Acad. Sci. 270B, 891-893.
12. Keszthelyi, L.:1981, Origins of Life 11, 9-21.
13. Dougherty, R.C.:1981, Origins of Life 11, 71-84.
14. Brack, A.:1982, BioSystems, in press.
15. Arnett, E.M., and Thompson, O.:1981, J. Am. Chem. Soc. 103, 968-970.
16. Arnett, E.M., and Gold, J.M.:1982, J. Am. Chem. Soc. 104, 636-639.
17. Lapicque, A., Cung, M.T., and Marraud, M.:1981, Tetrahedron 37, 891-
    898.
18. Cung, M.T., Marraud, M., and Néel, J.:1978, Biopolymers 17, 1149-
    1173.
19. Nelsestuen, G.L.:1978, J. Molecular Evol. 11, 109-120.
20. Tracey, A.S., and Diehl, P.:1975, FEBS Letters 59, 131-132.
21. Ponnamperuma, C., Shimoyama, A., and Friebele, E.:1982, Origins of
    Life 12, 9-40.

TOPOLOGICAL AND QUANTITATIVE RELATIONSHIPS IN EVOLVING GENOMES

Emile Zuckerkandl
Linus Pauling Institute of Science and Medicine
440 Page Mill Road, Palo Alto, California 94306

ABSTRACT

The network properties of the circuitry (topology) of gene interactions are to be distinguished from the quantitative aspects of these interactions. The difference between homologous and analogous morphological structures is best founded on this distinction. It is proposed that gene interaction topologies change very slowly during evolution, while quantitative aspects of gene action and interaction change faster. "Islands" of extremely old sectors of gene interaction topology may have been preserved. Cases of apparent convergent evolution of morphological features thus may contain a component of parallel evolution, which renders the independent appearance of such similar structures more plausible. Two types of gene interaction topology, a trans-spatiotemporal interaction topology and a stage-and-tissue-specific interaction topology need to be distinguished. Pathways for changes in gene interaction topology include gene duplication, heterochrony (the switch of the activity of a gene from one developmental time to another) and heterohistosis (the switch of the activity of a gene from one tissue to another). It is surmised that even higher organisms are built on the basis of a surprisingly small number of genes controlling polypeptides with significantly different functions, though in the same context the importance of the presence of a multiplicity of only slightly diverged gene duplicates is recognized. Even quantitative aspects of gene action and interaction are often conserved over rather long evolutionary periods. When quantitative relationships between gene activities change during evolution, they mostly are expected not to affect the network properties of the gene interaction system.

INTRODUCTION

The understanding of development and evolution is, I believe, in part contingent upon a simple distinction and the emphasis that we place on it. This distinction is that between the network properties

C. Hélène (ed.), Structure, Dynamics, Interactions and Evolution of Biological Macromolecules, 395–412.
Copyright © 1983 by D. Reidel Publishing Company.

of gene interaction, which can be referred to as gene interaction topology, and the various quantitative relationships in gene interaction
and gene expression within a same interaction topology.

To show that this distinction is meaningful, it may suffice, by
way of illustration, to point out that it offers the most rational
grounds on which to base the important classical distinction between
homologous and analogous structures. Homologous morphological structures can be defined as those that depend mostly on identical sets of
genes integrated into a nearly identical sector of gene interaction
topology, irrespective of differences in the quantitative aspect of
gene action and interaction. Such differences in quantitative aspects
make for the large morphological differences often observed between
homologous structures. Analogous morphological structures are then defined as structures based to a significant extent on different sectors
of the organism's gene interaction topology.

Admittedly this definition, like most biological definitions,
though potentially clarifying, leaves room for intermediate cases that
will be hard to classify when classifying eventually becomes possible
thanks to the elucidation of gene interaction patterns.

What are the quantitative aspects of gene action and interaction
that are being opposed here to the network properties of genomes?
These quantitative aspects relate to two functional characteristics of
genetic systems. On the one hand, structural genes are expressed at
certain rates. On the other hand, noncoding sequences linked to the
structural genes and instrumental for their expression (notably for
transcription and for the processing of the transcripts) are or are not
in a state that allows them to receive certain signals. Such signals
are surely themselves, for the most part, direct or indirect gene products. If the signals can be received, which in molecular terms means
bound, then the cell, in regard to these signals, will be said to be in
a state of competence. The reception of one signal implies a change in
the nature or rate of emission of another. With this emission we are
back to the first component of the dual system and into a new round of
signal expression - signal reception. The quantitative aspects of
signal reception are defined by the affinity of the signal for the target and by the concentration of units of signal and target. In addition there are the affinities and concentrations of other molecules
that, in combining with the signal molecule, change the latter's affinity for the target.

It is useful to formulate questions of gene interaction in terms
of controller node interaction. Let us first define the functional
unit of gene action, abbreviated fuga. It encompasses the coding
sequences of a structural gene and its noncoding dependencies, insofar
as they are involved in the regulation of the expression of the
structural gene (Zuckerkandl, 1978). Gene expression includes at least
transcription of a certain part of the fuga and the subsequent
processing of the transcript to messenger RNA. A controller node

(Zuckerkandl, 1978) is a set of molecular components directly involved
in the regulation of either transcription or processing
(transcriptional or processing controller node, respectively). The set
includes receptor sequences on the polynucleotide, regulator molecules
that combine with the receptor sequences and effectors that combine
with the regulators and change the latter's affinity for the receptors.

We shall refer to patterns of interaction between controller nodes
as controller node interaction topologies. Such topologies, again, do
not distinguish between quantitative aspects of the interaction, which
can be varied within the same topology, but deal exclusively with
interactive connections and the direction of the interactions. It is
important not to confuse the topological pattern of gene interactions
with the topographical pattern of the sequence of genes along the chromosomes (Fig. 1), even though the two patterns are probably interdepen-
dent to a significant degree.

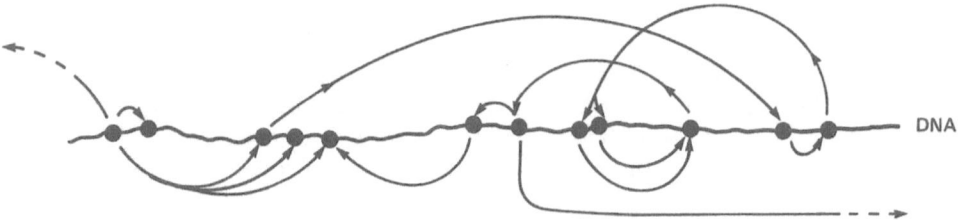

Fig. 1: Distinction between the topographical arrangement of
genes (fugas) along the chromosome and the topology of gene
interactions. The latter is represented by the set of arrows.
Each dot along the chromosome represents a fuga. It is not
specified whether the interactions are activating or repressing.
Both types of interaction are assumed to occur.

EVOLUTIONARY STABILITY AND CHANGE IN GENE INTERACTION TOPOLOGY

It seems to be accepted now that structural genes that correspond
to polypeptide domains are extremely old objects on earth (Zuckerkandl,
1975). This is so in spite of sequence changes during evolution that
are usually relatively superficial in their effects, though sometimes
numerous.

I wish to propose that some regulatory relationships between genes
are almost as old. Interaction patterns of genes defined by the connections between controller nodes probably remained near to constant
for certain sets of genes throughout the evolution of, say, the animal
kingdom, and for larger parts of the genome no doubt for only some
hundreds of millions of years. This is somewhat less old than are most
of the proteins or protein domains. Less old still are quantitative
relationships in gene interaction and ratios of rates in gene expression. This last category of quantitative relationships, though by far

the most variable, nevertheless in turn frequently displays long term evolutionary stability.

One general reason for expecting the topology of regulatory inter-relations between genes to be very stable in evolution hinges on the fact that the conservation of the structural genes is linked to their functions, which cannot be exercised independently the ones of the others. In their essential features these functions remain extra-ordinarily constant, even though, from time to time, offspring arise from gene duplicates that plow new functional ground. The decisive changes in these duplicates are mostly mutational events other than point mutations, namely events that lead to modifications in the ter-tiary structures of the proteins. When tertiary structure and function are significantly altered, a correlated evolution at the DNA level probably occurs in duplicated receptor sequences of regulator molecules. Meanwhile the established and essentially constant function of the preexisting protein cannot be carried out in a vacuum, but only in conjunction with other spatially and chronologically coordinated molecular functions. Thus proteins whose basic function is maintained will presumably also tend to conserve their regulatory relationships in the topological sense. Indeed, for each protein the correlated syn-thesis of other proteins creates the indispensable framework within which the function of the first protein can be carried out. Thus the very stability in protein function points to probable evolutionary stability in certain gene interaction patterns.

Let us define programs of gene action as the interactions over de-velopmental time within a certain set of controller nodes, a set that represents a certain sector of the topology of gene interactions. Each such sector is concretized by a mutational state, namely a certain total set of sequence characters that relate to the sector. These sequence characters, this mutational state vary from individual to individual and of course from species to species. Each mutational state (taking into account the coexistence of different alleles in diploid organisms) sets limits to the variation during development in rates of expression of the individual genes included in the topology sector. This variation and its limits are the results of the interactions between the mutational state, the developmentally changing internal milieu, and the external milieu. Natural selection may see to it that the variations in the limits of the rates of gene expression during development be themselves kept within certain bounds in spite of the inexorably progressing divergence of mutational states in isolated gene pools (distinct species).

Each program of gene action, we said, involves a gene interaction topology sector. Strong indications are already available about how ancient some such sectors are. As has been abundantly documented, often programs no longer obviously used during late developmental stages of organisms remain at least temporarily active during earlier stages. The set of controller node interactions that leads, say, to the formation of temporary gill slit-like structures in the human embryo is very likely to date back at least to the common ancestor of fish and man. The presence among man's gene programs of a partial fish program and many similar observations suggest a quasi-universal conservation of partial programs over much of evolution.

Another illustration of the persistence of such programs is the reawakening of atavistic gene programs in the adult population of a species. Examples are found in teratology, for instance the reappearance in rare individuals of a rudimentary tail in man (Warkany, 1971; Ledley, 1982), or the recurrence of an atavistic little toe in guinea pigs (Wright, 1934). In normal biology, an example is furnished by the reappearance of the second lower molar tooth in a lynx the ancestors of which had lost that tooth in the miocene about 20 million years ago (Kurten, 1963).

At least in the case of human tails and perhaps in most cases of evolutionary flashbacks it seems likely that the atavistic reversion is founded on heterochrony (Ledley, 1982), namely on a change in scheduling of gene action, a process likely to be predominantly based on events other than point mutations (Zuckerkandl, in prep.). In fact, the phenotypic end result of very old programs of gene action as known from fossils can be artificially reproduced in a contemporary adult. This has been accomplished through surgical intervention during development. By this type of intervention, indeed by the addition of supplemental embryonic material, Hampé (1959) produced in the adult chicken a disposition of lower-leg bones that is close to that of Archaeopterix (see Frazzetta, 1975, p. 225). This experiment demonstrates that the topological interrelations between controller nodes involved with programs for bones in the lower adult leg (and presumably by no means for this skeletal region alone!) has not essentially changed since the common ancestor of Archaeopterix and chicken. The alternative supposition, namely that the regulatory consequences of a surgical operation lead to the reestablishment of an ancestral gene topology, obviously cannot be considered, since this could be accomplished only through a series of specific mutational events.

Insofar as distantly related organisms have in common not only homologous structural genes, which we know to be the case, but also, to an important extent, the topology of functional connections between genes, the frequently emphasized opposition between parallelism and convergence may, in many cases, vanish. In evolutionary parallelism, two forms descending from a common ancestor have acquired independently a common feature that the ancestor did not possess. In convergent evolution a common feature has independently been acquired by two organisms or parts of organisms that are ancestrally "unrelated". Supposed instances of convergence may turn out to be instances of parallelism, at least in part. This would apply to homomorphosis, the evolution of similar organs in distantly related animals, such as the vertebrate and the cephalopod mollusc eye (Packard, 1972). Indeed, even in very distant organisms, such as cephalopods and vertebrates, whose common ancestor may have had a worm-like appearance, islands of gene programs, defined by groups of interconnected controller nodes and present in the worm-like ancestor, may have been preserved throughout subsequent evolution. Similar morphological structures may in part be the expression of parallel changes in the ratios of rates of expression of the different structural genes that compose a conserved island of connected controller nodes. It will eventually be possible to test this hypothesis through the techniques of molecular biology.

As organisms begin to diverge in gene regulation, they are expected to pass from an identical topology of controller nodes through a state describable in terms of continents of common controller node topology to that of ever smaller - yet not necessarily very small - islands of common controller node topology. The latter situation probably characterizes the degree of divergence between high categories of taxa, particularly phyla. Identical controller node topology is no doubt compatible with striking changes in growth and form through quantitative changes in gene interaction (see below).

TWO TYPES OF GENE INTERACTION TOPOLOGY

It will be necessary to distinguish two types of gene interaction topology. One, a virtual, multidimensional gene interaction topology may record all structural genes in a genome and each channel and direction of gene interaction extant at some organismal time and place, provided the interaction takes place during at least one developmental phase in at least one type of cells. The second kind would be a gene interaction topology as it is actually realized in a given tissue at a given developmental stage. There will be as many of the latter topologies as there are tissues and stages. The total genome topology is a trans-spatiotemporal interaction topology. Evolutionarily it is no doubt the most invariant. Stage-and-tissue-specific interaction topologies on the other hand are expected to change more frequently.

Topographical alterations in the linkage between genes (e.g. brought closer together or pulled further apart) will not by themselves bring about any alteration in the affinity between components of a controller node. A topographical reshuffling of genes thus does not imply - though it might eventually lead to - changes in trans-spatiotemporal topology. This statement is expected to hold only conditionally for transpositions. The condition is that the structural genes be transposed with their cis-acting regulatory sequences attached. In the case of insertions and deletions it is in turn required that these events do not destroy critical aspects of the relationship between the cis-acting regulatory sequences and the coding sequences (see Kolata, 1981).

The topographical reshuffling of genes may however, in a certain fraction of cases, lead to changes in stage-and-tissue-specific interaction topology and, via such an event, eventually to changes in

trans-spatiotemporal topology as well. This is expected to occur when
a transposition inserts a structural gene and its dependencies (a fuga)
into a region of chromatin that, as a region, is subject to a different
temporal and/or spatial control. Such a more radical regulatory effect
of a transposition may lead to a discontinuation in the interaction between certain components of a controller node. If this interaction
continues to take place at certain developmental stages, the integrity
of the controller node is likely to be maintained. If, on the contrary, any contact between certain controller node components is prevented throughout all tissues and developmental stages, the controller
node is likely to decay. In such a case the change in stage-and-
tissue-specific topology will have led to a change in trans-
spatiotemporal gene interaction topology (Fig. 2).

PATHWAYS FOR CHANGES IN GENE INTERACTION TOPOLOGY

     There are various pathways for changes in gene interaction topol-
ogy.

     Gene duplication is one of them. It is preferable to speak of
fuga duplication when one wishes to indicate that receptor sequences
close to the structural gene are also duplicated. Gene duplications
introduce additional topological connections when they include the
duplication of neighboring receptor sequences. Such gene duplications
as well as losses of duplicates no doubt represent the most frequent
changes in gene interaction topology. Their biological impact, which is
mostly modest, is expected to increase if and when the divergent
structural genes control proteins that evolve new functions.

     Several stages in gene topology change on the basis of fuga
duplication can be distinguished (Fig. 3). The initial stages,
topologically speaking, are events of little consequence. With progressing divergence of the duplicates, the change can be of increasing
scope. It is well known that the generation and loss of gene
duplicates is a very frequent evolutionary event (loss meaning either
deletion or reduction of the duplicate to a pseudogene), but in most
cases such an event, though it does represent a topological change,
can, in a general treatment of evolutionary topological constancy and
change, be discounted as mere topological "noise". This is the case
when the loss of gene duplicates occurs at stage A of Fig. 3 and at any
rate before stage D is reached.

     The process depicted in Fig. 3, founded on duplications of both
fugas controlling ordinary proteins (other than regulators) and of
fugas controlling the regulators themselves, may well represent a major
mode of evolution of gene regulation. This is suggested by the
homology between different regulators that is now being uncovered,
though not yet in eukaryotes (Matthews et al., 1982; Steitz et al.,
1982; Wilcken-Bergmann and Müller-Hill, 1982).

A second pathway of topology change involves heterochrony. Fig. 2 represents an example of how heterochrony, when sufficiently marked, will lead to significant changes in gene interaction topology.

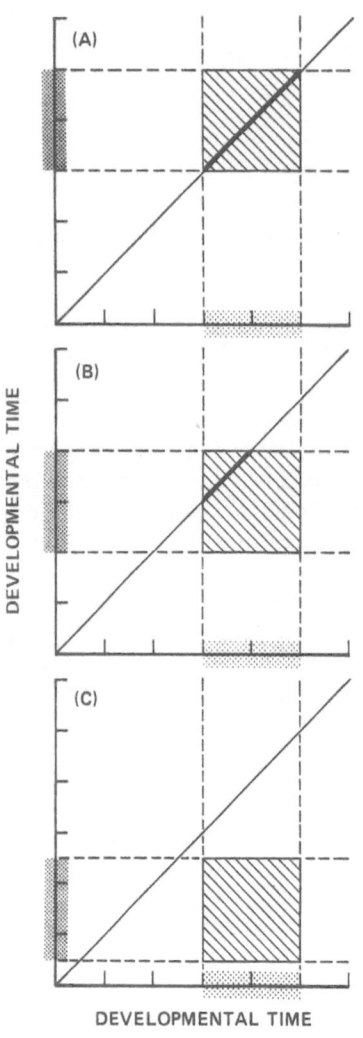

▨ DEVELOPMENTAL TIME
     AT WHICH RECEPTOR
     LINKED TO A GIVEN
     GENE IS ACCESSIBLE
     TO REGULATOR

▨ DEVELOPMENTAL TIME
     OF SYNTHESIS OF
     REGULATOR

Figure 2: Topology changes through heterochronies. It is assumed in these examples that the effectors necessary for effective regulator-receptor interaction are present at the same time as the regulators.
     Case A. The developmental time over which a regulator protein is synthesized (or, rather, present above a certain threshold value) and the time over which a given receptor for this regulator is accessible to the regulator are congruent in this example. In many real cases there may be accessibility some time before as well as after availability of sufficient quantities of "punctate" (Zuckerkandl, 1981) regulator, as suggested by DNAse I sensitivity and other studies (Weintraub and Groudine, 1976; Palmiter et al., 1978; Elgin et al., 1978; Weisbrod, 1982).
     Case B. A time shift has occurred through the integration of the structural gene under consideration into a gene complex whose members become accessible to regulators at an earlier developmental time than in case A. There remains some overlap in timing, albeit a reduced one, between regulator synthesis and receptor accessibility. Hence trans-spatiotemporal interaction topology of the system will not tend to change.
     Case C. The same structural gene is now in a gene complex whose members are accessible only during an early developmental period so that there is no time at which the regulator under consideration can be bound specifically. The time of synthesis of the regulator has indeed not changed in this example. At least one interaction in at least one controller node is thus abolished. The structural basis for this interaction may eventually disappear through lack of negative selection against certain mutations in controller node components. A modification in gene interaction topology has taken place, which may lead to other such modifications.

Heterohistosis, namely the switch of the activity of a gene from one tissue to another, probably as a consequence of a translocation (Zuckerkandl, in prep.), represents a third pathway of change in gene interaction topology, and could be represented in analogous fashion.

In many or most of these cases, especially in cases of heterochrony and heterohistosis, receptors near the structural gene may become inaccessible as a consequence of some higher level regulation whereby the "competence" of the cell is changed. Slight heterochronies, on the other hand, might be explainable by simple rate changes in the production of regulator and/or effector (Zuckerkandl, 1976, p. 400). Likewise certain types of heterohistoses, which are matters of proportion of proteins found in different tissues rather than all-or-none effects, might be interpreted in terms of situations such as the one represented on Fig. 3B and 3C (see also Fig. 1 and 2 in Zuckerkandl, 1978). This might be the case, for instance, if isozymes A and B are present in both tissues a and b, but in tissue a, if A+B represent 100%, there is only 10% of isozyme A, whereas there is 90% of it in tissue b. Higher level regulation would intervene in many of the all-or-none situations illustrated by myoglobin and hemoglobin (myoglobin, but no hemoglobin in muscle cells; hemoglobin, but no myoglobin in erythrocytes).

FROM TOPOLOGICAL TO QUANTITATIVE FEATURES OF GENE INTERACTION

If gene interaction topologies change only very slowly during evolution and do not significantly change over periods of time over which morphology is often drastically modified, one is led to assume that differences in morphological formations are primarily due to two factors, namely (1) to different quantitative relationships that involve component concentrations and/or affinity constants as they apply to the interaction of components of controller nodes, and (2) to different timing relationships between gene expressions or in the actual use (Davidson et al., 1982) by the cell of available gene products. Whatever the factors on which timing relationships may depend, the process, as mentioned, may be expected to occur mostly without a change in pattern of functional connections between controller nodes.

A general rationale can be given for this assumption. It depends, in turn, indirectly, on a generally accepted view, namely that "useless" components of the genome are progressively deleted (e.g. Koch, 1972; Petit and Zuckerkandl, 1976; Bailey et al., 1978; Takahata and Maruyama, 1979) or adulterated sequencewise to a nonfunctional state. The frequent occurrence of sequence adulteration has recently been confirmed by the discovery of numerous "pseudogenes" in the genome (Proudfoot, 1980; Leder et al, 1981). Pseudogenes have become useless as structural genes, even though in a number of cases they might not be useless otherwise, e.g. by providing functionally important distance and/or receptor sequences for proteins between other components of the genome.

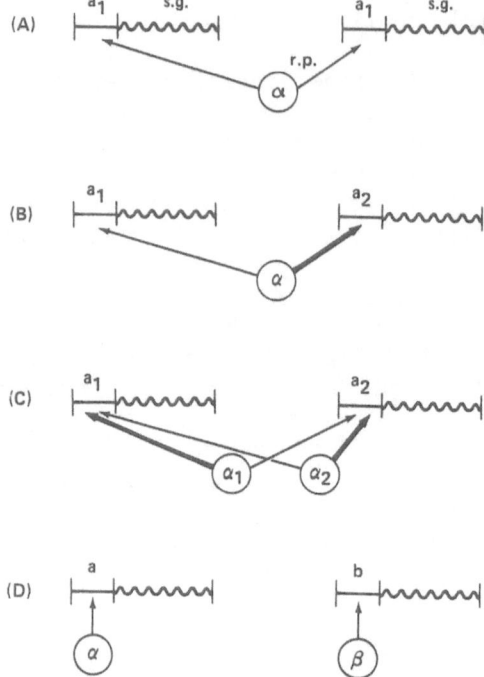

Fig. 3: <u>Stages in gene interaction topology changes consecutive to fuga
duplications.</u>

s.g. = structural gene
a, b = receptor sequences for regulator proteins (r.p.) α or β

Case A: a fuga has just been duplicated. There has been no time for
receptor sequence a to diverge in the two duplicates. Therefore regulator α
has the same affinity for the receptor sequences in the duplicate fugas.

Case B: like case A, but at a later evolutionary time. The receptor
sequences a_1 and a_2 have diverged slightly and the affinity of the
regulator protein for receptors a_1 and a_2 is no longer the same. The
heavier arrow indicates the stronger affinity. As a result there is a
difference in rate of expression between the duplicate genes (or this
difference is altered if it already existed for other reasons).

Case C: the fuga that controls the regulator protein now has also
duplicated. The structural genes for the two regulators $α_1$ and $α_2$ have
slightly diverged. Each of the two regulators still has a specific affinity
for both receptors a_1 and a_2, i.e. an affinity higher than background
nonspecific affinity for DNA. Though both regulators thus cross-react with
both receptors, the affinities of each regulator for the two receptors are
unequal. The heavier arrows indicate the stronger affinities between
regulator and receptor. Only one of the alternatives is represented.

Case D: the two regulators and the two receptors of case C have now
diverged to the extent that there no longer is any significant cross-
reaction. Receptors a and b combine specifically only with regulators α
and β respectively.

The molarless ancestor of the lynx that was mentioned above can be assumed to date back far enough in time for an irreversible decay in the gene program for "second lower molar tooth" to have occurred if the same program of gene action had not continued to be functional other- wise - no doubt in relation to all teeth and in part probably also to other morphological formations.

Thus, for the mere reason that, in its essential structure, the program for a lost morphological trait has been maintained, it may be concluded that the same gene topology must be used in other ways, prob- ably for the generation of some other morphological formations. If we find a single set of controller nodes at the origin of divergent morph- ological formations, the differences between these formations may well be due essentially to quantitative alterations in gene interactions within the same interaction pattern.

Just as it is turning out that the great diversity of living forms is based on a relatively very small number of types of proteins (De- rancourt et al, 1967; Dayhoff, 1974; Zuckerkandl, 1974; Doolittle, 1981), I anticipate that arrays of different morphological formations within the same organism will be found to be due mostly to the same[1] structural genes as well as to a rather limited array of partially overlapping blueprints for programs of gene action. The total number of such blueprints, when it will become possible to separate them by arbitrary boundaries for counting purposes, will presumably be found to be surprisingly restricted. Blueprints for gene programs are de- fined here by interaction topologies, not by interaction intensities. It is the latter that primarily vary, I believe, from organism to organism and from morphological feature to morphological feature of the same organism when these features involve the same genes. Thus many fewer different genes may be used for building the morphology of different parts of an organism than one might think. In fact, many fewer really different genes are present in higher organisms than believed, if one counts as units very closely related genes (whether contiguous or dispersed) rather than genes. Under these conditions, the fruit fly, Drosophila, may not have many more "different" genes than certain bacteria. It is apparently possible to put together a higher organism with rather few markedly different genes, with a mechanism for generating multiple specificities in cellular interaction (in part through an abundance of slightly different gene duplicates), with the laws of physics and physical chemistry, with the variable and yet reproducible environments that parts of organisms constitute for each other, and with another essential ingredient, evolutionary time.

The laws of physics and physical chemistry could not be omitted from this list. Genes do not by themselves generate morphologies. They only render them possible or impossible. They furnish building materials and determine their amount, time, and place. What happens next to these materials is no longer directly the business of the genes. It is the business of matter's self-organization. Such

self-organization may even lead to direct morphological heredity independent of the genes - as long as the genes continue to furnish the necessary building materials. There is an old experiment of Sonneborn's
(1963), one that did not keep on making as much noise among biologists
as it deserved to make. Sonneborn incorporated on the surface of a
Paramecium (a Ciliate protozoon) a piece of paroral cortex excised from
another individual. Over subsequent generations the offspring of the
recipient individual faithfully reproduced the same microsurgically -
not genetically - introduced aberrant morphological feature. This kind
of phenomenon may well be less peculiar and rare than we perhaps tend
to believe. It might, for instance, have some relationship with the
transmission from mother to daughter cell of specific local structural
features of chromatin that determine the transcribability of genes,
though admittedly there are other ways of attempting to explain this
transmission (Weintraub, 1979). The morphology of living systems
requires the presence of the right genes, to be sure, but it also owes
much to forces operating at other levels (e.g. Markiewiez et al.,
1982). Thanks to this circumstance the number of "really" different
genes can remain rather small even in a higher organism.

EVOLUTIONARY CONSERVATION OF QUANTITATIVE ASPECTS OF GENE ACTION AND
INTERACTION

    Even purely quantitative relationships between genes, within an
established gene interaction topology, may often be highly conserved by
evolution, as the long persistence of many allometric relationships
suggests, such as that between the length of the skull and the length
of the face in horses and in their ancestors. This relationship appears to have remained quantitatively identical for some 45 million
years (Simpson, 1944).

    Though stability in quantitative relationships between genes lasts
less long than the presumed stability in topological relationships,
both stabilities may nowadays seem surprising to molecular biologists.
Transposable elements (e.g. Calos and Miller, 1980) and translocations
appear in principle to be capable of reshuffling the genome extensively
and fast. At present we do not know whether their failure to do so in
functionally prominent ways is to be attributed primarily to stabilizing selection or largely to the fact that many insertions and deletions
between coding sequences have at best only minor effects (e.g. Bell et
al., 1981) or, also, to a significant extent, to internal restrictions
imposed by the organisms's mutational state. Thus, for example, in a
42 kilobasepair fragment of human DNA that includes the insulin gene,
there appears to be only a single region where insertion or deletion
events are very frequent (Bell et al., 1981). Certain insertion
sequences, such as phage μ appear to be capable of insertion almost
anywhere in the genome. But most transposable elements require that
DNA sequences fulfill certain conditions in order for insertion or deletion to occur (c.f. Calos and Miller, 1980). In chloroplast genomes
(Palmer and Thompson, 1982) and in the yeast mitochondrial genome

(Clark-Walker et al., 1981) large inverted repeats appear to inhibit
genome reshuffling, although this causal relation has yet to be proven.
Conceivably large inverted repeats have similar effects in nuclear
genomes. The distribution over the genome of sequence features that
inhibit sequence rearrangements and other sequence changes may have
been selected in such a way as to minimize drastic effects on the patterns of gene interaction. When relatively few insertion hot-spots are
distributed over the genome, a species might be intrinsically more
stable than when there are more. It will be worthwhile to verify
whether the genomes of so-called living fossils do or do not possess
this distinctive trait.

     During development, the constellations of quantitative aspects of
gene interaction change in a planified way. Some of the stages in
these constellations "recapitulate" phylogeny. (It does not matter
here that it is indeed the embryonic stages of ancestral forms that are
recapitulated [de Beer, 1930; Gould, 1976]). Such recapitulation is
probably due to the fact that 1) the topology of gene relations remains
constant, and 2) some of the discrete steps in quantitative
relationships that are observed within the established gene interaction
topology correspond to ancestral quantitative relationships. Such
constancy in quantitative relationships and the planified changes in
these relationships during development (either involving the same genes
or other gene duplicates; both probably occur) must be assumed to have
a structural basis. Several possibilities for such a basis can at
present be considered, but will not be discussed. One of the tasks of
molecular biology in the coming years will be the elucidation of the
molecular basis for these "quantomes," i.e., for defined, reproducible,
locked-in, if temporary, quantitative relationships between rates of
expression in a set of genes within a given genome.

EVOLUTION THROUGH QUANTITATIVE CHANGES IN GENE ACTION AND INTERACTION

     Even though molecular biology has recently shown genomes to be in
a state of turmoil in regard to DNA sequence, this turmoil,
evolutionarily, mostly appears to affect only the quantitative aspects
of gene interaction and gene expression. Most topographical changes
accepted by evolution may only lead to changes during development in
the concentration of certain controller node components in the vicinity
of target fugas or changes in the developmental time at which a component of the gene regulatory system reaches a given fuga. Thus, when
we deal with topographical rearrangements of genes, again many aspects
of the divergence between organisms probably are not due to changes in
the gene interaction topology, but to quantitative changes in the
acromolecular relationships within an established interaction topology.

     I called heterotachies (Zuckerkandl, in prep.) changes in rate of
expression of individual structural genes. Evolutionarily effective
heterotachies are frequent and increase with phylogenetic distance

Table 1.

STABILITY AND CHANGE IN ORGANIZATIONAL RELATIONSHIPS BETWEEN STRUCTURAL GENES

Item a	correlated with	Item b	Comments
Same (1) genes	-	Same (2) topology	This is thought to hold as a basic generality
Same (1) genes	-	Radically different topology	Deemed highly improbable. To be discounted
Additional genes generated by fuga duplication, controlling polypeptides that fill the same basic function	-	Marginally changed topology (changes can eventually become substantial by long-term divergent evolution)	Frequent evolutionary event, frequently reversed by loss (by either elimination or adulteration of the duplicated segment of DNA). Microevolution
Duplication of structural gene without concomitant duplication of cis-acting linked receptor sequences	-	No change in topology	Likely to occur at times
Duplication of structural gene with concomitant duplication of cis-acting receptor sequences, followed by the generation of a new polypeptide function from one of the duplicates	-	Concomitant evolution of regulatory relationship, leading to a significant change in topology	Rare event, important by its consequences for gene topology, evolutionary divergence, formation of a new type of tissue. Macroevolution
Same (1,3) genes	-	Heterotachy	No change in topology; microevolution
Same (1,3) genes	-	Heterochrony	According to degree of heterochrony, may or may not lead to change in topology; macroevolution
Same (1,3) genes	-	Heterohistosis	Change in topology; macroevolution
Same topology (at most modified by the presence of homofunctional fuga duplicates)	-	Same quantitative relationships in gene interaction and gene expression	Evolutionary stasis
Same topology (at most modified by the presence of homofunctional fuga duplicates)	-	Different quantitative relationships	Heterotachies; changes in allometric coefficients; microevolution and part of macroevolution; different phases of ontogeny

(1) "Same" includes genes whose protein product may have changed in amino acid sequence but not in overall tertiary structure nor in basic function

(2) In the sense of trans-spatiotemporal topology

(3) Includes gene duplicates that control similar proteins

(Ferris and Whitt, 1979; Philipp et al., 1979; Gregory S. Whitt, personal communication; Li, in press). Since they are unlikely to lead to modifications in gene interaction topology, they will not be further commented on. They appear in Table 1, which gives a summary of relationships that have been considered here from the point of view of their potential for evolutionary stability and change. Reference is made in the Table to the significance of the relationships considered for microevolution and macroevolution, a topic that is specifically addressed in another paper (Zuckerkandl, in prep.).

Even macroevolution, according to the view expressed here, is mostly limited to quantitative changes within essentially constant gene interaction topologies (trans-spatiotemporal interaction topologies). This statement includes adaptative radiation, however striking its results may be, such as the formation, within a rather short period of evolutionary time, of a number of mammalian orders about 80 million years ago.

To sum up the main point made in this paper: much of what strikes us as a major evolutionary change, for instance the difference between a mouse and a man, is probably due to quantitative changes in gene action and interaction taking place within the framework of a gene interaction circuitry that has itself been altered in only minor ways.

[1] The "same" structural gene may be a structurally and functionally similar duplicate of the gene. The concept of "same gene" is here extended to include such gene duplicates.

REFERENCES

Bailey, G.S., Poulter, R.T.M., Stockwell, P.A. (1978). Gene duplication in tetraploid fish: model for gene silencing at unlinked duplicated loci. Proc. Natl. Acad. Sci. 75, pp. 5575-5579.

Bell, G.I., Karam, J.H., Rutter, W.J. (1981). Polymorphic DNA region adjacent to the 5' end of the human insulin gene. Proc. Natl. Acad. Sci. 78, pp. 5759-5763.

Calos, M.P., Miller, J.H. (1980). Transposable elements. Cell 20, pp. 579-595.

Clark-Walker, G.D., McArthur, C.R., Sriprakash, K.S. (1981). Partial duplication of the large ribosomal RNA sequence in an inverted repeat in circular mitochondrial DNA from Kloeckera africana. J. Mol. Biol. 147, pp. 399-415.

Davidson, E.H., Hough-Evans, B.R., Britten, R.J. (1982). Molecular biology of the sea urchin embryo. Science 217, pp. 17-26.

Dayhoff, M.O. (1974).   Computer analysis of protein sequences.   Fed.
    Proc. 33, pp. 2314-2316

de Beer, G.R. (1930).   Embryology and Evolution.   Oxford Clarendon
    Press, 116 pp.

Derancourt, J., Lebor, A.S., Zuckerkandl, E. (1967).   Séquence des
    acides aminès, séquence des nuclèotides et èvolution.
    Bull. Soc. Chim. Biol. 49, pp. 577-607.

Doolittle, R. (1981).   Similar amino acid sequences : chance or common
    ancestry? Science 214, pp. 149-159.

Elgin, S.C.R., Serunian, L.A., Silver, L.M. (1978).   Distribution
    patterns of Drosophila nonhistone chromosomal proteins.   Cold
    Spring Harbor Symp. Quant. Biol.   "1977", 42, pp. 839-849.
    Followed by comment by E. Zuckerkandl.

Ferris, S.D., Whitt, G.S. (1979).   Evolution of the differential
    regulation of duplicate genes after polyploidization.   J. Mol.
    Evol. 12, pp. 267-317.

Frazzetta, T.H. (1975).   Complex Adaptations in Evolving Populations.
    Sinauer Assoc., Sunderland, Mass., 267 pp.

Gould, S.J. (1977).   Ontogeny and Phylogeny.   Belknap Press of Harvard
    University, Cambridge, Mass., 501 pp.

Hampè, A. (1959).   Contribution a l'ètude du dèveloppement et de
    la règulation des dèficiences et des excedents dans la patte
    de l'embryon de poulet.   Arch. Anat. Micr. Morph. Expèrim. 48,
    supp., pp. 345-478.

Koch, A.L. (1972).   Enzyme evolution: I.   The importance of metastable
    intermediates.   Genetics 72, pp. 297-316.

Kolata, G.B. (1981).   Genes regulated through chromatin structure.
    Science 214, pp. 775-776.

Kurten, B. (1963).   Return of a lost structure in the evolution of the
    felid dentition.   Commentationes Biologicae 26, No.4, Societas
    Scientiarum Fennica, 11 pp.

Leder, A., Swan, D., Ruddle, F., Eustachio, P. d', Leder, P. (1981).
    Dispersion of α-like globin genes of the mouse to three different
    chromosomes.   Nature 293, pp. 196-200.

Ledley, F.D. (1982).   Evolution of the human tail.   New Engl. J. Med.
    306, pp. 1212-1215.

Li, W.-H. (in press) Evolution of duplicate genes and pseudogenes. In: Evolution of Genes and Proteins, M. Nei and R.K. Koehn, eds., Sinauer, Mass.

Markiewicz, Z., Broome-Smith, J.K., Schwarz, U., Spratt, B.G. (1982). Spherical E. coli due to elevated levels of D-alanine carboxypeptidase. Nature 297, pp. 702-704.

Matthews, B.W., Ohlendorf, D.H., Anderson, W.F., Takeda, Y. (1982). Structure of the DNA-binding region of lac repressor inferred from its homology with cro repressor. Proc. Natl. Acad. Sci. 79, pp. 1428-1432.

Packard, A. (1972). Cephalopods and fish: the limits of convergence. Biol. Rev. 47, pp. 241-307.

Palmer, J.D., Thompson, W.F. (1982). Chloroplast DNA rearrangements are more frequent when a large inverted repeat sequence is lost. Cell 29, pp. 537-550

Palmiter, R.D., Mulvihill, E.R., McKnight, G.S., Senear, A.W. (1978). Regulation of gene expression in the chick oviduct by steroid hormones. Cold Spring Harbor Symp. Quant. Biol. "1977", 42, pp. 639-647.

Petit, C., Zuckerkandl, E. (1976). Evolution, Génétique des Populations, Evolution Moléculaire. Hermann, Paris, 278 pp.

Philipp, D.P., Childers, W.F., Whitt, G.S. (1979). Evolution of patterns of differential gene expression : a comparison of the temporal and spatial patterns of isozyme locus expression in two closely related fish species. J. Exp. Zool. 210, pp. 473-488.

Proudfoot, N. (1980). Pseudogenes. Nature 286, pp. 840-841.

Simpson, G.G. (1944). Tempo and Mode in evolution. Columbia University Press, New York, 237 pp.

Sonneborn, T.M. (1963). Does preformed cell structure play an essential role in cell heredity? In: The Nature of Biochemical Diversity, J.M. Allen, Ed., McGraw Hill, New York, p. 165.

Steitz, T.A., Ohlendorf, D.H., McKay, D.B., Anderson, W.F., Matthews, B.W. (1982). Structural similarity in the DNA-binding domains of catabolite gene activator and cro repressor proteins. Proc. Natl. Acad. Sci. 79, pp. 3097-3100.

Takahata, N., Maruyama, T. (1979). Polymorphism and loss of duplicate gene expression: a theoretical study with application to tetraploid fish. Proc. Natl. Acad. Sci. 76, pp. 4521-4525.

Warkany, J. (1971). Congenital Malformation. Year Book Medical Publ., Chicago, 1309 pp.

Weintraub, H. (1979). Assembly of an active chromatin structure during replication. Nucleic Acids Res. 7, pp. 781-792.

Weintraub, H., Groudine, M. (1976). Chromosomal subunits in active genes have an altered conformation. Science 193, pp. 848-856.

Weisbrod, S. (1982). Active chromatin. Nature 297, pp. 289-295.

Wilcken-Bergmann, B. von, Müller-Hill, B. (1982). Sequence of galR gene indicates a common evolutionary origin of lac and gal represssor in Escherichia coli. Proc. Natl. Acad. Sci. 79, pp. 2427-2431.

Wright, S. (1934). An analysis of variability in number of digits in an inbred strain of guinea pigs. Genetics 19, pp. 506-551.

Zuckerkandl, E. (1974). Accomplissements et perspectives de la paléogenetique chimique. Ecole de Roscoff, Editions du Centre National de la Recherche Scientifique, Paris.

Zuckerkandl, E. (1975). The appearance of new structures and functions in proteins during evolution. J. Mol. Evol. 7, pp. 1-57.

Zuckerkandl, E. (1976). Programs of gene action and progressive evolution. In: Molecular Anthropology - Genes and Proteins in the Evolutionary Ascent of the Primates. M. Goodman and R.E. Tashian, Eds., Plenum Press, New York, pp. 387-447.

Zuckerkandl, E. (1978). Multilocus enzymes, gene regulation, and genetic sufficiency. J. Mol. Evol. 12, pp. 57-89.

Zuckerkandl, E. (1981). A general function of noncoding polynucleotide sequences: mass binding of transconformational proteins. Molec. Biol. Rep. 7, pp. 149-158.